Ship Stability for Masters and Mates

Ship Stability for Masters and Mates

Sixth edition – Consolidated 2006

Revised by Dr C.B. Barrass M.Sc C.Eng FRINA CNI

By Captain D.R. Derrett

AMSTERDAM BOSTON HEIDELBERG LONDON
NEW YORK OXFORD PARIS SAN DIEGO
SAN FRANCISCO SINGAPORE SYDNEY TOKYO

Butterworth-Heinemann is an imprint of Elsevier
Linacre House, Jordan Hill, Oxford OX2 8DP, UK
30 Corporate Drive, Suite 400, Burlington, MA 01803, USA

First edition published by Stanford Maritime Ltd 1964
Third edition (metric) 1972
Reprinted 1973, 1975, 1977, 1979, 1982
Fourth edition 1984
Fourth edition by Butterworth-Heinemann Ltd 1990
Reprinted 1990 (twice), 1991, 1993, 1997, 1998, 1999
Fifth edition 1999
Reprinted 2000 (twice), 2001, 2002, 2003, 2004
Sixth edition 2006
Reprinted 2006, 2007

Notice
No responsibility is assumed by the publisher for any injury and/or damage to persons
or property as a matter of products liability, negligence or otherwise, or from any use
or operation of any methods, products, instructions or ideas contained in the material
herein. Because of rapid advances in the medical sciences, in particular, independent
verification of diagnoses and drug dosages should be made

British Library Cataloguing in Publication Data
A catalogue record for this book is available from the British Library

Library of Congress Cataloging-in-Publication Data
A catalog record for this book is available from the Library of Congress

ISBN: 978-0-7506-6784-5

For information on all Butterworth-Heinemann publications
visit our website at books.elsevier.com

Printed and bound in *Great Britain*

07 08 09 10 10 9 8 7 6 5 4 3

Working together to grow
libraries in developing countries

www.elsevier.com | www.bookaid.org | www.sabre.org

ELSEVIER   BOOK AID International   Sabre Foundation

Contents

Part 2 Linking Ship Stability and Ship Strength

Part 3 Endnotes

Part 4 Appendices

To my wife Hilary and our family

Acknowledgements

I gladly acknowledge with grateful thanks the help, comments and encouragement afforded to me by the following personnel in the Maritime Industry:

Captain D.R. Derrett, Author of 'Ship Stability for Masters and Mates', Third edition (metric) 1972, published by Stanford Maritime Ltd.

Captain Sergio Battera, Vice-Chief (Retired) Pilot, Co-operation of Venice Port and Estuary Authority.

Julian Parker, Secretariat, The Nautical Institute, London.

Tim Knaggs, Editor of the Naval Architect, Royal Institute of Naval Architects, London.

Gary Quinn, Head of Testing Services, Scottish Qualification Authority (SQA) Glasgow.

Roger Towner, Chief Examiner, Department for Transport/Maritime and Coastguard Agency (DfT/MCA), Southampton.

Captain G.C. Leggett, Area Operations Manager (Surveys and Inspections), Maritime and Coastguard Agency, Liverpool.

Captain Neil McQuaid, Chief Executive, Marcon Associates Ltd, Southport.

Malcolm Dann, Partner, Brookes Bell Jarrett Kirmann Ltd, Liverpool.

Captain I.C. Clark, Maritime Author for The Nautical Institute, London.

Darren Dodd, Managing Director, Saab Tank Control (UK), Wokingham.

Colin Jones, Stock Control Manager, DPM Ltd, Liverpool.

Preface

This book was written specifically to meet the needs of students studying for their Transport Certificates of Competency for Deck Officers and Engineering Officers, and STCW equivalent international qualifications. Several specimen examination questions, together with a marking scheme, have been kindly supplied by SQA/MCA.

The book will also prove to be extremely useful to Maritime Studies degree students and serve as a quick and handy reference for Shipboard Officers, Naval Architects, Ship Designers, Ship Classification Surveyors, Marine Consultants, Marine Instrument Manufactures, Drydock Personnel, Ship-owner Superintendents and Cargo-Handling Managers.

Stability can exist when a vessel is rolling or trimming – it is the ability to remain in stable equilibrium or otherwise. Hence there is a link between Ship Stability and Ship Motion. Stability can also exist in ship structures via the strength of the material from which they are built. A material may be stressed or strained and not return to its initial form, thereby losing its stability. Hence there is a link between Ship Stability and Ship Strength. Another type of Ship Stability exists when dealing with course-headings and course keeping. This is called Directional Stability. Examples of this are given in Chapter 38, Interaction.

Note
Throughout this book, when dealing with Transverse Stability, BM, GM and KM will be used. When dealing with Trim, i.e. Longitudinal Stability, then BM_L, GM_L and KM_L will be used to denote the longitudinal considerations.

Therefore, there will be no suffix 'T' for Transverse Stability, but there will be a suffix 'L' for the Longitudinal Stability text and diagrams.

C.B. Barrass

Introduction

In 1968, Captain D.R. Derrett wrote the highly acclaimed standard textbook *'Ship Stability for Masters and Mates'*. In 1999, for the Fifth edition, I revised several areas of the Fourth edition (1984) book and introduced new topics that were in keeping with examinations and developments within the shipping industry.

Changes to the Sixth edition

In 2004, the SQA/MCA made major changes to the syllabuses for the STCW 95 Chief Mate/Master Reg. 11/2 (Unlimited) Ship Stability course. Changes were also made to the STCW 95 Officer in Charge of Navigational Watch <500 gt Reg. 11/3 (Near Coastal) General Ship Knowledge and Operations syllabus.

Other key changes since the Fifth (1999) edition and this Sixth edition include the following:

- IMO Grain Rules and angle of list
- Floodable and permissible length curves
- Icing allowances – effects on trim and stability
- A Trim and Stability pro-forma sheet
- Tabular and assigned freeboard values
- Load lines and freeboard marks
- Effects of side winds on stability – wind levers and wind moments
- The calibration book
- Update of research into squat and interaction
- Air draft considerations
- Draft Surveys – procedures and calculations
- Synchronous rolling of ships and associated dangers
- Parametric rolling of ships and associated dangers
- Timber ship freeboard marks
- Trimming moments about the aft perpendicular
- Changes in lightweight and its KG over a period of time
- Recent SQA/MCA examination questions.

My main aims and objectives for this Sixth edition of the book are:

1. To help Masters, Mates and Engineering Officers prepare for their SQA/MCA written and oral examinations.
2. To provide knowledge at a basic level for those whose responsibilities include the loading and safe operation of ships.
3. To give Maritime students and Marine Officers an awareness of problems when dealing with stability and strength, and to suggest methods for solving these problems if they meet them in their day-to-day operation of ships.
4. To act as a good quick reference source for those officers who obtained their Certificates of Competency a few months/years prior to joining their ship, port or dry-dock.
5. To help students of naval architecture/ship technology in their studies on ONC, HNC, HND and initial years on undergraduate degree courses.
6. To assist dry-dock personnel, Ship-designers, Dft ship-surveyors, Port Authorities, Marine Consultants, Nautical Study Lecturers, Marine Superintendents, etc. in their Ship Stability deliberations.

There are 12 new chapters in this new edition. Also included are tabular presentations of several vessels delivered to their ship-owners in 2002–2005. They show the typical deadweight, lengths, breadths, depths, drafts and service speeds for 40 ships. These parameters give a good awareness of just how large and how fast merchant ships can be. Photographs of ships have been added to this edition.

Another addition to the 1999 edition is the nomenclature or glossary of ship terms. This will prove useful for the purpose of rapid consultation. The revision one-liners have been extended by 35 questions to bring the final total to 100. A case study, involving squat, interaction and the actual collision of two vessels is analysed in detail.

The discourse on how to pass Maritime exams has now been modified and expanded. A selection of SQA/MCA exam questions set recently, together with a marking scheme, has been incorporated into this new edition.

Maritime courts are continually dealing with ships that have grounded, collided or capsized. If this book helps to prevent further incidents of this sort, then the efforts of Captain D.R. Derrett and myself will have been worthwhile.

Finally, it only remains for me to wish the student every success in the exams, and to wish those working within the shipping industry continued success in your chosen career. I hope this book will be of interest and assistance.

C.B. Barrass

Part 1
Linking Ship Stability and Ship Motions

Chapter 1
Forces and moments

The solution of many of the problems concerned with ship stability involves an understanding of the resolution of forces and moments. For this reason a brief examination of the basic principles will be advisable.

Forces

A *force* can be defined as any push or pull exerted on a body. The S.I. unit of force is the Newton, one Newton being the force required to produce in a mass of one kilogram an acceleration of one metre per second. When considering a force the following points regarding the force must be known:

(a) The magnitude of the force.
(b) The direction in which the force is applied.
(c) The point at which the force is applied.

The resultant force. When two or more forces are acting at a point, their combined effect can be represented by one force which will have the same effect as the component forces. Such a force is referred to as the 'resultant force', and the process of finding it is called the 'resolution of the component forces'.

The resolution of forces. When resolving forces it will be appreciated that a force acting towards a point will have the same effect as an equal force acting away from the point, so long as both forces act in the same direction and in the same straight line. Thus a force of 10 Newtons (N) pushing to the right on a certain point can be substituted for a force of 10 Newtons (N) pulling to the right from the same point.

(a) *Resolving two forces which act in the same straight line*
If both forces act in the same straight line and in the same direction the resultant is their sum, but if the forces act in opposite directions the resultant is the difference of the two forces and acts in the direction of the larger of the two forces.

Example 1

Whilst moving an object one man pulls on it with a force of 200 Newtons, and another pushes in the same direction with a force of 300 Newtons. Find the resultant force propelling the object.

Component forces 300 N   A   200 N

The resultant force is obviously 500 Newtons, the sum of the two forces, and acts in the direction of each of the component forces.

Resultant force 500 N   A   or   A   500 N

Example 2

A force of 5 Newtons is applied towards a point whilst a force of 2 Newtons is applied at the same point but in the opposite direction. Find the resultant force.

Component forces 5 N   A   2 N

Since the forces are applied in opposite directions, the magnitude of the resultant is the difference of the two forces and acts in the direction of the 5 N force.

Resultant force 3 N   A   or   A   3 N

(b) *Resolving two forces which do not act in the same straight line*
When the two forces do not act in the same straight line, their resultant can be found by completing a parallelogram of forces.

Example 1

A force of 3 Newtons and a force of 5 N act towards a point at an angle of 120 degrees to each other. Find the direction and magnitude of the resultant.

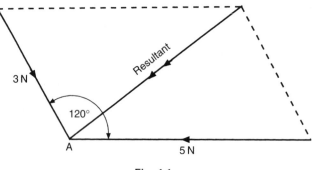

Fig. 1.1

Ans. Resultant 4.36 N at 36° 34½′ to the 5 N force.

Note. Notice that each of the component forces and the resultant all act towards the point A.

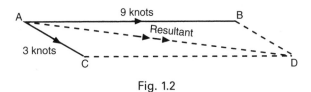

Fig. 1.2

Example 2

A ship steams due east for an hour at 9 knots through a current which sets 120 degrees (T) at 3 knots. Find the course and distance made good.

The ship's force would propel her from A to B in one hour and the current would propel her from A to C in one hour. The resultant is AD, $0.97\frac{1}{2}^\circ \times 11.6$ miles and this will represent the course and distance made good in one hour.

Note. In the above example both of the component forces and the resultant force all act away from the point A.

Example 3

A force of 3 N acts downwards towards a point whilst another force of 5 N acts away from the point to the right as shown in Figure 1.3. Find the resultant.

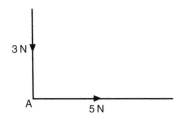

Fig. 1.3

In this example one force is acting towards the point and the second force is acting away from the point. Before completing the parallelogram, substitute either a force of 3 N acting away from the point for the force of 3 N towards the point as shown in Figure 1.4, or a force of 5 N towards the point

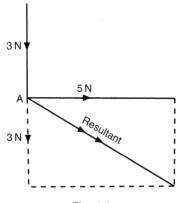

Fig. 1.4

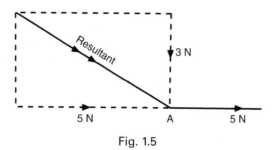

Fig. 1.5

for the force of 5 N away from the point as shown in Figure 1.5. In this way both of the forces act either towards or away from the point. The magnitude and direction of the resultant is the same whichever substitution is made; i.e. 5.83 N at an angle of 59° to the vertical.

(c) *Resolving two forces which act in parallel directions*
When two forces act in parallel directions, their combined effect can be represented by one force whose magnitude is equal to the algebraic sum of the two component forces, and which will act through a point about which their moments are equal.

The following two examples may help to make this clear.

Example 1
In Figure 1.6 the parallel forces W and P are acting upwards through A and B respectively. Let W be greater than P. Their resultant (W + P) acts upwards through the point C such that $P \times y = W \times x$. Since W is greater than P, the point C will be nearer to B than to A.

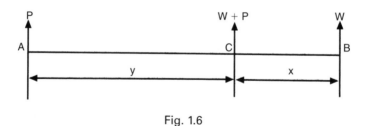

Fig. 1.6

Example 2
In Figure 1.7 the parallel forces W and P act in opposite directions through A and B respectively. If W is again greater than P, their resultant (W − P) acts through point C on AB produced such that $P \times y = W \times x$.

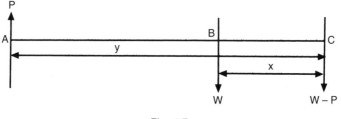

Fig. 1.7

Moments of forces

The moment of a force is a measure of the turning effect of the force about a point. The turning effect will depend upon the following:

(a) The magnitude of the force.
(b) The length of the lever upon which the force acts, the lever being the perpendicular distance between the line of action of the force and the point about which the moment is being taken.

The magnitude of the moment is the product of the force and the length of the lever. Thus, if the force is measured in Newtons and the length of the lever in metres, the moment found will be expressed in Newton-metres (Nm).

Resultant moment. When two or more forces are acting about a point their combined effect can be represented by one imaginary moment called the 'Resultant Moment'. The process of finding the resultant moment is referred to as the 'Resolution of the Component Moments'.

Resolution of moments. To calculate the resultant moment about a point, find the sum of the moments to produce rotation in a clockwise direction about the point, and the sum of the moments to produce rotation in an anti-clockwise direction. Take the lesser of these two moments from the greater and the difference will be the magnitude of the resultant. The direction in which it acts will be that of the greater of the two component moments.

Example 1

A capstan consists of a drum 2 metres in diameter around which a rope is wound, and four levers at right angles to each other, each being 2 metres long. If a man on the end of each lever pushes with a force of 500 Newtons, what strain is put on the rope? (See Figure 1.8(a).)

Moments are taken about O, the centre of the drum.

$$\text{Total moment in an anti-clockwise direction} = 4 \times (2 \times 500) \text{ Nm}$$
$$\text{The resultant moment} = 4000 \text{ Nm (Anti-clockwise)}$$
$$\text{Let the strain on the rope} = P \text{ Newtons}$$
$$\text{The moment about O} = (P \times 1) \text{ Nm}$$
$$\therefore P \times 1 = 4000$$
$$\text{or } P = 4000 \text{ N}$$

Ans. The strain is 4000 N.

Note. For a body to remain at rest, the resultant force acting on the body must be zero and the resultant moment about its centre of gravity must also be zero, if the centre of gravity be considered a fixed point.

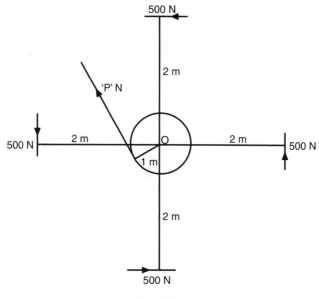

Fig. 1.8(a)

Mass

In the S.I. system of units it is most important to distinguish between the mass of a body and its weight. Mass is the fundamental measure of the quantity of matter in a body and is expressed in terms of the kilogram and the tonne, whilst the weight of a body is the force exerted on it by the Earth's gravitational force and is measured in terms of the Newton (N) and kilo-Newton (kN).

Weight and mass are connected by the formula:

$$\text{Weight} = \text{Mass} \times \text{Acceleration}$$

Example 2

Find the weight of a body of mass 50 kilograms at a place where the acceleration due to gravity is 9.81 metres per second per second.

$$\text{Weight} = \text{Mass} \times \text{Acceleration}$$
$$= 50 \times 9.81$$

Ans. Weight = 490.5 N

Moments of mass

If the force of gravity is considered constant then the weight of bodies is pro-portional to their mass and the resultant moment of two or more weights about a point can be expressed in terms of their mass moments.

Example 3

A uniform plank is 3 metres long and is supported at a point under its mid-length. A load having a mass of 10 kilograms is placed at a distance of 0.5 metres from one end and a second load of mass 30 kilograms is placed at a distance of one metre from the other end. Find the resultant moment about the middle of the plank.

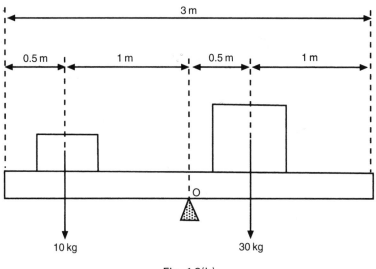

Fig. 1.8(b)

Moments are taken about O, the middle of the plank.

$$\text{Clockwise moment} = 30 \times 0.5$$
$$= 15 \text{ kg m}$$
$$\text{Anti-clockwise moment} = 10 \times 1$$
$$= 10 \text{ kg m}$$
$$\text{Resultant moment} = 15 - 10$$

Ans. Resultant moment = 5 kg m clockwise

Exercise 1

1   A capstan bar is 3 metres long. Two men are pushing on the bar, each with a force of 400 Newtons. If one man is placed half-way along the bar and the other at the extreme end of the bar, find the resultant moment about the centre of the capstan.

2   A uniform plank is 6 metres long and is supported at a point under its mid-length. A 10 kg mass is placed on the plank at a distance of 0.5 metres from one end and a 20 kg mass is placed on the plank 2 metres from the other end. Find the resultant moment about the centre of the plank.

3   A uniform plank is 5 metres long and is supported at a point under its mid-length. A 15 kg mass is placed 1 metre from one end and a 10 kg mass is placed 1.2 metres from the other end. Find where a 13 kg mass must be placed on the plank so that the plank will not tilt.

4   A weightless bar 2 metres long is suspended from the ceiling at a point which is 0.5 metres in from one end. Suspended from the same end is a mass of 110 kg. Find the mass which must be suspended from a point 0.3 metres in from the other end of the bar so that the bar will remain horizontal.

5   Three weights are placed on a plank. One of 15 kg mass is placed 0.6 metres in from one end, the next of 12 kg mass is placed 1.5 metres in from the same end and the last of 18 kg mass is placed 3 metres from this end. If the mass of the plank be ignored, find the resultant moment about the end of the plank.

Chapter 2
Centroids and the centre of gravity

The centroid of an area is situated at its geometrical centre. In each of the following figures 'G' represents the centroid, and if each area was suspended from this point it would balance.

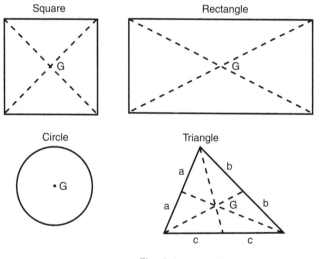

Fig. 2.1

The centre of gravity of a body is the point at which all the mass of the body may be assumed to be concentrated and is the point through which the force of gravity is considered to act vertically downwards, with a force equal to the weight of the body. It is also the point about which the body would balance.

The centre of gravity of a homogeneous body is at its geometrical centre. Thus the centre of gravity of a homogeneous rectangular block is half-way along its length, half-way across its breadth and at half its depth.

Let us now consider the effect on the centre of gravity of a body when the distribution of mass within the body is changed.

Effect of removing or discharging mass

Consider a rectangular plank of homogeneous wood. Its centre of gravity will be at its geometrical center: – i.e., half-way along its length, half-way across its breadth, and at half depth. Let the mass of the plank be W kg and let it be supported by means of a wedge placed under the centre of gravity as shown in Figure 2.2. The plank will balance.

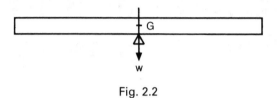

Fig. 2.2

Now let a short length of the plank, of mass w kg, be cut from one end such that its centre of gravity is d metres from the centre of gravity of the plank. The other end, now being of greater mass, will tilt downwards. Figure 2.3(a) shows that by removing the short length of plank a resultant moment of w × d kg m has been created in an anti-clockwise direction about G.

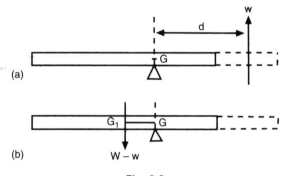

Fig. 2.3

Now consider the new length of plank as shown in Figure 2.3(b). The centre of gravity will have moved to the new half-length indicated by the distance G to G_1. The new mass, (W – w) kg, now produces a tilting moment of (W – w) × GG_1 kg m about G.

Since these are simply two different ways of showing the same effect, the moments must be the same, i.e.

$$(W - w) \times GG_1 = w \times d$$

or

$$GG_1 = \frac{w \times d}{W - w} \text{ metres}$$

From this it may be concluded that when mass is removed from a body, the centre of gravity of the body will move directly away from the centre of gravity of the mass removed, and the distance it moves will be given by the formula:

$$GG_1 = \frac{w \times d}{\text{Final mass}} \text{ metres}$$

where GG_1 is the shift of the centre of gravity of the body, w is the mass removed, and d is the distance between the centre of gravity of the mass removed and the centre of gravity of the body.

Application to ships

In each of the above figures, G represents the centre of gravity of the ship with a mass of w tonnes on board at a distance of d metres from G. G to G_1 represents the shift of the ship's centre of gravity due to discharging the mass.

In Figure 2.4(a), it will be noticed that the mass is vertically below G, and that when discharged G will move vertically upwards to G_1.

In Figure 2.4(b), the mass is vertically above G and the ship's centre of gravity will move directly downwards to G_1.

In Figure 2.4(c), the mass is directly to starboard of G and the ship's centre of gravity will move directly to port from G to G_1.

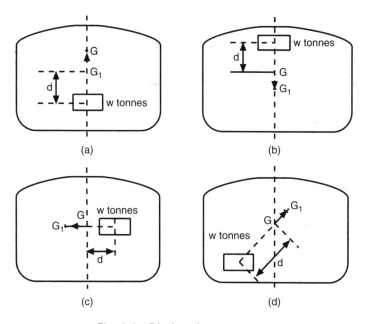

Fig. 2.4   Discharging a mass w.

In Figure 2.4(d), the mass is below and to port of G, and the ship's centre of gravity will move upwards and to starboard.

In each case:

$$GG_1 = \frac{w \times d}{\text{Final displacement}} \text{ metres}$$

Effect of adding or loading mass

Once again consider the plank of homogeneous wood shown in Figure 2.2. Now add a piece of plank of mass w kg at a distance of d metres from G as shown in Figure 2.5(a).

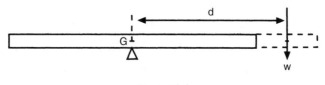

Fig. 2.5(a)

The heavier end of the plank will again tilt downwards. By adding a mass of w kg at a distance of d metres from G a tilting moment of $w \times d$ kg m about G has been created.

Now consider the new plank as shown in Figure 2.5(b). Its centre of gravity will be at its new half-length (G_1), and the new mass, (W + w) kg, will produce a tilting moment of $(W + w) \times GG_1$ kg m about G.

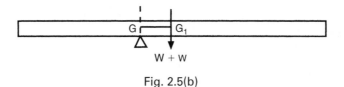

Fig. 2.5(b)

These tilting moments must again be equal, i.e.

$$(W + w) \times GG_1 = w \times d$$

or

$$GG_1 = \frac{w \times d}{W + w} \text{ metres}$$

From the above it may be concluded that when mass is added to a body, the centre of gravity of the body will move directly towards the centre of gravity of the mass added, and the distance which it moves will be given by the formula:

$$GG_1 = \frac{w \times d}{\text{Final mass}} \text{ metres}$$

where GG_1 is the shift of the centre of gravity of the body, w is the mass added, and d is the distance between the centres of gravity.

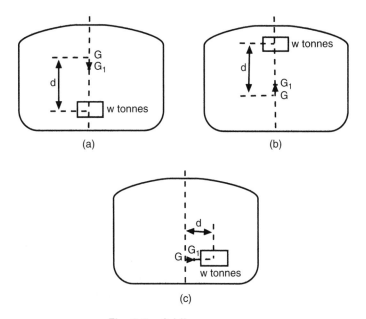

Fig. 2.6   Adding a mass w.

Application to ships

In each of the above figures, G represents the position of the centre of gravity of the ship before the mass of w tonnes has been loaded. After the mass has been loaded, G will move directly towards the centre of gravity of the added mass (i.e. from G to G_1).

Also, in each case:

$$GG_1 = \frac{w \times d}{\text{Final displacement}} \text{ metres}$$

Effect of shifting weights

In Figure 2.7, G represents the original position of the centre of gravity of a ship with a weight of 'w' tonnes in the starboard side of the lower hold having its centre of gravity in position g_1. If this weight is now discharged the ship's centre of gravity will move from G to G_1 directly away from g_1. When

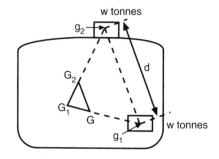

Fig. 2.7   Discharging, adding and moving a mass w.

the same weight is reloaded on deck with its centre of gravity at g_2 the ship's centre of gravity will move from G_1 to G_2.

From this it can be seen that if the weight had been shifted from g_1 to g_2 the ship's centre of gravity would have moved from G to G_2.

It can also be shown that GG_2 is parallel to g_1g_2 and that

$$GG_2 = \frac{w \times d}{W} \text{ metres}$$

where w is the mass of the weight shifted, d is the distance through which it is shifted and W is the ship's displacement.

The centre of gravity of the body will always move parallel to the shift of the centre of gravity of any weight moved within the body.

Effect of suspended weights

The centre of gravity of a body is the point through which the force of gravity may be considered to act vertically downwards. Consider the centre of gravity of a weight suspended from the head of a derrick as shown in Figure 2.8.

It can be seen from Figure 2.8 that whether the ship is upright or inclined in either direction, the point in the ship through which the force of gravity may be considered to act vertically downwards is g_1, the point of suspension. Thus the centre of gravity of a suspended weight is considered to be at the point of suspension.

Conclusions

1. The centre of gravity of a body will move directly *towards* the centre of gravity of any *weight added*.
2. The centre of gravity of a body will move directly *away* from the centre of gravity of any *weight removed*.

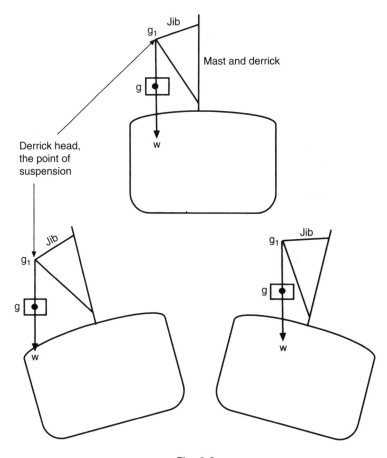

Fig. 2.8

3. The centre of gravity of a body will *move parallel* to the shift of the centre of gravity of any *weight moved* within the body.
4. No matter where the weight 'w' was initially in the ship relative to G, when this weight is moved *downwards* in the ship, then the ship's overall G will also be moved *downwards* to a *lower* position. Consequently, the ship's stability will be *improved*.
5. No matter where the weight 'w' was initially in the ship relative to G, when this weight is moved *upwards* in the ship, then the ship's overall G will also be moved *upwards* to a *higher* position. Consequently, the ship's stability will be *decreased*.
6. The *shift of the centre of gravity* of the body in each case is given by the formula:

$$GG_1 = \frac{w \times d}{W} \text{ metres}$$

where w is the mass of the weight added, removed or shifted, W is the *final* mass of the body, and d is, in 1 and 2, the distance between the centres of gravity, and in 3, the distance through which the weight is shifted.

7. When a weight is *suspended* its centre of gravity is considered to be at the *point of suspension.*

Example 1

A hold is partly filled with a cargo of bulk grain. During the loading, the ship takes a list and a quantity of grain shifts so that the surface of the grain remains parallel to the waterline. Show the effect of this on the ship's centre of gravity.

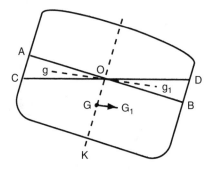

Fig. 2.9

In Figure 2.9, G represents the original position of the ship's centre of gravity when upright. AB represents the level of the surface of the grain when the ship was upright and CD the level when inclined. A wedge of grain AOC with its centre of gravity at g has shifted to ODB with its centre of gravity at g_1. The ship's centre of gravity will shift from G to G_1, such that GG_1 is parallel to gg_1, and the distance

$$GG_1 = \frac{w \times d}{W} \text{ metres}$$

Example 2

A ship is lying starboard side to a quay. A weight is to be discharged from the port side of the lower hold by means of the ship's own derrick. Describe the effect on the position of the ship's centre of gravity during the operation.

Note. When a weight is suspended from a point, the centre of gravity of the weight appears to be at the point of suspension regardless of the distance between the point of suspension and the weight. Thus, as soon as the weight is clear of the deck and is being borne at the derrick head, the centre of gravity of the weight appears to move from its original position to the derrick head. For example, it does not matter whether the weight is 0.6 metres or 6.0 metres above the deck, or whether it is being raised or lowered; its centre of gravity will appear to be at the derrick head.

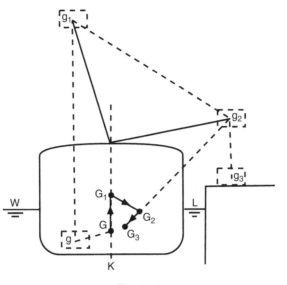

Fig. 2.10

In Figure 2.10, G represents the original position of the ship's centre of gravity, and g represents the centre of gravity of the weight when lying in the lower hold. As soon as the weight is raised clear of the deck, its centre of gravity will appear to move vertically upwards to g_1. This will cause the ship's centre of gravity to move upwards from G to G_1, parallel to gg_1. The centres of gravity will remain at G_1 and g_1 respectively during the whole of the time the weight is being raised. When the derrick is swung over the side, the derrick head will move from g_1 to g_2, and since the weight is suspended from the derrick head, its centre of gravity will also appear to move from g_1 to g_2. This will cause the ship's centre of gravity to move from G_1 to G_2. If the weight is now landed on the quay it is in effect being discharged from the derrick head and the ship's centre of gravity will move from G_2 to G_3 in a direction directly away from g_2. G_3 is therefore the final position of the ship's centre of gravity after discharging the weight.

From this it can be seen that the net effect of discharging the weight is a shift of the ship's centre of gravity from G to G_3, directly away from the centre of gravity of the weight discharged. This would agree with the earlier conclusions which have been reached in Figure 2.4.

Note. The only way in which the position of the centre of gravity of a ship can be altered is by changing the distribution of the weights within the ship, i.e. by *adding removing* or *shifting* weights.

Students find it hard sometimes to accept that the weight, when suspended from the derrick, acts at its point of suspension.

However, it can be proved, by experimenting with ship models or observing full-size ship tests. The final angle of heel when measured verifies that this assumption is indeed correct.

Exercise 2

1 A ship has displacement of 2400 tonnes and KG = 10.8 metres. Find the new KG if a weight of 50 tonnes mass already on board is raised 12 metres vertically.

2 A ship has displacement of 2000 tonnes and KG = 10.5 metres. Find the new KG if a weight of 40 tonnes mass already on board is shifted from the 'tween deck to the lower hold, through a distance of 4.5 metres vertically.

3 A ship of 2000 tonnes displacement has KG = 4.5 metres. A heavy lift of 20 tonnes mass is in the lower hold and has KG = 2 metres. This weight is then raised 0.5 metres clear of the tank top by a derrick whose head is 14 metres above the keel. Find the new KG of the ship.

4 A ship has a displacement of 7000 tonnes and KG = 6 metres. A heavy lift in the lower hold has KG = 3 metres and mass 40 tonnes. Find the new KG when this weight is raised through 1.5 metres vertically and is suspended by a derrick whose head is 17 metres above the keel.

5 Find the shift in the centre of gravity of a ship of 1500 tonnes displacement when a weight of 25 tonnes mass is shifted from the starboard side of the lower hold to the port side on deck through a distance of 15 metres.

Chapter 3
Density and specific gravity

Density is defined as 'mass per unit volume'. For example, the mass density of

$$FW = 1000 \text{ kg per cubic metre or } 1.000 \text{ tonne/m}^3$$

$$SW = 1025 \text{ kg per cubic metre or } 1.025 \text{ tonne/m}^3$$

The specific gravity (SG) or relative density of a substance is defined as the ratio of the weight of the substance to the weight of an equal volume of fresh water.

If a volume of one cubic metre is considered, then the SG or relative density of a substance is the ratio of the density of the substance to the density of fresh water; i.e.

$$\text{SG or relative density of a substance} = \frac{\text{Density of the substance}}{\text{Density of fresh water}}$$

$$\text{The density of FW} = 1000 \text{ kg per cu. m}$$

$$\therefore \text{SG of a substance} = \frac{\text{Density of the substance in kg per cu. m}}{1000}$$

or

$$\text{Density in kg per cu. m} = 1000 \times \text{SG}$$

Example 1

Find the relative density of salt water whose density is 1025 kg per cu. m.

$$\text{Relative density} = \frac{\text{Density of SW in kg per cu. m}}{1000}$$

$$= \frac{1025}{1000}$$

$$\therefore \text{Relative density of salt water} = 1.025$$

Example 2

Find the density of a fuel oil whose relative density is 0.92.

$$\text{Density in kg per cu. m} = 1000 \times \text{SG}$$
$$= 1000 \times 0.92$$
$$\therefore \text{Density} = 920 \text{ kg per cu. m}$$

Example 3

When a double-bottom tank is full of fresh water it holds 120 tonnes. Find how many tonnes of oil of relative density 0.84 it will hold.

$$\text{Relative density} = \frac{\text{Mass of oil}}{\text{Mass of FW}}$$

or

$$\text{Mass of oil} = \text{Mass of FW} \times \text{relative density}$$
$$= 120 \times 0.84 \text{ tonnes}$$
$$\text{Mass of oil} = 100.8 \text{ tonnes}$$

Example 4

A tank measures 20 m × 24 m × 10.5 m and contains oil of relative density 0.84. Find the mass of oil it contains when the ullage is 2.5 m. An ullage is the distance from the surface of the liquid in the tank to the top of the tank. A sounding is the distance from the surface of the liquid to the base of the tank or sounding pad.

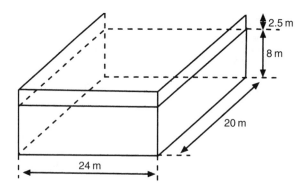

Fig. 3.1

$$\text{Volume of oil} = L \times B \times D$$
$$= 20 \times 24 \times 8 \text{ cu. m}$$
$$\text{Density of oil} = \text{SG} \times 1000$$
$$= 840 \text{ kg per cu. m or } 0.84 \text{ t/m}^3$$
$$\text{Mass of oil} = \text{Volume} \times \text{density}$$
$$= 20 \times 24 \times 8 \times 0.84$$
$$\text{Mass of oil} = 3225.6 \text{ tonnes}$$

Example 5

A tank will hold 153 tonnes when full of fresh water. Find how many tonnes of oil of relative density 0.8 it will hold allowing 2% of the oil loaded for expansion.

$$\text{Mass of freshwater} = 153 \text{ tonnes}$$
$$\therefore \text{Volume of the tank} = 153 \text{ m}^3$$
$$\text{Volume of oil} + 2\% \text{ of volume of oil} = \text{Volume of the tank}$$
$$\text{or } 102\% \text{ of volume of the oil} = 153 \text{ m}^3$$
$$\therefore \text{volume of the oil} = 153 \times \frac{100}{102} \text{ m}^3$$
$$= 150 \text{ m}^3$$
$$\text{Mass of the oil} = \text{Volume} \times \text{Density}$$
$$= 150 \times 0.8 \text{ tonnes}$$

Ans. = 120 tonnes

In Load Lines (2002 Edition), the IMO suggests that weights shall be calculated on the basis of the following values for specific gravities and densities of liquids:

Salt water = 1.025   Fresh water = 1.000   Oil fuel = 0.950
Diesel fuel = 0.900   Lubricating oil = 0.900

In addition to these values, many river authorities use for river water and dock water allowances a range of densities from 1.005 to 1.025 t/m³. This range exists because the density alters with state of tide and also after heavy rainfall.

Exercise 3

1 A tank holds 120 tonnes when full of fresh water. Find how many tonnes of diesel oil of relative density 0.880 it will hold, allowing 2% of the volume of the tank for expansion in the oil.
2 A tank when full will hold 130 tonnes of salt water. Find how many tonnes of oil fuel relative density 0.940 it will hold, allowing 1% of the volume of the tank for expansion.
3 A tank measuring 8 m × 6 m × 7 m is being filled with diesel oil of relative density 0.9. Find how many tonnes of diesel oil in the tank when the ullage is 3 metres.
4 Oil fuel of relative density 0.925 is run into a tank measuring 6 m × 4 m × 8 m until the ullage is 2 metres. Calculate the number of tonnes of oil the tank then contains.
5 A tank will hold 100 tonnes when full of fresh water. Find how many tonnes of oil of relative density 0.880 may be loaded if 2% of the volume of the oil loaded is to be allowed for expansion.
6 A deep tank 10 metres long, 16 metres wide and 6 metres deep has a coaming 4 metres long, 4 metres wide and 25 cm deep. (Depth of tank does not include depth of coaming). How may tonnes of oil, of relative density 0.92, can it hold if a space equal to 3% of the oil loaded is allowed for expansion?

Chapter 4
Laws of flotation

Archimedes' Principle states that when a body is wholly or partially immersed in a fluid it appears to suffer a loss in mass equal to the mass of the fluid it displaces.

The mass density of fresh water is 1000 kg per cu. m. Therefore, when a body is immersed in fresh water it will appear to suffer a loss in mass of 1000 kg for every 1 cu. m of water it displaces.

When a box measuring 1 cu. m and of 4000 kg mass is immersed in fresh water it will appear to suffer a loss in mass of 1000 kg. If suspended from a spring balance the balance would indicate a mass of 3000 kg.

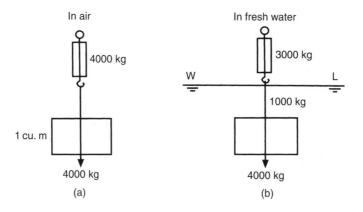

Fig. 4.1

Since the actual mass of the box is not changed, there must be a force acting vertically upwards to create the apparent loss of mass of 1000 kg. This force is called the *force of buoyancy,* and is considered to act vertically upwards through a point called the *centre of buoyancy*. The centre of buoyancy is the centre of gravity of the underwater volume.

Now consider the box shown in Figure 4.2(a) which also has a mass of 4000 kg, but has a volume of 8 cu. m. If totally immersed in fresh water it will displace 8 cu. m of water, and since 8 cu. m of fresh water has a mass of 8000 kg, there will be an upthrust or force of buoyancy causing an apparent loss of mass of 8000 kg. The resultant apparent loss of mass is 4000 kg. When released, the box will rise until a state of equilibrium is reached, i.e. when the buoyancy is equal to the mass of the box. To make the buoyancy produce a loss of mass of 4000 kg the box must be displacing 4 cu. m of water. This will occur when the box is floating with half its volume immersed, and the resultant force then acting on the box will be zero. This is shown in Figure 4.2(c).

Now consider the box to be floating in fresh water with half its volume immersed as shown in Figure 4.2(c). If a mass of 1000 kg be loaded on deck as shown in Figure 4.3(a) the new mass of the body will be 5000 kg, and since this exceeds the buoyancy by 1000 kg, it will move downwards.

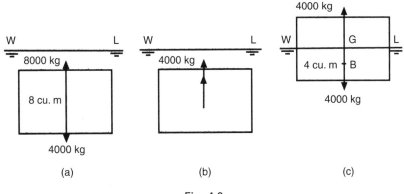

Fig. 4.2

The downwards motion will continue until buoyancy is equal to the mass of the body. This will occur when the box is displacing 5 cu. m of water and the buoyancy is 5000 kg, as shown in Figure 4.3(b).

The conclusion which may be reached from the above is that for a body to float at rest in still water, it must be displacing its own weight of water and the centre of gravity must be vertically above or below the centre of buoyancy.

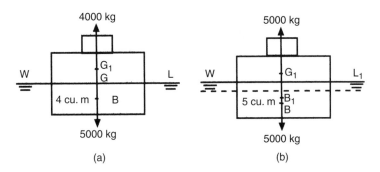

Fig. 4.3

The variable immersion hydrometer

The variable immersion hydrometer is an instrument, based on the Law of Archimedes, which is used to determine the density of liquids. The type of hydrometer used to find the density of the water in which a ship floats is usually made of a non-corrosive material and consists of a weighted bulb with a narrow rectangular stem which carries a scale for measuring densities between 1000 and 1025 kilograms per cubic metre, i.e. 1.000 and 1.025 t/m³.

The position of the marks on the stem are found as follows. First let the hydrometer, shown in Figure 4.4, float upright in fresh water at the mark X. Take the hydrometer out of the water and weigh it. Let the mass be M_x kilograms. Now replace the hydrometer in fresh water and add lead shot in the bulb until it floats with the mark Y, at the upper end of the stem, in the waterline. Weigh the hydrometer again and let its mass now be M_y kilograms.

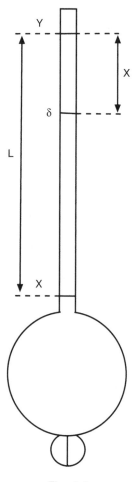

Fig. 4.4

The mass of water displaced by the stem between X and Y is therefore equal to $M_y - M_x$ kilograms. Since 1000 kilograms of fresh water occupy one cubic metre, the volume of the stem between X and Y is equal to $\dfrac{M_y - M_x}{1000}$ cu. m.

Let L represent the length of the stem between X and Y, and let 'a' represent the cross-sectional area of the stem:

$$a = \frac{\text{Volume}}{\text{Length}}$$

$$= \frac{M_y - M_x}{1000\ L}\ \text{sq m}$$

Now let the hydrometer float in water of density $\delta\ kg/m^3$ with the water-line 'x' metres below Y:

$$\text{Volume of water displaced} = \frac{M_y}{1000} - x\,a$$

$$= \frac{M_y}{1000} - x\left(\frac{M_y - M_x}{1000\ L}\right) \text{--------- (I)}$$

But

$$\text{Volume of water displaced} = \frac{\text{Mass of water displaced}}{\text{Density of water displaced}}$$

$$= \frac{M_y}{1000\ \delta} \text{-------------------------- (II)}$$

$$\text{Equate (I) and (II)} \quad \therefore\ \frac{M_y}{1000\ \delta} = \frac{M_y}{1000} - x\left(\frac{M_y - M_x}{1000\ L}\right)$$

or

$$\delta = \frac{M_y}{M_y - x\left(\dfrac{M_y - M_x}{L}\right)}$$

In this equation, M_y, M_x and L are known constants whilst δ and x are variables. Therefore, to mark the scale it is now only necessary to select various values of δ and to calculate the corresponding values of x.

Tonnes per centimetre immersion (TPC)

The TPC for any draft is the mass which must be loaded or discharged to change a ship's mean draft in salt water by one centimetre, where:

$$\text{TPC} = \frac{\text{Water-plane area}}{100} \times \text{Density of water}$$

$$\therefore \text{TPC} = \frac{\text{WPA}}{100} \times \rho$$

WPA is in m^2.
ρ is in t/m^3.

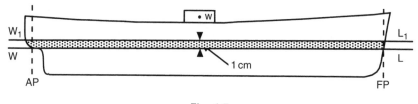

Fig. 4.5

Consider a ship floating in *salt water* at the waterline WL as shown in Figure 4.5. Let 'A' be the area of the water-plane in square metres.

Now let a mass of 'w' tonnes be loaded so that the mean draft is increased by one centimetre. The ship then floats at the waterline $W_1 L_1$. Since the draft has been increased by one centimetre, the mass loaded is equal to the TPC for this draft. Also, since an extra mass of water equal to the mass loaded must be displaced, then the mass of water in the layer between WL and $W_1 L_1$ is also equal to the TPC:

$$\text{Mass} = \text{Volume} \times \text{Density}$$

$$= A \times \frac{1}{100} \times \frac{1025}{1000} \text{ tonnes} = \frac{1.025\,A}{100} \text{ tonnes}$$

$$\therefore \text{TPC}_{sw} = \frac{1.025\,A}{100} = \frac{\text{WPA}}{97.56}. \quad \text{Also, TPC}_{fw} = \frac{\text{WPA}}{100}$$

TPC in dock water

Note. When a ship is floating in dock water of a relative density other than 1.025 the weight to be loaded or discharged to change the mean draft by 1 centimetre (TPC_{dw}) may be found from the TPC in salt water (TPC_{sw}) by simple proportion as follows:

$$\frac{\text{TPC}_{dw}}{\text{TPC}_{sw}} = \frac{\text{Relative density of dock water (RD}_{dw})}{\text{Relative density of salt water (RD}_{sw})}$$

or

$$TPC_{dw} = \frac{RD_{dw}}{1.025} \times TPC_{sw}$$

Reserve buoyancy

It has already been shown that a floating vessel must displace its own weight of water. Therefore, it is the submerged portion of a floating vessel which provides the buoyancy. The volume of the enclosed spaces above the waterline are not providing buoyancy but are being held in reserve. If extra weights are loaded to increase the displacement, these spaces above the waterline are there to provide the extra buoyancy required. Thus, *reserve buoyancy* may be defined as the volume of the enclosed spaces above the waterline. It may be expressed as a volume or as a percentage of the total volume of the vessel.

Example 1

A box-shaped vessel 105 m long, 30 m beam, and 20 m deep, is floating upright in fresh water. If the displacement is 19 500 tonnes, find the volume of reserve buoyancy.

$$\text{Volume of water displaced} = \frac{\text{Mass}}{\text{Density}} = 19\,500 \text{ cu. m}$$

$$\text{Volume of vessel} = 105 \times 30 \times 20 \text{ cu. m}$$
$$= 63\,000 \text{ cu. m}$$
$$\text{Reserve buoyancy} = \text{Volume of vessel} - \text{Volume of water displaced}$$

Ans. Reserve buoyancy = 43 500 cu. m

Example 2

A box-shaped barge 16 m × 6 m × 5 m is floating alongside a ship in fresh water at a mean draft of 3.5 m. The barge is to be lifted out of the water and loaded on to the ship with a heavy-lift derrick. Find the load in tonnes borne by the purchase when the draft of the barge has been reduced to 2 metres.

Note. By Archimedes' Principle the barge suffers a loss in mass equal to the mass of water displaced. The mass borne by the purchase will be the difference between the actual mass of the barge and the mass of water displaced at any draft, or the difference between the mass of water originally displaced by the barge and the new mass of water displaced.

$$\text{Mass of the barge} = \text{Original mass of water displaced}$$
$$= \text{Volume} \times \text{Density}$$
$$= 16 \times 6 \times 3.5 \times 1 \text{ tonnes}$$
$$\text{Mass of water displace at 2 m draft} = 16 \times 6 \times 2 \times 1 \text{ tonnes}$$
$$\therefore \text{ Load borne by the purchase} = 16 \times 6 \times 1 \times (3.5 - 2) \text{ tonnes}$$

Ans. = 144 tonnes

Example 3

A cylindrical drum 1.5 m long and 60 cm in diameter has mass 20 kg when empty. Find its draft in water of density 1024 kg per cu. m if it contains 200 litres of paraffin of relative density 0.6, and is floating with its axis perpendicular to the waterline (Figure 4.6).

Note. The drum must displace a mass of water equal to the mass of the drum plus the mass of the paraffin.

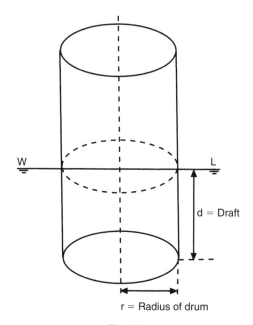

W ⎯ L

d = Draft

r = Radius of drum

Fig. 4.6

$$\text{Density of the paraffin} = \text{SG} \times 1000 \text{ kg per cu. m}$$
$$= 600 \text{ kg per cu. m}$$
$$\text{Mass of the paraffin} = \text{Volume} \times \text{Density} = 0.2 \times 600 \text{ kg}$$
$$= 120 \text{ kg}$$
$$\text{Mass of the drum} = 20 \text{ kg}$$
$$\text{Total mass} = 140 \text{ kg}$$

Therefore the drum must displace 140 kg of water.

$$\text{Volume of water displaced} = \frac{\text{Mass}}{\text{Density}} = \frac{140}{1024} \text{ cu. m}$$
$$\text{Volume of water displaced} = 0.137 \text{ cu. m}$$

Let d = draft, and r = radius of the drum, where $r = \dfrac{60}{2} = 30 \text{ cm} = 0.3 \text{ m}$.
Volume of water displaced $(V) = \pi r^2 d$

or

$$d = \frac{V}{\pi r^2}$$

$$= \frac{0.137}{\frac{22}{7} \times 0.3 \times 0.3} \text{ m}$$

$$= 0.484 \text{ m}$$

Ans. Draft = 0.484 m

Homogeneous logs of rectangular section

The draft at which a rectangular homogeneous log will float may be found as follows:

$$\text{Mass of log} = \text{Volume} \times \text{Density}$$
$$= L \times B \times D \times SG \text{ of log} \times 1000 \text{ kg}$$
$$\text{Mass of water displaced} = \text{Volume} \times \text{Density}$$
$$= L \times B \times d \times SG \text{ of water} \times 1000 \text{ kg}$$
$$\text{But Mass of water displaced} = \text{Mass of log}$$
$$\therefore L \times B \times d \times SG \text{ of water} \times 1000 = L \times B \times D \times SG \text{ of log} \times 1000$$

or

$$d \times SG \text{ of water} = D \times SG \text{ of log}$$

$$\frac{\text{Draft}}{\text{Depth}} = \frac{SG \text{ of log}}{SG \text{ of water}} \quad \text{or} \quad \frac{\text{Relative density of log}}{\text{Relative density of water}}$$

Example 1

Find the distance between the centres of gravity and buoyancy of a rectangular log 1.2 m wide, 0.6 m deep, and of relative density 0.8 when floating in fresh water with two of its sides parallel to the waterline.

If BM is equal to $\dfrac{b^2}{12\,d}$ determine if this log will float with two of its sides parallel to the waterline.

Note. The centre of gravity of a homogeneous log is at its geometrical centre (see Figure 4.7).

$$\frac{\text{Draft}}{\text{Depth}} = \frac{\text{Relative density of log}}{\text{Relative density of water}}$$

$$\text{Draft} = \frac{0.6 \times 0.8}{1}$$

$$\left.\begin{array}{l} \text{Draft} = 0.48 \text{ m} \\ \text{KB} = 0.24 \text{ m} \\ \text{KG} = 0.30 \text{ m} \end{array}\right\} \text{ see Figure 4.8}$$

Ans. BG = 0.06 m

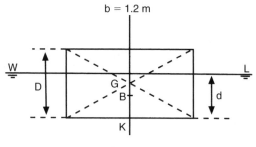

Fig. 4.7

$$BM = \frac{b^2}{12 \times d} = \frac{1.2^2}{12 \times 0.48} = 0.25 \text{ m}$$

$$KB \text{ as above} = +\underline{0.24} \text{ m}$$

$$KM = KB + BM = +0.49 \text{ m}$$

$$-KG = -\underline{0.30} \text{ m}$$

$$GM = 0.19 \text{ m}$$

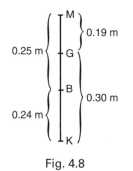

Fig. 4.8

Conclusion

Because G is below M, this homogeneous log is in stable equilibrium. Consequently, it *will* float with two of its sides parallel to the waterline.

Exercise 4

1   A drum of mass 14 kg when empty, is 75 cm long, and 60 cm in diameter. Find its draft in salt water if it contains 200 litres of paraffin of relative density 0.63.

2   A cube of wood of relative density 0.81 has sides 30 cm long. If a mass of 2 kg is placed on the top of the cube with its centre of gravity vertically over that of the cube, find the draft in salt water.

3   A rectangular tank (3 m × 1.2 m × 0.6 m) has no lid and is floating in fresh water at a draft of 15 cm. Calculate the minimum amount of fresh water which must be poured into the tank to sink it.

4   A cylindrical salvage buoy is 5 metres long, 2.4 metres in diameter, and floats on an even keel in salt water with its axis in the water-plane. Find the upthrust which this buoy will produce when fully immersed.

5   A homogeneous log of rectangular cross-section is 30 cm wide and 25 cm deep. The log floats at a draft of 17 cm. Find the reserve buoyancy and the distance between the centre of buoyancy and the centre of gravity.

6   A homogeneous log of rectangular cross-section is 5 m. long, 60 cm wide, 40 cm deep, and floats in fresh water at a draft of 30 cm. Find the mass of the log and its relative density.

7   A homogeneous log is 3 m long, 60 cm wide, 60 cm deep, and has relative density 0.9. Find the distance between the centres of buoyancy and gravity when the log is floating in fresh water.

8   A log of square section is 5 m × 1 m × 1 m. The relative density of the log is 0.51 and it floats half submerged in dock water. Find the relative density of the dock water.

9   A box-shaped vessel 20 m × 6 m × 2.5 m floats at a draft of 1.5 m in water of density 1013 kg per cu. m. Find the displacement in tonnes, and the height of the centre of buoyancy above the keel.

10   An empty cylindrical drum 1 metre long and 0.6 m. in diameter has mass 20 kg. Find the mass which must be placed in it so that it will float with half of its volume immersed in (a) salt water and (b) fresh water.

11   A lifeboat, when fully laden, displaces 7.2 tonnes. Its dimensions are 7.5 m × 2.5 m × 1 m, and its block coefficient 0.6. Find the percentage of its volume under water when floating in fresh water.

12   A homogeneous log of relative density 0.81 is 3 metres long, 0.5 metres square cross-section, and is floating in fresh water. Find the displacement of the log, and the distance between the centres of gravity and buoyancy.

13   A box-shaped barge 55 m × 10 m × 6 m. is floating in fresh water on an even keel at 1.5 m draft. If 1800 tonnes of cargo is now loaded, find the difference in the height of the centre of buoyancy above the keel.

14   A box-shaped barge 75 m × 6 m × 4 m displaces 180 tonnes when light. If 360 tonnes of iron are loaded while the barge is floating in fresh water, find her final draft and reserve buoyancy.

15   A drum 60 cm in diameter and 1 metre long has mass 30 kg when empty. If this drum is filled with oil of relative density 0.8, and is floating in fresh water, find the percentage reserve buoyancy.

Chapter 5
Group weights, water draft, air draft and density

Group weights in a ship

The first estimate that the Naval Architect makes for a new ship is to estimate the lightweight.

Lightweight: This is the weight of the ship itself when completely *empty*, with boilers topped up to working level. It is made up of steel weight, wood and outfit weight, and the machinery weight. This lightweight is evaluated by conducting an inclining experiment normally just prior to delivery of the new vessel. Over the years, this value will change (see Chapter 33).

Deadweight: This is the weight that a ship *carries*. It can be made up of oil fuel, fresh water, stores, lubricating oil, water ballast, crew and effects, cargo and passengers. This deadweight will vary, depending on how much the ship is loaded between light ballast and fully-loaded departure conditions.

Displacement: This is the weight of the volume of water that the ship displaces:

$$\text{Displacement} = \text{Lightweight} + \text{Deadweight}$$

Hence

$$W = Lwt + Dwt$$

Water draft: This is the vertical distance from the waterline down to the keel. If it is to the top of the keel, then it is draft moulded. If it is to the bottom of the keel, then it is draft extreme. Draft moulded is used mainly by Naval Architects. Draft extreme is used mainly by masters, mates, port authorities and dry-dock personnel.

Air draft: This is the quoted vertical distance from the waterline to the highest point on the ship when at *zero forward speed*. It indicates the ability of a ship to pass under a bridge spanning a waterway that forms part of the intended route.

For the Panama Canal, this air draft is to be no greater than 57.91 m. For the St Lawrence Seaway the maximum air draft is to be 35.5 m.

In order to go beneath bridges over rivers, some vessels have telescopic masts. Others have hinged masts to lessen the chances of contact with the bridge under which they are sailing.

Occasionally, the master or mate needs to calculate the maximum cargo to discharge or the minimum ballast to load to safely pass under a bridge. This will involve moment of weight estimates relating to the final end drafts, as detailed in Chapter 16.

What must be remembered is that the vessel is at a forward speed. Therefore, allowances have to be made for the squat components of mean bodily sinkage and trim ratio forward and aft (see Chapter 37).

Effect of change of density when the displacement is constant

When a ship moves from water of one density to water of another density, without there being a change in her mass, the draft will change. This will happen because the ship must displace the same mass of water in each case. Since the density of the water has changed, the volume of water displaced must also change. This can be seen from the formula:

$$\text{Mass} = \text{Volume} \times \text{Density}$$

If the density of the water increases, then the volume of water displaced must decrease to keep the mass of water displaced constant, and vice versa.

The effect on box-shaped vessels

$$\text{New mass of water displaced} = \text{Old mass of water displaced}$$
$$\therefore \text{New volume} \times \text{New density} = \text{Old volume} \times \text{Old density}$$
$$\frac{\text{New volume}}{\text{Old volume}} = \frac{\text{Old density}}{\text{New density}}$$
$$\text{But volume} = L \times B \times \text{Draft}$$
$$\therefore \frac{L \times B \times \text{New draft}}{L \times B \times \text{Old draft}} = \frac{\text{Old density}}{\text{New density}}$$

or

$$\frac{\text{New draft}}{\text{Old draft}} = \frac{\text{Old density}}{\text{New density}}$$

Example 1

A box-shaped vessel floats at a mean draft of 2.1 metres, in dock water of density 1020 kg per cu. m. Find the mean draft for the same mass displacement in salt water of density 1025 kg per cubic metre.

$$\frac{\text{New draft}}{\text{Old draft}} = \frac{\text{Old density}}{\text{New density}}$$

$$\text{New draft} = \frac{\text{Old density}}{\text{New density}} \times \text{Old draft}$$

$$= \frac{1020}{1025} \times 2.1\,\text{m}$$

$$= 2.09\,\text{m}$$

Ans. <u>New draft = 2.09 m</u>

Example 2

A box-shaped vessel floats upright on an even keel as shown in fresh water of density 1000 kg per cu. m, and the centre of buoyancy is 0.50 m above the keel. Find the height of the centre of buoyancy above the keel when the vessel is floating in salt water of density 1025 kg per cubic metre.

Note. The centre of buoyancy is the geometric centre of the underwater volume and for a box-shaped vessel must be at half draft, i.e. KB = $\frac{1}{2}$ draft.

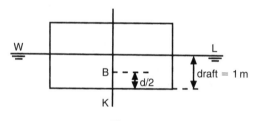

Fig. 5.1

In Fresh Water

$$\text{KB} = 0.5 \text{ m, and since KB} = \tfrac{1}{2} \text{ draft, then draft} = 1\,\text{m}$$

In Salt Water

$$\frac{\text{New draft}}{\text{Old draft}} = \frac{\text{Old density}}{\text{New density}}$$

$$\text{New draft} = \text{Old draft} \times \frac{\text{Old density}}{\text{New density}}$$

$$= 1 \times \frac{1000}{1025}$$

$$\text{New draft} = 0.976\,\text{m}$$

$$\text{New KB} = \tfrac{1}{2} \text{ new draft}$$

Ans. <u>New KB = 0.488 m, say 0.49 m.</u>

The effect on ship-shaped vessels

It has already been shown that when the density of the water in which a vessel floats is changed the draft will change, but the mass of water in kg or tonnes displaced will be unchanged; i.e.

$$\text{New displacement} = \text{Old displacement}$$

or

$$\text{New volume} \times \text{New density} = \text{Old volume} \times \text{Old density}$$

$$\therefore \frac{\text{New volume}}{\text{Old volume}} = \frac{\text{Old density}}{\text{New density}}$$

With ship shapes this formula should not be simplified further as it was in the case of a box shape because the underwater volume is not rectangular. To find the change in draft of a ship shape due to change of density a quantity known as the 'Fresh Water Allowance' must be known.

The *Fresh Water Allowance* is the number of millimetres by which the mean draft changes when a ship passes from salt water to fresh water, or vice versa, whilst floating at the loaded draft. It is found by the formula:

$$\text{FWA (in mm)} = \frac{\text{Displacement (in tonnes)}}{4 \times \text{TPC}}$$

The proof of this formula is as follows:

$$\text{To show that FWA (in mm)} = \frac{\text{Displacement (in tonnes)}}{4 \times \text{TPC}}$$

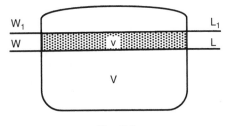

Fig. 5.2

Consider the ship shown in Figure 5.2 to be floating at the load Summer draft in salt water at the waterline, WL. Let V be the volume of salt water displaced at this draft.

Now let W_1L_1 be the waterline for the ship when displacing the same mass of fresh water. Also, let 'v' be the extra volume of water displaced in fresh water.

The total volume of fresh water displaced is then V + v.

$$\text{Mass} = \text{Volume} \times \text{Density}$$
$$\therefore \text{ Mass of SW displaced} = 1025\,V$$
$$\text{and mass of FW displaced} = 1000\,(V + v)$$
$$\text{but mass of FW displaced} = \text{Mass of SW displaced}$$
$$\therefore \quad 1000\,(V + v) = 1025\,V$$
$$1000\,V + 1000\,v = 1025\,V$$
$$1000\,v = 25\,V$$
$$v = V/40$$

Now let w be the mass of salt water in volume v, in tonnes and let W be the mass of salt water in volume V, in tonnes,

$$\therefore \ w = W/40$$
$$\text{but } w = \frac{FWA}{10} \times TPC$$
$$\frac{FWA}{10} \times TPC = W/40$$

or

$$FWA = \frac{W}{4 \times TPC} \ \text{mm}$$

where

W = Loaded salt water displacement in tonnes

Figure 5.3 shows a ship's load line marks. The centre of the disc is at a distance below the deck line equal to the ship's statutory freeboard. Then 540 mm forward of the disc is a vertical line 25 mm thick, with horizontal lines measuring 230×25 mm on each side of it. The upper edge of the one marked 'S' is in line with the horizontal line through the disc and indicates the draft to which the ship may be loaded when floating in salt water in a Summer Zone. Above this line and pointing aft is another line marked 'F', the upper edge of which indicates the draft to which the ship may be loaded when floating in fresh water in a Summer Zone. If loaded to this draft in fresh water the ship will automatically rise to 'S' when she passes into salt water. The perpendicular distance in millimetres between the upper edges of these two lines is therefore the ship's Fresh Water Allowance.

When the ship is loading in dock water which is of a density between these two limits 'S' may be submerged such a distance that she will automatically rise to 'S' when the open sea and salt water is reached. The

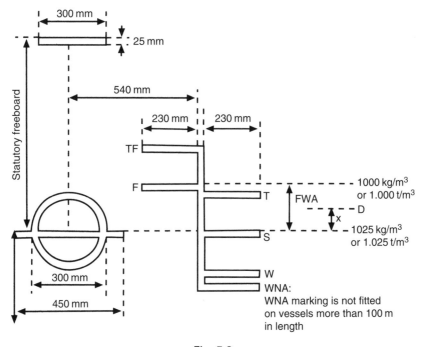

Fig. 5.3

distance by which 'S' can be submerged, called the *Dock Water Allowance*, is found in practice by simple proportion as follows:

$$\text{Let } x = \text{The Dock Water Allowance}$$
$$\text{Let } \rho_{DW} = \text{Density of the dock water}$$

Then

$$\frac{x \text{ mm}}{\text{FWA mm}} = \frac{1025 - \rho_{DW}}{1025 - 1000}$$

or

$$\text{Dock Water Allowance} = \frac{\text{FWA}(1025 - \rho_{DW})}{25}$$

Example 3

A ship is loading in dock water of density 1010 kg per cu. m. FWA = 150 mm. Find the change in draft on entering salt water.

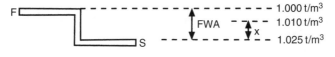

Fig. 5.4

$$\text{Let } x = \text{The change in draft in millimetres}$$

$$\text{Then } \frac{x}{\text{FWA}} = \frac{1025 - 1010}{25}$$

$$x = 150 \times \frac{15}{25}$$

$$x = 90 \, \text{mm}$$

Ans. Draft will decrease by 90 mm, i.e. 9 cm.

Example 4

A ship is loading in a Summer Zone in dock water of density 1005 kg per cu. m. FWA = 62.5 mm, TPC = 15 tonnes. The lower edge of the Summer load line is in the waterline to port and is 5 cm above the waterline to starboard. Find how much more cargo may be loaded if the ship is to be at the correct load draft in salt water.

Note. This ship is obviously listed to port and if brought upright the lower edge of the 'S' load line on each side would be 25 mm above the waterline. Also, it is the upper edge of the line which indicates the 'S' load draft and, since the line is 25 mm thick, the ship's draft must be increased by 50 mm to bring her to the 'S' load line in dock water. In addition 'S' may be submerged by x mm.

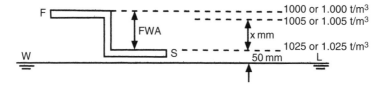

Fig. 5.5

$$\frac{x}{\text{FWA}} = \frac{1025 - \rho_{DW}}{25}$$

$$x = 62.5 \times \frac{20}{25}$$

$$x = 50 \, \text{mm}$$

$$\therefore \text{ Total increase in draft required} = 100 \text{ mm or } 10 \text{ cm}$$

and

$$\text{Cargo to load} = \text{Increase in draft} \times \text{TPC}$$
$$= 10 \times 15$$

Ans. Cargo to load = 150 tonnes

Effect of density on displacement when the draft is constant

Should the density of the water in which a ship floats be changed without the ship altering her draft, then the mass of water displaced must have changed. The change in the mass of water displaced may have been brought about by bunkers and stores being loaded or consumed during a sea passage, or by cargo being loaded or discharged. In all cases:

New volume of water displaced = Old volume of water displaced

or

$$\frac{\text{New displacement}}{\text{New density}} = \frac{\text{Old displacement}}{\text{Old density}}$$

or

$$\frac{\text{New displacement}}{\text{Old displacement}} = \frac{\text{New density}}{\text{Old density}}$$

Example 1

A ship displaces 7000 tonnes whilst floating in fresh water. Find the displacement of the ship when floating at the same draft in water of density 1015 kg per cubic metre, i.e. 1.015 t/m^3.

$$\frac{\text{New displacement}}{\text{Old displacement}} = \frac{\text{New density}}{\text{Old density}}$$

$$\text{New displacement} = \text{Old displacement} \times \frac{\text{New density}}{\text{Old density}}$$

$$= 7000 \times \frac{1015}{1000}$$

Ans. New displacement = 7105 tonnes

Example 2

A ship of 6400 tonnes displacement is floating in salt water. The ship has to proceed to a berth where the density of the water is 1008 kg per cu. m. Find how much cargo must be discharged if she is to remain at the salt water draft.

$$\frac{\text{New displacement}}{\text{Old displacement}} = \frac{\text{New density}}{\text{Old density}}$$

or

$$\text{New displacement} = \text{Old displacement} \times \frac{\text{New density}}{\text{Old density}}$$

$$= 6400 \times \frac{1008}{1025}$$

New displacement = 6293.9 tonnes
Old displacement = 6400.0 tonnes

Ans. Cargo to discharge = 106.1 tonnes

Example 3

A ship 120 m × 17 m × 10 m has a block coefficient 0.800 and is floating at the load Summer draft of 7.2 metres in fresh water. Find how much more cargo can be loaded to remain at the same draft in salt water.

$$\text{Old displacement} = L \times B \times \text{Draft} \times C_b \times \text{Density}$$
$$= 120 \times 17 \times 7.2 \times 0.800 \times 1000 \text{ tonnes}$$
$$\text{Old displacement} = 11\,750 \text{ tonnes}$$
$$\frac{\text{New displacement}}{\text{Old displacement}} = \frac{\text{New density}}{\text{Old density}}$$
$$\text{New displacement} = \text{Old displacement} \times \frac{\text{New density}}{\text{Old density}}$$
$$= 11\,750 \times \frac{1025}{1000}$$
$$\text{New displacement} = 12\,044 \text{ tonnes}$$
$$\text{Old displacement} = 11\,750 \text{ tonnes}$$

Ans. Cargo to load = 294 tonnes

Note. This problem should not be attempted as one involving TPC and FWA.

Exercise 5

Density and draft

1   A ship displaces 7500 cu. m. of water of density 1000 kg per cu. m. Find the displacement in tonnes when the ship is floating at the same draft in water of density 1015 kg per cu. m.

2   When floating in fresh water at a draft of 6.5 m a ship displaces 4288 tonnes. Find the displacement when the ship is floating at the same draft in water of density 1015 kg per cu. m.

3   A box-shaped vessel 24 m × 6 m × 3 m displaces 150 tonnes of water. Find the draft when the vessel is floating in salt water.

4   A box-shaped vessel draws 7.5 m in dock water of density 1006 kg per cu. m. Find the draft in salt water of density 1025 kg per cu. m.

5   The KB of a rectangular block which is floating in fresh water is 50 cm. Find the KB in salt water.

6   A ship is lying at the mouth of a river in water of density 1024 kg per cu. m and the displacement is 12 000 tonnes. The ship is to proceed up river and to berth in dock water of density 1008 kg per cu. m with the same draft as at present. Find how much cargo must be discharged.

7   A ship arrives at the mouth of a river in water of density 1016 kg per cu. m with a freeboard of 'S' m. She then discharges 150 tonnes of cargo, and proceeds up river to a second port, consuming 14 tonnes of bunkers. When she arrives at the second port the freeboard is again 'S' m, the density of the

water being 1004 kg per cu. m. Find the ship's displacement on arrival at the second port.

8 A ship loads in fresh water to her salt water marks and proceeds along a river to a second port consuming 20 tonnes of bunkers. At the second port, where the density is 1016 kg per cu. m, after 120 tonnes of cargo have been loaded, the ship is again at the load salt water marks. Find the ship's load displacement in salt water.

The TPC and FWA, etc.

9 A ship's draft is 6.40 metres forward and 6.60 metres aft. FWA = 180 mm. Density of the dock water is 1010 kg per cu. m. If the load mean draft in salt water is 6.7 metres, find the final drafts F and A in dock water if this ship is to be loaded down to her marks and trimmed 0.15 metres by the stern. (Centre of flotation is amidships.)

10 A ship floating in dock water of density 1005 kg per cu. m has the lower edge of her Summer load line in the waterline to starboard and 50 mm above the waterline to port. FWA = 175 mm and TPC = 12 tonnes. Find the amount of cargo which can yet be loaded in order to bring the ship to the load draft in salt water.

11 A ship is floating at 8 metres mean draft in dock water of relative density 1.01. TPC = 15 tonnes and FWA = 150 mm. The maximum permissible draft in salt water is 8.1 m. Find the amount of cargo yet to load.

12 A ship's light displacement is 3450 tonnes and she has on board 800 tonnes of bunkers. She loads 7250 tonnes of cargo, 250 tonnes of bunkers and 125 tonnes of fresh water. The ship is then found to be 75 mm from the load draft. TPC = 12 tonnes. Find the ship's deadweight and load displacement.

13 A ship has a load displacement of 5400 tonnes, TPC = 30 tonnes. If she loads to the Summer load line in dock water of density 1010 kg per cu. m, find the change in draft on entering salt water of density 1025 kg per cu. m.

14 A ship's FWA is 160 mm, and she is floating in dock water of density 1012 kg per cu. m. Find the change in draft when she passes from dock water to salt water.

Chapter 6
Transverse statical stability

Recapitulation

1. The centre of gravity of a body 'G' is the point through which the force of gravity is considered to act vertically downwards with a force equal to the weight of the body. KG is VCG of the ship.
2. The centre of buoyancy 'B' is the point through which the force of buoyancy is considered to act vertically upwards with a force equal to the weight of water displaced. It is the centre of gravity of the underwater volume. KB is VCB of the ship.
3. To float at rest in still water, a vessel must displace her own weight of water, and the centre of gravity must be in the same vertical line as the centre of buoyancy.
4. KM = KB + BM. Also KM = KG + GM.

Definitions

1. *Heel*. A ship is said to be heeled when she is inclined by an external force. For example, when the ship is inclined by the action of the waves or wind.
2. *List*. A ship is said to be listed when she is inclined by forces within the ship. For example, when the ship is inclined by shifting a weight transversely within the ship. This is a fixed angle of heel.

The metacentre

Consider a ship floating upright in still water as shown in Figure 6.1(a). The centres of gravity and buoyancy are at G and B, respectively. Figure 6.1(c) shows the righting couple. GZ is the righting lever.

Now let the ship be inclined by an external force to a small angle (θ) as shown in Figure 6.1(b). Since there has been no change in the distribution of weights the centre of gravity will remain at G and the weight of the ship (W) can be considered to act vertically downwards through this point.

When heeled, the wedge of buoyancy WOW_1 is brought out of the water and an equal wedge LOL_1 becomes immersed. In this way a wedge of buoyancy having its centre of gravity at g is transferred to a position with its centre of gravity at g_1. The centre of buoyancy, being the centre of gravity of the underwater volume, must shift from B to the new position B_1, such that BB_1 is parallel to gg_1, and $BB_1 = \dfrac{v \times gg_1}{V}$ where v is the volume of the transferred wedge, and V is the ship's volume of displacement.

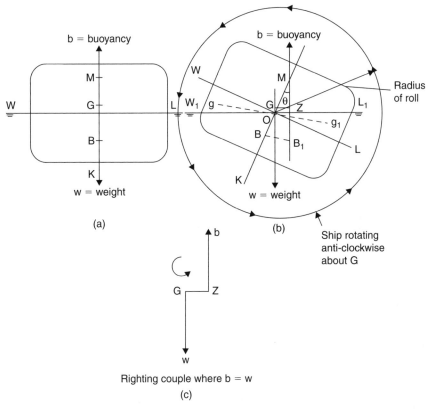

(a)

(b)

Ship rotating
anti-clockwise
about G

Righting couple where b = w

(c)

Fig. 6.1   Stable equilibrium.

The verticals through the centres of buoyancy at two consecutive angles of heel intersect at a point called the *metacentre*. For angles of heel up to about 15° the vertical through the centre of buoyancy may be considered to cut the centre line at a fixed point called the initial metacentre (M in Figure 6.1(b)). The height of the initial metacentre above the keel (KM) depends upon a ship's underwater form. Figure 6.2 shows a typical curve of KM's for a ship plotted against draft.

The vertical distance between G and M is referred to as the *metacentric height*. If G is below M the ship is said to have positive metacentric height, and if G is above M the metacentric height is said to be negative.

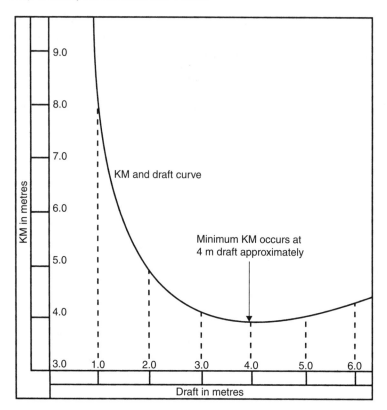

Fig. 6.2

Equilibrium

Stable equilibrium

A ship is said to be in stable equilibrium if, when inclined, she tends to return to the initial position. For this to occur the centre of gravity must be below the metacentre, that is, the ship must have positive initial metacentric height. Figure 6.1(a) shows a ship in the upright position having a positive GM. Figure 6.1(b) shows the same ship inclined to a small angle. The position of G remains unaffected by the heel and the force of gravity is considered to act vertically downwards through this point. The centre of buoyancy moves out to the low side from B to B_1 to take up the new centre of gravity of the underwater volume, and the force of buoyancy is considered to act vertically upwards through B_1 and the metacentre M. If moments are taken about G there is a moment to return the ship to the upright. This moment is referred to as the *Moment of Statical Stability* and is equal to the product of the force 'W' and the length of the lever GZ; i.e.

Moment of Statical Stability $= W \times GZ$ *tonnes metres*

The lever GZ is referred to as the *righting lever* and is the perpendicular distance between the centre of gravity and the vertical through the centre of buoyancy.

At a small angle of heel (less than 15°):

$$GZ = GM \times \sin\theta \text{ and Moment of Statical Stability} = W \times GM \times \sin\theta$$

Unstable equilibrium

When a ship which is inclined to a small angle tends to heel over still further, she is said to be in unstable equilibrium. For this to occur the ship must have a negative GM. Note how G is above M.

Figure 6.3(a) shows a ship in unstable equilibrium which has been inclined to a small angle. The moment of statical stability, W × GZ, is clearly a capsizing moment which will tend to heel the ship still further.

Note. A ship having a very small negative initial metacentric height GM need not necessarily capsize. This point will be examined and explained later. This situation produces an angle of loll.

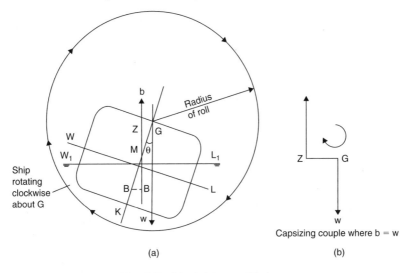

Capsizing couple where b = w

(a)   (b)

Fig. 6.3   Unstable equilibrium.

Neutral equilibrium

When G coincides with M as shown in Figure 6.4(a), the ship is said to be in neutral equilibrium, and if inclined to a small angle she will tend to remain at that angle of heel until another external force is applied. The ship has zero GM. Note that KG = KM.

Moment of Statical Stability = W × GZ, but in this case GZ = 0
∴ Moment of Statical Stability = 0 see Figure 6.4(b)

Therefore there is no moment to bring the ship back to the upright or to heel her over still further. The ship will move vertically up and down in the water at the fixed angle of heel until further external or internal forces are applied.

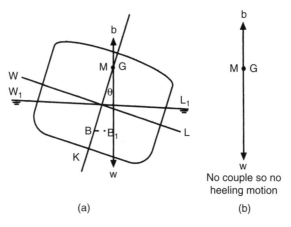

(a)   (b)

Fig. 6.4   Neutral equilibrium.

Correcting unstable and neutral equilibrium

When a ship in unstable or neutral equilibrium is to be made stable, the
effective centre of gravity of the ship should be lowered. To do this one or
more of the following methods may be employed:

1. Weights already in the ship may be lowered.
2. Weights may be loaded below the centre of gravity of the ship.
3. Weights may be discharged from positions above the centre of gravity.
4. Free surfaces within the ship may be removed.

The explanation of this last method will be found in Chapter 7.

Stiff and tender ships

The *time period* of a ship is the time taken by the ship to roll from one side to
the other and back again to the initial position.

When a ship has a comparatively large GM, for example 2 m to 3 m, the
righting moments at small angles of heel will also be comparatively large.
It will thus require larger moments to incline the ship. When inclined she
will tend to return more quickly to the initial position. The result is that the
ship will have a comparatively short time period, and will roll quickly –
and perhaps violently – from side to side. A ship in this condition is said to
be 'stiff', and such a condition is not desirable. The time period could be as
low as 8 seconds. The effective centre of gravity of the ship should be raised
within that ship (see Figure 40.2).

When the GM is comparatively small, for example 0.16 m to 0.20 m the
righting moments at small angles of heel will also be small. The ship will
thus be much easier to incline and will not tend to return so quickly to the
initial position. The time period will be comparatively long and a ship, for
example 25 to 35 seconds, in this condition is said to be 'tender'. As before,
this condition is not desirable and steps should be taken to increase the GM
by lowering the effective centre of gravity of the ship (see Figure 40.2).

The officer responsible for loading a ship should aim at a happy medium between these two conditions whereby the ship is neither too stiff nor too tender. A time period of 15 to 25 seconds would generally be acceptable for those on board a ship at sea (see Figure 40.2).

Negative GM and angle of loll

It has been shown previously that a ship having a negative initial metacentric height will be unstable when inclined to a small angle. This is shown in Figure 6.5(a).

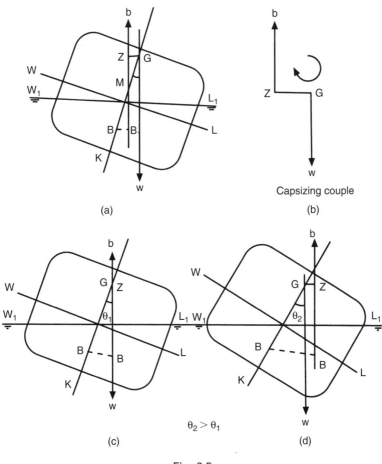

Fig. 6.5

As the angle of heel increases, the centre of buoyancy will move out still further to the low side. If the centre of buoyancy moves out to a position vertically under G, the capsizing moment will have disappeared as shown in Figure 6.5(b). The angle of heel at which this occurs is called the *angle of loll*. It will be noticed that at the angle of loll, the GZ is zero. G remains on the centre line.

If the ship is heeled beyond the angle of loll from θ_1 to θ_2, the centre of buoyancy will move out still further to the low side and there will be a moment to return her to the angle of loll as shown in Figure 6.5(c).

From this it can be seen that the ship will oscillate about the angle of loll instead of about the vertical. If the centre of buoyancy does not move out far enough to get vertically under G, the ship will capsize.

The angle of loll will be to port or starboard and back to port depending on external forces such as wind and waves. One minute it may flop over to $3°$ P and then suddenly flop over to $3°$ S.

There is always the danger that G will rise above M and create a situation of unstable equilibrium. This will cause capsizing of the ship.

The GM value

GM is crucial to ship stability. The table below shows *typical* working values for GM for several ship-types all at *fully-loaded* drafts.

Ship type	GM at fully-loaded condition
General cargo ships	0.30–0.50 m
Oil tankers	0.50–2.00 m
Double-hull supertankers	2.00–5.00 m
Container ships	1.50–2.50 m
Ro-Ro vessels	1.50 m approximately
Bulk ore carriers	2–3 m

At drafts below the fully-loaded draft, due to KM tending to be larger in value it will be found that corresponding GM values will be *higher* than those listed in the table above. For all conditions of loading the Department for Transport (DfT) stipulate that the GM must never be less than 0.15 m.

Exercise 6

1   Define the terms 'heel', 'list', 'initial metacentre' and 'initial metacentric height'.
2   Sketch transverse sections through a ship, showing the positions of the centre of gravity, centre of buoyancy, and initial metacentre, when the ship is in (a) Stable equilibrium, (b) Unstable equilibrium, and (c) Neutral equilibrium.
3   Explain what is meant by a ship being (a) tender and, (b) stiff.
4   With the aid of suitable sketches, explain what is meant by 'angle of loll'.
5   A ship of 10 000 t displacement has an initial metacentric height of 1.5 m. What is the moment of statical stability when the ship is heeled 10 degrees?

Chapter 7
Effect of free surface of liquids on stability

When a tank is completely filled with a liquid, the liquid cannot move within the tank when the ship heels. For this reason, as far as stability is concerned, the liquid may be considered as a static weight having its centre of gravity at the centre of gravity of the liquid within the tank.

Figure 7.1(a) shows a ship with a double-bottom tank filled with a liquid having its centre of gravity at g. The effect when the ship is heeled to a small angle θ is shown in Figure 7.1(b). No weights have been moved within the ship, therefore the position of G is not affected. The centre of buoyancy will move out to the low side indicated by BB_1.

$$\text{Moment of statical stability} = W \times GZ$$
$$= W \times GM \times \sin \theta$$

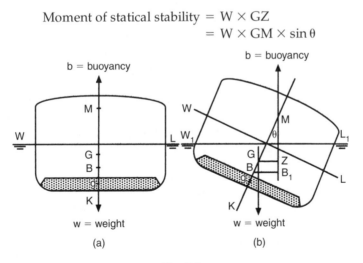

Fig. 7.1

Now consider the same ship floating at the same draft and having the same KG, but increase the depth of the tank so that the liquid now only partially fills it as shown in Figures 7.1(c) and (d).

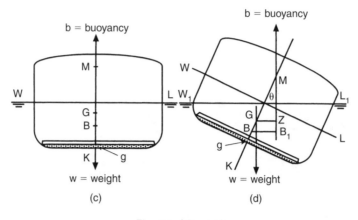

Fig. 7.1 (Contd.)

When the ship heels, as shown in Figure 7.2, the liquid flows to the low side of the tank such that its centre of gravity shifts from g to g_1. This will cause the ship's centre of gravity to shift from G to G_1, parallel to gg_1.

$$\text{Moment of statical stability} = W \times G_1Z_1$$
$$= W \times G_vZ_v$$
$$= W \times G_vM \times \sin\theta$$

This indicates that the effect of the free surface is to reduce the effective metacentric height from GM to G_vM. GG_v is therefore the virtual loss of GM due to the free surface. Any loss in GM is a loss in stability.

If free surface be created in a ship with a small initial metacentric height, the virtual loss of GM due to the free surface may result in a negative metacentric height. This would cause the ship to take up an angle of loll which may be dangerous and in any case is undesirable. This should be borne in mind when considering whether or not to run water ballast into tanks to correct an angle of loll, or to increase the GM. Until the tank is full there will be a virtual loss of GM due to the free surface effect of the liquid. It should also be noted from Figure 7.2 that even though the distance GG_1 is fairly small it produces a relatively large virtual loss in GM (GG_v).

Correcting an angle of loll

If a ship takes up an angle of loll due to a very small negative GM it should be corrected as soon as possible. GM may be, for example -0.05 to $-0.10\,\text{m}$, well below the DfT minimum stipulation of $0.15\,\text{m}$.

First make sure that the heel is due to a negative GM and not due to uneven distribution of the weights on board. For example, when bunkers are burned from one side of a divided double bottom tank it will obviously cause G to move to G_1, away from the centre of gravity of the burned bunkers, and will result in the ship listing as shown in Figure 7.3.

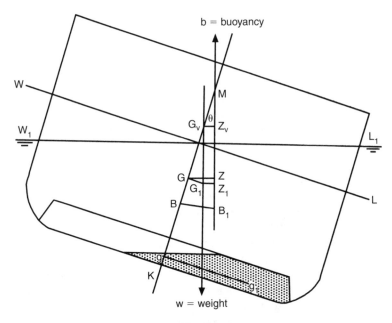

Fig. 7.2

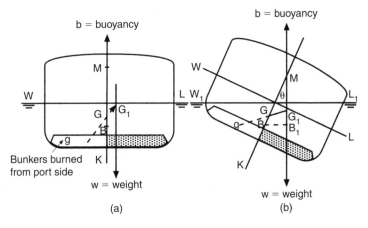

Fig. 7.3

Having satisfied oneself that the weights within the ship are uniformly distributed, one can assume that the list is probably due to a very small negative GM. To correct this it will be necessary to lower the position of the effective centre of gravity sufficiently to bring it below the initial meta-centre. Any slack tanks should be topped up to eliminate the virtual rise of G due to free surface effect. If there are any weights which can be lowered within the ship, they should be lowered. For example, derricks may be lowered if any are topped; oil in deep tanks may be transferred to double bottom tanks, etc.

Assume now that all the above action possible has been taken and that the ship is still at an angle of loll. Assume also that there are double-bottom tanks which are empty. Should they be filled, and if so, which ones first?

Before answering these questions consider the effect on stability during the filling operation. Free surfaces will be created as soon as liquid enters an empty tank. This will give a virtual rise of G which in turn will lead to an increased negative GM and an increased angle of loll. Therefore, if it is decided that it is safe to use the tanks, those which have the smallest area can be filled first so that the increase in list is cut to a minimum. Tanks should be filled one at a time.

Next, assume that it is decided to start by filling a tank which is divided at the centre line. Which side is to be filled first? If the high side is filled first the ship will start to right herself but will then roll suddenly over to take up a larger angle of loll on the other side, or perhaps even capsize. Now consider filling the low side first. Weight will be added low down in the vessel and G will thus be lowered, but the added weight will also cause G to move out of the centre line to the low side, increasing the list. Free surface is also being created and this will give a virtual rise in G, thus causing a loss in GM, which will increase the list still further.

Figure 7.4(a) shows a ship at an angle of loll with the double-bottom tanks empty and in Figure 7.4(b) some water has been run into the low side. The shift of the centre of gravity from G to G_v is the virtual rise of G due to the free surface, and the shift from G_v to G_1 is due to the weight of the added water.

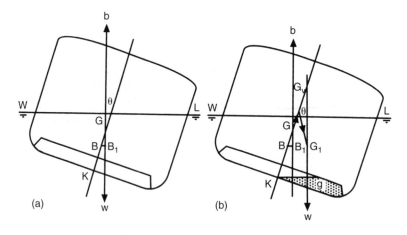

Fig. 7.4

It can be seen from the figure that the net result is a moment to list the ship over still further, but the increase in list is a gradual and controlled increase. When more water is now run into the tank the centre of gravity of the ship will gradually move downwards and the list will start to decrease. As the list decreases, water may be run into the other side of the tank. The water will then be running in much more quickly, causing G to move downwards more quickly. The ship cannot roll suddenly over to the other side as there is more water in the low side than in the high side. If sufficient weight of water is loaded to bring G on the centre line below M, the ship should complete the operation upright.

To summarize:

(a) Check that the list is due to a very small negative GM, for example −0.05 to −0.10 m.
(b) Top up any slack tanks and lower weights within the ship if possible.
(c) If the ship is still listed and it is decided to fill double-bottom tanks, start by filling the low side of a tank which is adequately sub-divided.
(d) The list is bound to be increased in the initial stages.
(e) Never start by filling tanks on the high side first.
(f) Always *calculate the effects first* before *authorizing action* to be taken to ballast any tanks.

Exercise 7

1 With the aid of suitable sketches, show the effect of slack tanks on a ship's stability.
2 A ship leaves port upright with a full cargo of timber, and with timber on deck. During the voyage, bunkers, stores and fresh water are consumed evenly from each side. If the ship arrives at her destination with a list, explain the probable cause of the list and how this should be remedied.
3 A ship loaded with timber and with timber on deck, berths with an angle of loll away from the quay. From which side should the timber on deck be discharged first and why?

Chapter 8
TPC and displacement curves

Recapitulation

The TPC is the mass which must be loaded or discharged to change the ship's mean draft by 1 cm. When the ship is floating in salt water it is found by using the formula:

$$\mathrm{TPC_{SW}} = \frac{\mathrm{WPA}}{97.56}$$

where

$$\mathrm{WPA} = \text{the area of the water-plane in sq. metres}$$

The area of the water-plane of a box-shaped vessel is the same for all drafts if the trim be constant, and so the TPC will also be the same for all drafts.

In the case of a ship the area of the water-plane is not constant for all drafts, and therefore the TPC will reduce at lower drafts, as shown in Figure 8.1. The TPCs are calculated for a range of drafts extending beyond the light and loaded drafts, and these are then tabulated or plotted on a graph. From the table or graph the TPC at intermediate drafts may be found.

TPC curves

When constructing a TPC curve the TPCs are plotted against the corresponding drafts. It is usually more convenient to plot the drafts on the vertical axis and the TPCs on the horizontal axis.

Example

(a) Construct a graph from the following information:

Mean draft (m)	3.0	3.5	4.0	4.5
TPC (tonnes)	8.0	8.5	9.2	10.0

(b) From this graph find the TPCs at drafts of 3.2 m; 3.7 m; and 4.3 m.

(c) If the ship is floating at a mean draft of 4 m and then loads 50 tonnes of cargo, 10 tonnes of fresh water and 25 tonnes of bunkers, whilst 45 tonnes of ballast are discharged, find the final mean draft.

(a) For the graph see Figure 8.1
(b) TPC at 3.2 m draft = 8.17 tonnes
     TPC at 3.7 m draft = 8.77 tonnes
     TPC at 4.3 m draft = 9.68 tonnes
(c) TPC at 4 m draft   = 9.2 tonnes

Loaded cargo	50 tonnes
Fresh water	10 tonnes
Bunkers	25 tonnes
Total	85 tonnes

Discharged ballast	45 tonnes
Net loaded	40 tonnes

$$\text{Increase in draft} = \frac{w}{\text{TPC}}$$
$$= \frac{40}{9.2}$$
$$= 4.35 \text{ cm}$$

Increase in draft   = 0.044 m
Original draft   = 4.000 m
New mean draft = 4.044 m

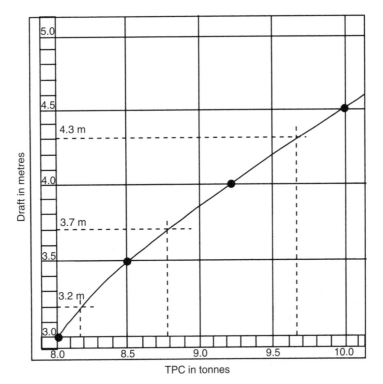

Fig. 8.1

Note. If the net weight loaded or discharged is very large, there is likely to be a considerable difference between the TPCs at the original and the new drafts, in which case to find the change in draft the procedure is as follows.

First find an approximate new draft using the TPC at the original draft, then find the TPC at the approximate new draft. Using the mean of these two TPCs find the actual increase or decrease in draft.

Displacement curves

A displacement curve is one from which the displacement of the ship at any particular draft can be found, and vice versa. The draft scale is plotted on the vertical axis and the scale of displacements on a horizontal axis. As a general rule the largest possible scale should be used to ensure reasonable accuracy. When the graph paper is oblong in shape, the length of the paper should be used for the displacement scale and the breadth for the drafts. It is quite unnecessary in most cases to start the scale from zero as the information will only be required for drafts between the light and load displacements (known as the boot-topping area).

Example

(a) Construct a displacement curve from the following data:

Draft (m)	3	3.5	4	4.5	5.0	5.5
Displacement (tonnes)	2700	3260	3800	4450	5180	6060

(b) If this ship's light draft is 3 m, and the load draft is 5.5 m, find the deadweight.
(c) Find the ship's draft when there are 500 tonnes of bunkers, and 50 tonnes of fresh water and stores on board.
(d) When at 5.13 m mean draft the ship discharges 2100 tonnes of cargo and loads 250 tonnes of bunkers. Find the new mean draft.
(e) Find the approximate TPC at 4.4 m mean draft.
(f) If the ship is floating at a mean draft of 5.2 m, and the load mean draft is 5.5 m, find how much more cargo may be loaded.

(a) See Figure 8.2 for the graph

(b) Load	Draft 5.5 m	Displacement	6060 tonnes
Light	Draft 3.0 m	Displacement	2700 tonnes

Deadweight = 3360 tonnes

(c) Light displacement	2700 tonnes
Bunkers	+500 tonnes
Fresh water and stores	+50 tonnes
New displacement	3250 tonnes

∴ Draft = 3.48 m

(d) Displacement at 5.13 m	5380 tonnes
Cargo discharged	−2100 tonnes
	3280 tonnes
Bunkers loaded	+ 250 tonnes
New displacement	3530 tonnes

∴ New draft = 3.775 m

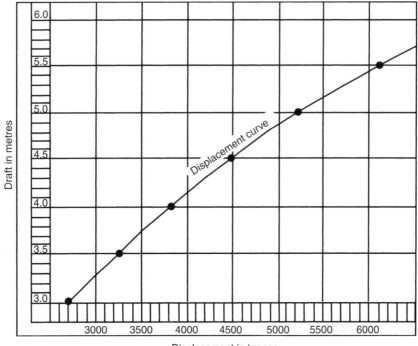

Fig. 8.2

(e) At 4.5 m draft the displacement is   4450 tonnes
    At 4.3 m draft the displacement is   − 4175 tonnes
    Difference to change the draft 0.2 m   275 tonnes

    Difference to change the draft 1 cm $\dfrac{275}{20}$ tonnes

∴ TPC = 13.75 tonnes

(f) Load draft 5.5 m   Displacement   6060 tonnes
    Present draft 5.2 m   Displacement   −5525 tonnes
                          Difference   535 tonnes

∴ Load = 535 tonnes

In fresh water, the TPC is calculated as follows:

$$\text{TPC}_{FW} = \frac{\text{WPA}}{100}$$

$$\text{or} \quad \text{TPC}_{FW} = \text{TC}_{SW} \times \frac{1.000}{1.025}$$

Exercise 8

TPC curves

1 (a) Construct a TPC curve from the following data:

Mean draft (m)	1	2	3	4	5
TPC (tonnes)	3.10	4.32	5.05	5.50	5.73

   (b) From this curve find the TPC at drafts of 1.5 m and 2.1 m.

   (c) If this ship floats at 2.2 m mean draft and then discharges 45 tonnes of ballast, find the new mean draft.

2 (a) From the following information construct a TPC curve:

Mean draft (m)	1	2	3	4	5
Area of water-plane (sq. m)	336	567	680	743	777

   (b) From this curve find the TPCs at mean drafts of 2.5 m and 4.5 m.

   (c) If, while floating at a draft of 3.8 m, the ship discharges 380 tonnes of cargo and loads 375 tonnes of bunkers, 5 tonnes of stores and 125 tonnes of fresh water, find the new mean draft.

3 From the following information construct a TPC curve:

Mean draft (m)	1	3	5	7
TPC (tonnes)	4.7	10.7	13.6	15.5

Then find the new mean draft if 42 tonnes of cargo is loaded whilst the ship is floating at 4.5 m mean draft.

Displacement curves

4 (a) From the following information construct a displacement curve:

Displacement (tonnes)	376	736	1352	2050	3140	4450
Mean draft (m)	1	2	3	4	5	6

   (b) From this curve find the displacement at a draft of 2.3 m.

   (c) If this ship floats at 2.3 m mean draft and then loads 850 tonnes of cargo and discharges 200 tonnes of cargo, find the new mean draft.

   (d) Find the approximate TPC at 2.5 m mean draft.

5 The following information is taken from a ship's displacement scale:

Displacement (tonnes)	335	1022	1949	2929	3852	4841
Mean draft (m)	1	1.5	2	2.5	3	3.5

   (a) Construct the displacement curve for this ship and from it find the draft when the displacement is 2650 tonnes.

   (b) If this ship arrived in port with a mean draft of 3.5 m, discharged her cargo, loaded 200 tonnes of bunkers and completed with a mean draft of 2 m, find how much cargo she discharged.

   (c) Assuming that the ship's light draft is 1 m, find the deadweight when the ship is floating in salt water at a mean draft of 1.75 m.

6 (a) From the following information construct a displacement curve:

Displacement (tonnes)	320	880	1420	2070	2800	3680
Draft (m)	1	1.5	2	2.5	3	3.5

(b) If this ship's draft light is 1.1 m, and the load draft 3.5 m, find the deadweight.

(c) If the vessel had on board 300 tonnes of cargo, 200 tonnes of ballast and 60 tonnes of fresh water and stores, what would be the salt water mean draft.

7 (a) Construct a displacement curve from the following data:

Draft (m)	1	2	3	4	5	6
Displacement (tonnes)	335	767	1270	1800	2400	3100

(b) The ship commenced loading at 3 m mean draft and, when work ceased for the day, the mean draft was 4.2 m. During the day 85 tonnes of salt water ballast had been pumped out. Find how much cargo had been loaded.

(c) If the ship's light draft was 2 m find the mean draft after she had taken in 870 tonnes of water ballast and 500 tonnes of bunkers.

(d) Find the TPC at 3 m mean draft.

8 (a) From the following information construct a displacement curve:

Draft (m)	1	2	3	4	5	6
Displacement (tonnes)	300	1400	3200	5050	7000	9000

(b) If the ship is floating at a mean draft of 3.2 m, and then loads 1800 tonnes of cargo and 200 tonnes of bunkers and also pumps out 450 tonnes of water ballast, find the new displacement and final mean draft.

(c) At a certain draft the ship discharged 1700 tonnes of cargo and loaded 400 tonnes of bunkers. The mean draft was then found to be 4.5 m. Find the original mean draft.

Chapter 9
Form coefficients

The coefficient of fineness of the water-plane area (C$_w$)

The coefficient of fineness of the water-plane area is the ratio of the area of the water-plane to the area of a rectangle having the same length and maximum breadth.

In Figure 9.1 the area of the ship's water-plane is shown shaded and ABCD is a rectangle having the same length and maximum breadth:

$$\text{Coefficient of fineness } (C_w) = \frac{\text{Area of water-plane}}{\text{Area of rectangle ABCD}}$$

$$= \frac{\text{Area of water-plane}}{L \times B}$$

$$\therefore \text{Area of the water-plane} = L \times B \times C_w$$

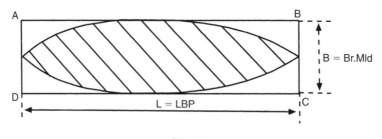

Fig. 9.1

Example 1

Find the area of the water-plane of a ship 36 metres long, 6 metres beam, which has a coefficient of fineness of 0.8.

$$\text{Area of water-plane} = L \times B \times C_w$$
$$= 36 \times 6 \times 0.8$$

Ans. Area of water-plane = 173 sq.m

Example 2

A ship 128 metres long has a maximum beam of 20 metres at the waterline, and coefficient of fineness of 0.85. Calculate the TPC at this draft.

$$\text{Area of water-plane} = L \times B \times C_w$$
$$= 128 \times 20 \times 0.85$$
$$= 2176 \text{ sq. metres}$$
$$TPC_{SW} = \frac{WPA}{97.56}$$
$$= \frac{2176}{97.56}$$

Ans. TPC$_{SW}$ = 22.3 tonnes

The block coefficient of fineness of displacement (C$_b$)

The block coefficient of a ship at any particular draft is the ratio of the volume of displacement at that draft to the volume of a rectangular block having the same overall length, breadth and depth.

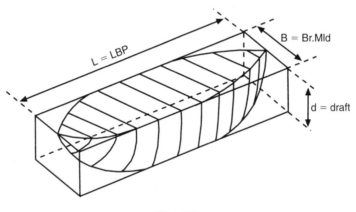

Fig. 9.2

In Figure 9.2 the shaded portion represents the volume of the ship's displacement at the draft concerned, enclosed in a rectangular block having the same overall length, breadth and depth.

$$\text{Block coefficient (C}_b) = \frac{\text{Volume of displacement}}{\text{Volume of the block}}$$
$$= \frac{\text{Volume of displacement}}{L \times B \times \text{draft}}$$

$\therefore$ Volume of displacement = L $\times$ B $\times$ draft $\times$ C$_b$

Ship's lifeboats

The cubic capacity of a lifeboat should be determined by Simpson's rules or by any other method giving the same degree of accuracy. The accepted C_b for a ship's lifeboat constructed of wooden planks is 0.6 and this is the figure to be used in calculations unless another specific value is given. Thus, the cubic capacity of a wooden lifeboat can be found using the formula:

$$\text{Volume} = (L \times B \times \text{Depth} \times 0.6) \text{ cubic metres}$$

The number of persons which a lifeboat may be certified to carry is equal to the greatest whole number obtained by the formula V/x where 'V' is the cubic capacity of the lifeboat in cubic metres and 'x' is the volume in cubic metres for each person. 'x' is 0.283 for a lifeboat 7.3 metres in length or over, and 0.396 for a lifeboat 4.9 metres in length. For intermediate lengths of lifeboats, the value of 'x' is determined by interpolation.

Example 1

Find the number of persons which a wooden lifeboat 10 metres long, 2.7 metres wide and 1 metre deep may be certified to carry.

$$
\begin{aligned}
\text{Volume of the boat} &= (L \times B \times D \times 0.6) \text{ cu. m} \\
&= 10 \times 2.7 \times 1 \times 0.6 \\
&= 16.2 \text{ cu. m} \\
\text{Number of persons} &= V/x \\
&= 16.2/0.283
\end{aligned}
$$

Ans. Number of persons = 57

Example 2

A ship 64 metres long, 10 metres maximum beam, has a light draft of 1.5 metres and a load draft of 4 metres. The block coefficient of fineness is 0.600 at the light draft and 0.750 at the load draft. Find the deadweight.

$$
\begin{aligned}
\text{Light displacement} &= (L \times B \times \text{draft} \times C_b) \text{ cu. m} \\
&= 64 \times 10 \times 1.5 \times 0.6 \\
&= 576 \text{ cu. m} \\
\text{Load displacement} &= (L \times B \times \text{draft} \times C_b) \text{ cu. m} \\
&= 64 \times 10 \times 4 \times 0.75 \\
&= 1920 \text{ cu. m} \\
\text{Deadweight} &= \text{Load displacement} - \text{Light displacement} \\
&= (1920 - 576) \text{ cu. m} \\
\text{Deadweight} &= 1344 \text{ cu. m} \\
&= 1344 \times 1.025 \text{ tonnes}
\end{aligned}
$$

Ans. Deadweight = 1378 tonnes

The midships coefficient (C_m)

The midships coefficient to any draft is the ratio of the transverse area of the midships section (A_m) to a rectangle having the same breadth and depths.

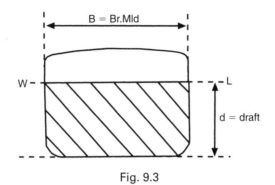

Fig. 9.3

In Figure 9.3 the shaded portion represents the area of the midships section to the waterline WL, enclosed in a rectangle having the same breadth and depth.

$$\text{Midships coefficient } (C_m) = \frac{\text{Midships area } (A_m)}{\text{Area of rectangle}}$$

$$= \frac{\text{Midships area } (A_m)}{B \times d}$$

or

$$\text{Midships area } (A_m) = B \times d \times C_m$$

The prismatic coefficient (C_p)

The prismatic coefficient of a ship at any draft is the ratio of the volume of displacement at that draft to the volume of a prism having the same length as the ship and the same cross-sectional area as the ship's midships area. The prismatic coefficient is used mostly by ship-model researchers.

In Figure 9.4 the shaded portion represents the volume of the ship's displacement at the draft concerned, enclosed in a prism having the same length as the ship and a cross-sectional area equal to the ship's midships area (A_m).

$$\text{Prismatic coefficient } (C_p) = \frac{\text{Volume of ship}}{\text{Volume of prism}}$$

$$= \frac{\text{Volume of ship}}{L \times A_m}$$

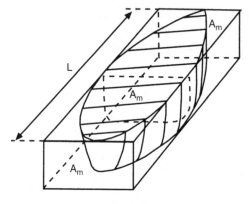

Fig. 9.4

or

$$\text{Volume of ship} = L \times A_m \times C_p$$

Note

$$C_m \times C_p = \frac{A_m}{B \times d} \times \frac{\text{Volume of ship}}{L \times A_m}$$

$$= \frac{\text{Volume of ship}}{L \times B \times d}$$

$$= C_b$$

$$\therefore C_m \times C_p = C_b \quad \text{or}$$

$$C_p = \frac{C_b}{C_m}$$

Note. C_p is always slightly higher than C_b at each waterline.

Having described exactly what C_w, C_b, C_m and C_p are, it would be useful to know what their values would be for several ship types.

First of all it must be remembered that all of these form coefficients will never be more than unity. To be so is not physically possible.

For the C_b values at *fully loaded drafts* the following table gives good typical values:

Ship type	Typical C_b fully loaded	Ship type	Typical C_b fully loaded
ULCC	0.850	General cargo ship	0.700
Supertanker	0.825	Passenger liner	0.575 to 0.625
Oil tanker	0.800	Container ship	0.575
Bulk carrier	0.775 to 0.825	Coastal tug	0.500

Medium form ships (C_b approx. 0.700), full-form ships ($C_b > 0.700$), fine-form ships ($C_b < 0.700$).

To estimate a value for C_w for these ship types at their *fully loaded* drafts, it is useful to use the following rule-of-thumb approximation:

$$C_w = \left(\tfrac{2}{3} \times C_b\right) + \tfrac{1}{3} \ \underline{@ \ \text{Draft Mld only!}}$$

Hence, for the oil tanker, C_w would be 0.867, for the general cargo ship C_w would be 0.800 and for the tug C_w would be 0.667 in fully loaded conditions.

For merchant ships, the midships coefficient or midship area coefficient is 0.980 to 0.990 at fully loaded draft. It depends on the rise of floor and the bilge radius. Rise of floor is almost obsolete nowadays.

As shown before,

$$C_p = \frac{C_b}{C_m}$$

Hence for a bulk carrier, if C_b is 0.750 with a C_m of 0.985, the C_p will be:

$$C_p = \frac{0.750}{0.985} = 0.761 \ \underline{@ \ \text{Draft Mld}}$$

C_p is used mainly by researchers at ship-model tanks carrying out tests to obtain the least resistance for particular hull forms of prototypes.

C_b and C_w change as the drafts move from fully loaded to light-ballast to lightship conditions. The diagram (Figure 9.5) shows the curves at drafts below the fully loaded draft for a general cargo ship of 135.5 m LBP.

'K' is calculated for the fully loaded condition and is *held constant* for all remaining drafts down to the ship's lightship (empty ship) waterline.

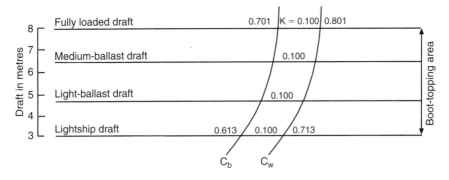

Fig. 9.5   Variation of C_b and C_w values with draft. (Note how the two curves are parallel at a distance of 0.100 apart.)

Exercise 9

1 (a) Define 'coefficient of fineness of the water-plane'.

   (b) The length of a ship at the waterline is 100 m, the maximum beam is 15 m and the coefficient of fineness of the water-plane is 0.8. Find the TPC at this draft.

2 (a) Define 'block coefficient of fineness of displacement'.

   (b) A ship's length at the waterline is 120 m when floating on an even keel at a draft of 4.5 m. The maximum beam is 20 m. If the ship's block coefficient is 0.75, find the displacement in tonnes at this draft in salt water.

3 A ship is 150 m long, has 20 m beam, load draft 8 m, light draft 3 m. The block coefficient at the load draft is 0.766, and at the light draft is 0.668. Find the ship's deadweight.

4 A ship 120 m long × 15 m beam has a block coefficient of 0.700 and is floating at the load draft of 7 m in fresh water. Find how much more cargo can be loaded if the ship is to float at the same draft in salt water.

5 A ship 100 m long, 15 m beam and 12 m deep is floating on an even keel at a draft of 6 m, block coefficient 0.8. The ship is floating in salt water. Find the cargo to discharge so that the ship will float at the same draft in fresh water.

6 A ship's lifeboat is 10 m long, 3 m beam and 1.5 m deep. Find the number of persons which may be carried.

7 A ship's lifeboat measures 10 m × 2.5 m × 1 m. Find the number of persons which may be carried.

Chapter 10
Simpson's Rules for areas and centroids

Areas and volumes

Simpson's Rules may be used to find the areas and volumes of irregular figures. The rules are based on the assumption that the boundaries of such figures are curves which follow a definite mathematical law. When applied to ships they give a good approximation of areas and volumes. The accuracy of the answers obtained will depend upon the spacing of the ordinates and upon how near the curve follows the law.

Simpson's First Rule

This rule assumes that the curve is a parabola of the second order. A parabola of the second order is one whose equation, referred to co-ordinate axes, is of the form $y = a_0 + a_1x + a_2x^2$, where a_0, a_1 and a_2 are constants.

Let the curve in Figure 10.1 be a parabola of the second order. Let y_1, y_2 and y_3 be three ordinates equally spaced at 'h' units apart.

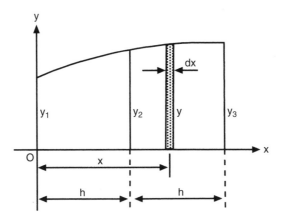

Fig. 10.1

The area of the elementary strip is y dx. Then the area enclosed by the curve and the axes of reference is given by:

$$\text{Area of figure} = \int_{O}^{2h} y\,dx$$

But

$$y = a_0 + a_1x + a_2x^2$$

$$\therefore \text{Area of figure} = \int_{O}^{2h} (a_0 + a_1x + a_2x^2)\,dx$$

$$= \left[a_0x + \frac{a_1x^2}{2} + \frac{a_2x^3}{3} \right]_{O}^{2h}$$

$$= 2a_0h + 2a_1h^2 + \tfrac{8}{3}a_2h^3$$

Assume that the area of figure $= Ay_1 + By_2 + Cy_3$

Using the equation of the curve and substituting 'x' for O, h and 2h respectively:

$$
\begin{aligned}
\text{Area of figure} &= Aa_0 + B(a_0 + a_1h + a_2h^2) \\
&\quad + C(a_0 + 2a_1h + 4a_2h^2) \\
&= a_0(A + B + C) + a_1h(B + 2C) \\
&\quad + a_2h^2(B + 4C)
\end{aligned}
$$

$$
\begin{aligned}
\therefore\ 2a_0h + 2a_1h^2 + \tfrac{8}{3}a_2h^3 &= a_0(A + B + C) + a_1h(B + 2C) \\
&\quad + a_2h^2(B + 4C)
\end{aligned}
$$

Equating coefficients:

$$A + B + C = 2h, \quad B + 2C = 2h, \quad B + 4C = \tfrac{8}{3}h$$

From which:

$$A = \frac{h}{3} \qquad B = \frac{4h}{3} \qquad C = \frac{h}{3}$$

$$\therefore \text{Area of figure} = \frac{h}{3}(y_1 + 4y_2 + y_3)$$

This is *Simpson's First Rule*.

It should be noted that Simpson's First Rule can also be used to find the area under a curve of the third order, i.e., a curve whose equation, referred to the co-ordinate axes, is of the form $y = a_0 + a_1x + a_2x^2 + a_3x^3$, where a_0, a_1, a_2 and a_3 are constants.

Summary

A coefficient of $\frac{1}{3}$ with multipliers of 1, 4, 1, etc.

Simpson's Second Rule

This rule assumes that the equation of the curve is of the third order, i.e. of a curve whose equation, referred to the co-ordinate axes, is of the form $y = a_0 + a_1x + a_2x^2 + a_3x^3$, where a_0, a_1, a_2 and a_3 are constants.

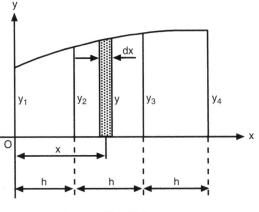

Fig. 10.2

In Figure 10.2:

$$\text{Area of elementary strip} = y\,dx$$

$$\text{Area of the figure} = \int_O^{3h} y\,dx$$

$$= \int_O^{3h} (a_0 + a_1x + a_2x^2 + a_3x^3)\,dx$$

$$= \left[a_0x + \tfrac{1}{2}a_1x^2 + \tfrac{1}{3}a_2x^3 + \tfrac{1}{4}a_3x^4 \right]_O^{3h}$$

$$= 3a_0h + \tfrac{9}{2}a_1h^2 + 9a_2h^3 + \tfrac{81}{4}a_3h^4$$

$$\begin{aligned}
\text{Let the area of the figure} &= Ay_1 + By_2 + Cy_3 + Dy_4 \\
&= Aa_0 + B(a_0 + a_1h + a_2h^2 + a_3h^3) \\
&\quad + C(a_0 + 2a_1h + 4a_2h^2 + 8a_3h^3) \\
&\quad + D(a_0 + 3a_1h + 9a_2h^2 + 27a_3h^3) \\
&= a_0(A + B + C + D) + a_1h(B + 2C + 3D) \\
&\quad + a_2h^2(B + 4C + 9D) + a_3h^3(B + 8C + 27D)
\end{aligned}$$

Equating coefficients:

$$A + B + C + D = 3h$$
$$B + 2C + 3D = \tfrac{9}{2}h$$
$$B + 4C + 9D = 9h$$
$$B + 8C + 27D = \tfrac{81}{4}h$$

From which:

$$A = \tfrac{3}{8}h, \quad B = \tfrac{9}{8}h, \quad C = \tfrac{9}{8}h, \quad D = \tfrac{3}{8}h$$

$$\therefore \text{ Area of figure } = \tfrac{3}{8}hy_1 + \tfrac{9}{8}hy_2 + \tfrac{9}{8}hy_3 + \tfrac{3}{8}hy_4$$

or

$$\text{Area of figure } = \tfrac{3}{8}h(y_1 + 3y_2 + 3y_3 + y_4)$$

This is *Simpson's Second Rule*.

Summary
A coefficient of $\tfrac{3}{8}$ with multipliers of 1, 3, 3, 1, etc.

Simpson's Third Rule
In Figure 10.3:

$$\text{Area of the elementary strip } = y \, dx$$

$$\text{Area between } y_1 \text{ and } y_2 \text{ in figure } = \int_O^h y \, dx$$

$$= a_0h + \tfrac{1}{2}a_1h^2 + \tfrac{1}{3}a_2h^3$$

$$\text{Let the area between } y_1 \text{ and } y_2 = Ay_1 + By_2 + Cy_3$$

$$\begin{aligned}
\text{Then area } &= Aa_0 + B(a_0 + a_1h + a_2h^2) \\
&\quad + C(a_0 + 2a_1h + 4a_2h^2) \\
&= a_0(A + B + C) + a_1h(B + 2C) \\
&\quad + a_2h^2(B + 4C)
\end{aligned}$$

Equating coefficients:

$$A + B + C = h, \quad B + 2C = h/2, \quad B + 4C = h/3$$

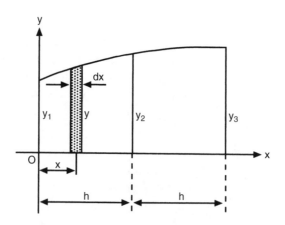

Fig. 10.3

From which:

$$A = \frac{5h}{12}, \quad B = \frac{8h}{12}, \quad C = -\frac{h}{12}$$

$\therefore$ Area of figure between y_1 and y_2 = $\frac{5}{12} hy_1 + \frac{8}{12} hy_2 + (-\frac{1}{12} hy_3)$

or

$$\text{Area} = \frac{h}{12}(5y_1 + 8y_2 - y_3)$$

This is the *Five/eight (or Five/eight minus one) rule, and is used to find the area between two consecutive ordinates when three consecutive ordinates are known.*

Summary
A coefficient of $\frac{1}{12}$ with multipliers of 5, 8, −1, etc.

Areas of water-planes and similar figures using extensions of Simpson's Rules
Since a ship is uniformly built about the centre line it is only necessary to calculate the area of half the water-plane and then double the area found to obtain the area of the whole water-plane.

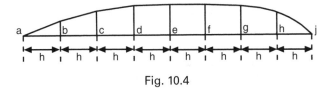

Fig. 10.4

Figure 10.4 represents the starboard side of a ship's water-plane area. To find the area, the centre line is divided into a number of equal lengths each 'h' m long. The length 'h' is called the *common interval*. The half-breadths, a, b, c, d, etc., are then measured and each of these is called a *half-ordinate*.

Using Simpson's First Rule
This rule can be used to find areas when there are an odd number of ordinates.

$$\text{Area of Figure 10.5(a)} = \frac{h}{13}(a + 4b + c)$$

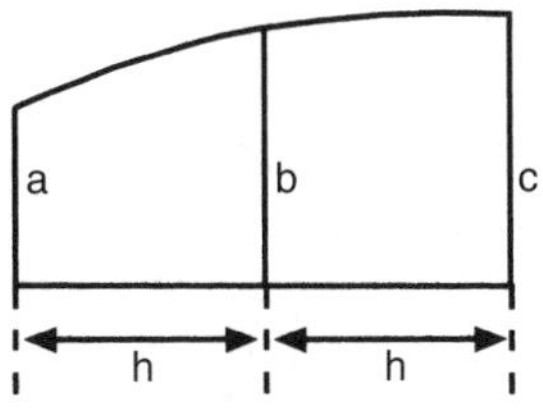

Fig. 10.5(a)

If the common interval and the ordinates are measured in metres, the area found will be in square metres.

Let this rule now be applied to a water-plane area such as that shown in Figure 10.5(b).

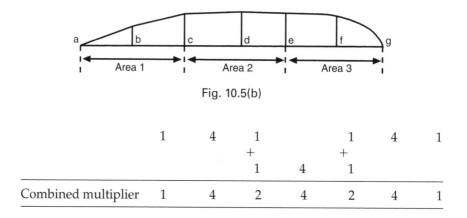

Fig. 10.5(b)

	1	4	1			1	4	1
			+			+		
			1	4	1			
Combined multiplier	1	4	2	4	2	4	1

The water-plane is divided into three separate areas and Simpson's First Rule is used to find each separate area:

$$\text{Area 1} = h/3(a + 4b + c)$$
$$\text{Area 2} = h/3(c + 4d + e)$$
$$\text{Area 3} = h/3(e + 4f + g)$$
$$\text{Area of } \tfrac{1}{2} \text{WP} = \text{Area 1} + \text{Area 2} + \text{Area 3}$$
$$\therefore \text{Area of } \tfrac{1}{2} \text{WP} = h/3(a + 4b + c) + h/3(c + 4d + e)$$
$$+ h/3(e + 4f + g)$$

or

$$\text{Area of } \tfrac{1}{2} \text{ WP} = h/3(a + 4b + 2c + 4d + 2e + 4f + g)$$

This is the form in which the formula should be used. Within the brackets the half-ordinates appear in their correct sequence from forward to aft. The coefficients of the half-ordinates are referred to as Simpson's Multipliers and they are in the form: 1424241. Had there been nine half-ordinates, the Multipliers would have been: 142424241. It is usually much easier to set out that part of the problem within the brackets in tabular form. Note how the Simpson's multipliers begin and end with 1, as shown in Figure 10.5(b).

Example

A ship 120 metres long at the waterline has equidistantly spaced half-ordinates commencing from forward as follows:

0, 3.7, 5.9, 7.6, 7.5, 4.6 and 0.1 metres, respectively.

Find the area of the water-plane and the TPC at this draft.

Note. There is an odd number of ordinates in the water-plane and therefore the First Rule can be used.

No.	$\frac{1}{2}$ ord.	SM	Area function
a	0	1	0
b	3.7	4	14.8
c	5.9	2	11.8
d	7.6	4	30.4
e	7.5	2	15.0
f	4.6	4	18.4
g	0.1	1	0.1
			$90.5 = \Sigma_1$

Σ_1 is used because it is a total; using Simpson's First Rule:

$$h = \frac{120}{6} = \text{the common interval CI}$$
$$\therefore CI = 20 \text{ metres}$$
$$\text{Area of WP} = \tfrac{1}{3} \times CI \times \Sigma_1 \times 2 \text{ (for both sides)}$$
$$= \tfrac{1}{3} \times 20 \times 90.5 \times 2$$

Ans. Area of WP = 1207 sq. m

$$TPC_{SW} = \frac{WPA}{97.56} = \frac{1207}{97.56}$$

Ans. TPC = 12.37 tonnes

Note. If the half-ordinates are used in these calculations, then the area of half the water-plane is found. If however the whole breadths are used, the total area of the water-plane will be found. If half-ordinates are given and the WPA is requested, simply multiply by 2 at the end of the formula as shown above.

Using the extension of Simpson's Second Rule

This rule can be used to find the area when the number of ordinates is such that if one be subtracted from the number of ordinates the remainder is divisible by 3.

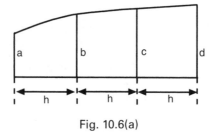

Fig. 10.6(a)

$$\text{Area of Figure 10.6(a)} = \tfrac{3}{8}\,h(a + 3b + 3c + d)$$

Now consider a water-plane area which has been divided up using seven half-ordinates as shown in Figure 10.6(b).

The water-plane can be split into two sections as shown, each section having four ordinates.

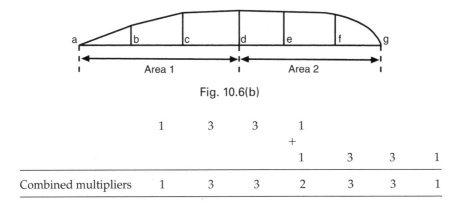

Fig. 10.6(b)

	1	3	3	1			
				+			
				1	3	3	1
Combined multipliers	1	3	3	2	3	3	1

$$\text{Area 1} = \tfrac{3}{8}\,h(a + 3b + 3c + d)$$
$$\text{Area 2} = \tfrac{3}{8}\,h(d + 3e + 3f + g)$$
$$\text{Area of } \tfrac{1}{2}\,WP = \text{Area 1} + \text{Area 2}$$
$$\therefore \quad \text{Area of } \tfrac{1}{2}\,WP = \tfrac{3}{8}\,h(a + 3b + 3c + d) + \tfrac{3}{8}\,h(d + 3e + 3f + g)$$

or

$$\text{Area of } \tfrac{1}{2}\,WP = \tfrac{3}{8}\,h(a + 3b + 3c + 2d + 3e + 3f + g)$$

This is the form in which the formula should be used. As before, all of the ordinates appear in their correct order within the brackets. The multipliers are now 1332331. Had there been 10 ordinates the multipliers would have been 1332332331. Note how the Simpson's multipliers begin and end with 1, as shown in Figure 10.6(b).

Example
Find the area of the water-plane described in the first example using Simpson's Second Rule.

No.	$\frac{1}{2}$ ord.	SM	Area function
a	0	1	0
b	3.7	3	11.1
c	5.9	3	17.7
d	7.6	2	15.2
e	7.5	3	22.5
f	4.6	3	13.8
g	0.1	1	0.1
			$80.4 = \Sigma_2$

Σ_2 is used because it is a total; using Simpson's Second Rule:

$$\text{Area of WP} = \tfrac{3}{8} \times CI \times \Sigma_2 \times 2 \text{ (for both sides)}$$
$$= \tfrac{3}{8} \times 20 \times 80.4 \times 2$$

Ans. Area WP = 1206 sq. m (compared to 1207 sq. m previous answer)
    The small difference in the two answers shows that the area found can only be a close approximation to the correct area.

The Five/Eight Rule (Simpson's Third Rule)
This rule may be used to find the area between two consecutive ordinates when three consecutive ordinates are known.
    The rule states that the area between two consecutive ordinates is equal to five times the first ordinate plus eight times the middle ordinate minus the external ordinate, all multiplied by $\frac{1}{12}$ of the common interval.

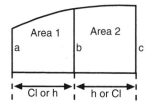

Fig. 10.6(c)

$$\text{Thus: Area } 1 = \frac{h}{12}(5a + 8b - c) \text{ or } \frac{1}{12} \times CI \times \Sigma_3$$
$$\text{Also: Area } 2 = \frac{h}{12}(5c + 8b - a) \text{ or } \frac{1}{12} \times CI \times \Sigma_3$$

Σ_3 is used because it is a total; using Simpson's Third Rule. Consider the next example.

Example

Three consecutive ordinates in a ship's water-plane, spaced 6 metres apart, are 14, 15 and 15.5 m, respectively. Find the area between the last two ordinates.

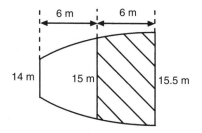

Fig. 10.6(d)

$$\text{Shaded area} = \frac{h}{12}(5a + 8b - c)$$

$$= \frac{6}{12}(77.5 + 120 - 14)$$

Ans. Area = 91.75 sq. m

Volumes of ship shapes and similar figures

Let the area of the elementary strip in Figures 10.7(a) and (b) be 'Y' square metres. Then the volume of the strip in each case is equal to Y dx and the volume of each ship is equal to $\int_{O}^{4h} Y\,dx$.

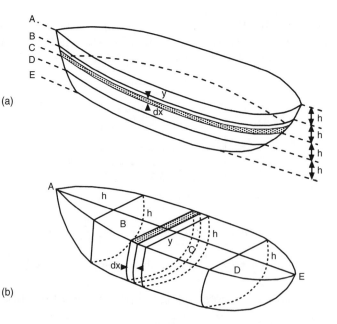

Fig. 10.7

The value of the integral in each case is found by Simpson's Rules using the areas at equidistant intervals as ordinates; i.e.

$$\text{Volume} = \frac{h}{3} (A + 4B + 2C + 4D + E)$$

or

$$\frac{\text{CI}}{3} \times \Sigma_1$$

Thus the volume of displacement of a ship to any particular draft can be found first by calculating the areas of water-planes or transverse areas at equidistant intervals and then using these areas as ordinates to find the volume by Simpson's Rules.

Example

The areas of a ship's water-planes are as follows:

Draft (m)	0	1	2	3	4
Area of WP (sq. m)	650	660	662	661	660

Calculate the ship's displacement in tonnes when floating in salt water at 4 metres draft. Also, if the ship's load draft is 4 metres. Find the FWA.

Draft (m)	Area	SM	Volume function
0	650	1	650
1	660	4	2640
2	662	2	1324
3	661	4	2644
4	660	1	660
			$7918 = \Sigma_1$

Σ_1 is used because it is a total; using Simpson's First Rule:

$$\text{Underwater volume} = \tfrac{1}{3} \times \text{CI} \times \Sigma_1 = \tfrac{1}{3} \times 1.0 \times 7918$$
$$= 2639 \tfrac{1}{3} \text{ cu. m}$$
$$\text{SW displacement} = 2639 \tfrac{1}{3} \times 1.025 \text{ tonnes}$$

Ans. SW displacement = 2705.3 tonnes

$$\text{Load TPC}_{SW} = \frac{\text{WPA}}{97.56}$$
$$= \frac{660}{97.56}$$
$$= 6.77 \text{ tonnes}$$
$$\text{FWA} = \frac{\text{Displacement}}{4 \times \text{TPC}}$$
$$= \frac{2705.3}{4 \times 6.77}$$

Ans. FWA = 99.9 mm or 9.99 cm

Appendages and intermediate ordinates

Appendages

It has been mentioned previously that areas and volumes calculated by the use of Simpson's Rules depend for their accuracy on the curvature of the sides following a definite mathematical law. It is very seldom that the ship's sides do follow one such curve. Consider the ship's water-plane area shown in Figure 10.8. The sides from the stem to the quarter form one curve but from this point to the stern is part of an entirely different curve. To obtain an answer which is as accurate as possible, the area from the stem to the quarter may be calculated by the use of Simpson's Rules and then the remainder of the area may be found by a second calculation. The remaining area mentioned above is referred to as an *appendage*.

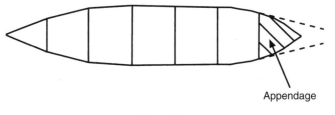

Appendage

Fig. 10.8

Similarly, in Figure 10.9 the side of the ship forms a reasonable curve from the waterline down to the turn of the bilge, but below this point the curve is one of a different form.

In this case the volume of displacement between the waterline (WL) and the water-plane XY could be found by use of Simpson's Rules and then the volume of the appendage found by means of a second calculation.

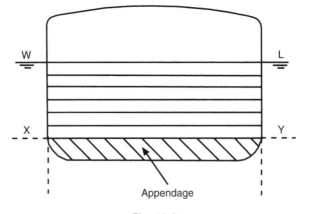

Appendage

Fig. 10.9

Example

A ship's breadths, at 9 m intervals commencing from forward are as follows:

0, 7.6, 8.7, 9.2, 9.5, 9.4 and 8.5 metres, respectively.

Abaft the last ordinate is an appendage of 50 sq. m. Find the total area of the water-plane.

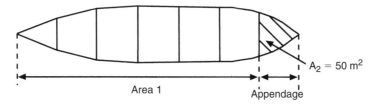

Fig. 10.10

Ord.	SM	Product for area
0	1	0
7.6	4	30.4
8.7	2	17.4
9.2	4	36.8
9.5	2	19.0
9.4	4	37.6
8.5	1	8.5
		$149.7 = \Sigma_1$

$$\text{Area 1} = \tfrac{1}{3} \times CI \times \Sigma_1$$
$$\text{Area 1} = \tfrac{9}{3} \times 149.7$$
$$\text{Area 1} = 449.1 \text{ sq. m}$$
$$\text{Appendage} = 50.0 \text{ sq. m} = \text{Area 2}$$

Ans. Area of WP = 499.1 sq. m

Subdivided common intervals

The area or volume of an appendage may be found by the introduction of intermediate ordinates. Referring to the water-plane area shown in Figure 10.11, the length has been divided into seven equal parts and the half ordinates have been set up. Also, the side is a smooth curve from the stem to the ordinate 'g'.

If the area of the water-plane is found by putting the eight half-ordinates directly through the rules, the answer obtained will obviously be an

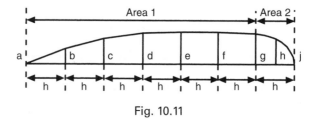

Fig. 10.11

erroneous one. To reduce the error the water-plane may be divided into two parts as shown in the figure.

Then,

$$\text{Area No. } 1 = h/3\,(a + 4b + 2c + 4d + 2e + 4f + g)$$

To find Area No. 2, an intermediate semi-ordinate is set up midway between the semi-ordinates g and j. The common interval for this area is $h/2$.

Then,

$$\text{Area No. } 2 = h/2 \times \tfrac{1}{3} \times (g + 4h + j)$$

or

$$\text{Area No. } 2 = h/3(\tfrac{1}{2}g + 2h + \tfrac{1}{2}j)$$

If CI is halved, then multipliers are halved, i.e. from 1, 4, 1 etc. to $\tfrac{1}{2}, 2, \tfrac{1}{2}$.

$$
\begin{aligned}
\text{Area of } \tfrac{1}{2}\text{WP} &= \text{Area } 1 + \text{Area } 2 \\
&= h/3(a + 4b + 2c + 4d + 2e + 4f + g) \\
&\quad + h/3(\tfrac{1}{2}g + 2h + \tfrac{1}{2}j) \\
&= h/3(a + 4b + 2c + 4d + 2e + 4f + g + \tfrac{1}{2}g \\
&\quad + 2h + \tfrac{1}{2}j) \\
\therefore \ \text{Area of } \tfrac{1}{2}\text{WP} &= h/3(a + 4b + 2c + 4d + 2e + 4f + 1\tfrac{1}{2}g \\
&\quad + 2h + \tfrac{1}{2}j) \quad \text{or} \quad \tfrac{1}{3} \times h \times \Sigma_1
\end{aligned}
$$

Example 1

The length of a ship's water-plane is 100 metres. The lengths of the half-ordinates commencing from forward are as follows:

0, 3.6, 6.0, 7.3, 7.7, 7.6, 4.8, 2.8 and 0.6 metres, respectively.

Midway between the last two half-ordinates is one whose length is 2.8 metres. Find the area of the water-plane.

$\frac{1}{2}$ ord.	SM	Area function
0	1	0
3.6	4	14.4
6.0	2	12.0
7.3	4	29.2
7.7	2	15.4
7.6	4	30.4
4.8	$1\frac{1}{2}$	7.2
2.8	2	5.6
0.6	$\frac{1}{2}$	0.3
		$114.5 = \Sigma_1$

$$\text{Area} = \tfrac{2}{3} \times \text{CI} \times \Sigma_1$$
$$\text{CI} = 100/7 = 14.29 \text{ m}$$
$$\text{Area of WP} = \tfrac{2}{3} \times 14.29 \times 114.5$$

Ans. Area of WP = 1,090.5 sq. m

Note how the CI used was the *Largest* CI in the ship's water-plane.

In some cases an even more accurate result may be obtained by dividing the water-plane into three separate areas as shown in Figure 10.12 and introducing intermediate semi-ordinates at the bow and the stern.

$$\text{Area 1} = h/2 \times \tfrac{1}{3} \times (a + 4b + c)$$
$$= h/3(\tfrac{1}{2}a + 2b + \tfrac{1}{2}c)$$
$$\text{Area 2} = h/3(c + 4d + 2e + 4f + g)$$
$$\text{Area 3} = h/2 \times \tfrac{1}{3}(g + 4h + j)$$
$$= h/3(\tfrac{1}{2}g + 2h + \tfrac{1}{2}j)$$
$$\text{Area } \tfrac{1}{2}\text{WP} = \text{Area 1} + \text{Area 2} + \text{Area 3}$$
$$= h/3(\tfrac{1}{2}a + 2b + \tfrac{1}{2}c) + c + 4d + 2e + 4f + g$$
$$+ \tfrac{1}{2}g + 2h + \tfrac{1}{2}j$$
$$\text{Area } \tfrac{1}{2}\text{WP} = h/3(\tfrac{1}{2}a + 2b + 1\tfrac{1}{2}c + 4d + 2e + 4f + 1\tfrac{1}{2}g$$
$$+ 2h + \tfrac{1}{2}j) \quad \text{or} \quad h/3 \times \Sigma_1$$

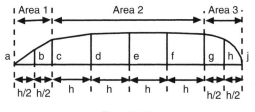

Fig. 10.12

$\frac{1}{2}$	2	$\frac{1}{2}$		1	4	1		
+	+			+				
		1	4	1		$\frac{1}{2}$	2	$\frac{1}{2}$

Combined multipliers	$\frac{1}{2}$	2	$1\frac{1}{2}$	4	2	4	$1\frac{1}{2}$	2	$\frac{1}{2}$

Example 2

A ship's water-plane is 72 metres long and the lengths of the half-ordinates commencing from forward are as follows:

0.2, 2.2, 4.4, 5.5, 5.8, 5.9, 5.9, 5.8, 4.8, 3.5 and 0.2 metres, respectively.

$\frac{1}{2}$ ord.	SM	Area function
0.2	$\frac{1}{2}$	0.1
2.2	2	4.4
4.4	$1\frac{1}{2}$	6.6
5.5	4	22.0
5.8	2	11.6
5.9	4	23.6
5.9	2	11.8
5.8	4	23.2
4.8	$1\frac{1}{2}$	7.2
3.5	2	7.0
0.2	$\frac{1}{2}$	0.1
		$117.6 = \Sigma_1$

The spacing between the first three and the last three half-ordinates is half of the spacing between the other half-ordinates. Find the area of the water-plane.

$$\text{Area} = \tfrac{1}{3} \times \text{CI} \times \Sigma_1 \times 2$$
$$\text{CI} = 72/8 = 9 \text{ m}$$
$$\text{Area of WP} = \tfrac{1}{3} \times 9 \times 117.6 \times 2$$

Ans. Area of WP = 705.6 sq. m

Note. It will be seen from this table that the effect of halving the common interval is to halve the Simpson's Multipliers.

Σ_1 is because it is using Simpson's First Rule.

Areas and volumes having an awkward number of ordinates

Occasionally the number of ordinates used is such that the area or volume concerned cannot be found directly by use of either the First or the Second

Rule. In such cases the area or volume should be divided into two parts, the area of each part being calculated separately, and the total area found by adding the areas of the two parts together.

Example 1

Show how the area of a water-plane may be found when using six semi-ordinates. Neither the First nor the Second Rule can be applied directly to the whole area but the water-plane can be divided into two parts as shown in Figure 10.13 Area No. 1 can be calculated using the First Rule and area No. 2 by the Second Rule. The areas of the two parts may then be added together to find the total area.

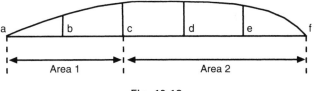

Fig. 10.13

An alternative method would be to find the area between the half-ordinates a and e by the First Rule and then find the area between the half-ordinates e and f by the 'five/eight' Rule.

Example 2

Show how the area may be found when using eight semi-ordinates.

Divide the area up as shown in Figure 10.14. Find area No. 1 using the Second Rule and area No. 2 using the First Rule.

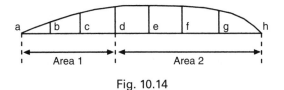

Fig. 10.14

An alternative method is again to find the area between the half-ordinates a and g by the first rule and the area between the half-ordinates g and h by the 'five/eight' rule.

In practice, the naval architect divides the ship's length into 10 stations and then subdivides the forward and aft ends in order to obtain extra accuracy with the calculations.

In doing so, the calculations can be made using Simpson's First and Second Rules, perhaps as part of a computer package.

Centroids and centres of gravity

To find the centre of flotation

The *centre of flotation* is the centre of gravity or *centroid* of the water-plane area, and is the point about which a ship heels and trims. It must lie on the longitudinal centre line but may be slightly forward or aft of amidships (from say 3 per cent L forward of amidships for oil tankers to say 3 per cent L aft of amidships for container ships).

To find the area of a water-plane by Simpson's Rules, the half-breadths are used as ordinates. If the moments of the half-ordinates about any point are used as ordinates, then the total moment of the area about that point will be found. If the total moment is now divided by the total area, the quotient will give the distance of the centroid of the area from the point about which the moments were taken. This may be shown as follows:

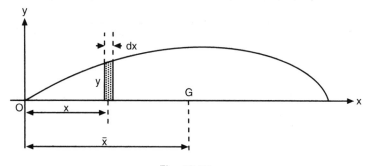

Fig. 10.15

In Figure 10.15:

$$\text{Area of strip} = y\,dx$$
$$\text{Area of the } \tfrac{1}{2}\,WP = \int_{O}^{L} y\,dx$$
$$\text{Area of the WP} = 2 \times \int_{O}^{L} y\,dx$$

The value of the integral is found using the formula:

$$\int_{O}^{L} y\,dx = \frac{h}{3}(a + 4b + 2c + 4d + e)$$

Thus, the value of the integral is found by Simpson's Rules using values of the variable y as ordinates.

$$\text{Moment of strip about OY} = x\,y\,dx$$
$$\text{Moment of } \tfrac{1}{2}\,WP \text{ about OY} = \int_{O}^{L} x\,y\,dx$$
$$\text{Moment of WP about OY} = 2 \times \int_{O}^{L} x\,y\,dx$$

The value of this integral is found by Simpson's Rules using values of the product x y as ordinates.

Let the distance of the centre of flotation be $\bar{X}$ from OY, then:

$$\bar{X} = \frac{\text{Moment}}{\text{Area}}$$

$$= \frac{2 \times \int_O^L x\,y\,dx}{2 \times \int_O^L y\,dx} = \frac{\Sigma_2}{\Sigma_1} \times CI$$

Example 1

A ship 150 metres long has half-ordinate commencing from aft as follows:

0, 5, 9, 9, 9, 7 and 0 metres, respectively.

Find the distance of the centre of flotation from forward (see Fig. 10.16).

Note. To avoid using large numbers the levers used are in terms of CI the common interval. It is more efficient than using levers in metres (see table on facing page).

$$\text{Area of the water-plane} = \tfrac{2}{3} \times CI \times \Sigma_1$$
$$= \tfrac{2}{3} \times 25 \times 376 \text{ sq. m}$$
$$\text{Distance of C.F. from aft} = \frac{\Sigma_2}{\Sigma_1} \times CI$$
$$= \frac{376}{120} \times CI$$
$$= 78.33 \text{ m}$$

Ans. C.F. is 78.33 m from aft

½ ord.		SM	Products for area	Levers from A*	Moment function
Aft	0	1	0	0	0
	5	4	20	1	20
	9	2	18	2	36
⊗	9	4	36	3	108
	9	2	18	4	72
	7	4	28	5	140
Forward	0	1	0	6	0
			$120 = \Sigma_1$		$376 = \Sigma_2$

* The Levers are in terms of the number of CI from the aft ordinate through to the foremost ordinate.
Σ_1, because it is the first total.
Σ_2, because it is the second total.

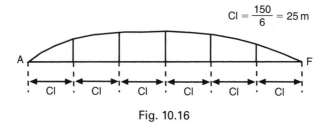

$$CI = \frac{150}{6} = 25\,m$$

Fig. 10.16

This problem can also be solved by taking the moments about amidships as in the following example:

Example 2

A ship 75 m long has half-ordinates at the load water-plane commencing from aft as follows:

0, 1, 2, 4, 5, 5, 5, 4, 3, 2 and 0 metres, respectively.

The spacing between the first three semi-ordinates and the last three semi-ordinates is half of that between the other semi-ordinates. Find the position of the Centre of Flotation relative to amidships.

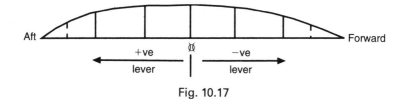

Fig. 10.17

Use +ve the sign for levers and moments AFT of amidships (⊗).
Use −ve sign for levers and moments FORWARD of amidships (⊗).

$\frac{1}{2}$ ord.		SM	Area function	Levers	Moment function
Aft	0	$\frac{1}{2}$	0	+4	0
	1	2	2	$+3\frac{1}{2}$	+7
	2	$1\frac{1}{2}$	3	+3	+9
	4	4	16	+2	+32
	5	2	10	+1	+10
⊗	5	4	20	0	0
	5	2	10	−1	−10
	4	4	16	−2	−32
	3	$1\frac{1}{2}$	4.5	−3	−13.5
	2	2	4.0	$-3\frac{1}{2}$	−14
Forward	0	$\frac{1}{2}$	0	−4	0
			$85.5 = \Sigma_1$		$-11.5 = \Sigma_2$

$$CI = \frac{75}{8} = 9.375 \text{ m}$$

Σ_1 denotes first total.
Σ_2 denotes second algebraic total.
The point having a lever of zero is the fulcrum point. All other levers +ve and −ve are then relative to this point.

$$\text{Distance of C.F. from amidships} = \frac{\Sigma_2}{\Sigma_1} \times CI$$

$$= \frac{-11.5}{85.5} \times 9.375$$

$$= -1.26 \text{ m}$$

The −ve sign shows it is forward of ⚓.

Ans. C.F. is 1.26 metres forward of amidships

To find the KB
The centre of buoyancy is the three-dimensional centre of gravity of the underwater volume and Simpson's Rules may be used to determine its height above the keel.

First, the areas of water-planes are calculated at equidistant intervals of draft between the keel and the waterline. Then the volume of displacement is calculated by using these areas as ordinates in the rules. The moments of these areas about the keel are then taken to find the total moment of the underwater volume about the keel. The KB is then found by dividing the total moment about the keel by the volume of displacement.

It will be noted that this procedure is similar to that for finding the position of the Centre of Flotation, which is the two-dimensional centre of gravity of each water-plane.

Example 1
A ship is floating upright on an even keel at 6.0 m draft F and A. The areas of the water-planes are as follows:

Draft (m)	0	1	2	3	4	5	6
Area (sq.m)	5000	5600	6020	6025	6025	6025	6025

Find the ship's KB at this draft.

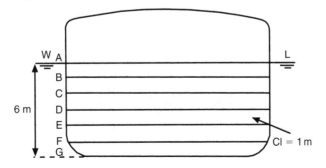

Fig. 10.18

Water-plane	Area	SM	Volume function	Levers	Moment function
A	6025	1	6025	6	36150
B	6025	4	24100	5	120500
C	6025	2	12050	4	48200
D	6025	4	24100	3	72300
E	6020	2	12040	2	24080
F	5600	4	22400	1	22400
G	5000	1	5000	0	0
			$105\,715 = \Sigma_1$		$323\,630 = \Sigma_2$

$$KB = \frac{\text{Moment about keel}}{\text{Volume of displacement}}$$

$$\therefore KB = \frac{\Sigma_2}{\Sigma_1} \times CI$$

$$= \frac{323\,630}{105\,715} \times 1.0$$

Ans. $\underline{KB} = \underline{3.06 \text{ metres}}$

$$= 0.51 \times d \text{ approximately}$$

The lever of zero was at the keel so the final answer was relative to this point, i.e. above base.

If we Simpsonize $\frac{1}{2}$ ords we will obtain *areas*.

If we Simpsonize areas we will obtain *volumes*.

Example 2

A ship is floating upright in S.W. on an even keel at 7 m draft F and A. The TPCs are as follows:

Draft (m)	1	2	3	4	5	6	7
TPC (tonnes)	60	60.3	60.5	60.5	60.5	60.5	60.5

The volume between the outer bottom and 1 m draft is 3044 cu. m, and its centre of gravity is 0.5 m above the keel. Find the ship's KB.

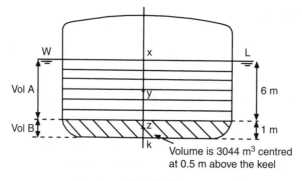

Fig. 10.19

In Figure 10.19:

Let KY represent the height of the centre of gravity of volume A above the keel, and KZ represent the height of the centre of gravity of volume B above the keel.

Let $X = \dfrac{100}{1.025}$ then the area of each water-plane is equal to TPC $\times$ X sq. m.

Σ_1 denotes the first total
Σ_2 denotes the second total $\left.\right\}$ see table below.

Draft	Area	SM	Volume function	Levers 1 m	Moment function
7	60.5	1	60.5	0	0
6	60.5	4	242.0	1	242.0
5	60.5	2	121.0	2	242.0
4	60.5	4	242.0	3	726.0
3	60.5	2	121.0	4	484.0
2	60.3	4	241.2	5	1206.0
1	60.0	1	60.0	6	360.0
			$1087.7 = \Sigma_1$		$3260.0 = \Sigma_2$

$$\text{Volume A} = \frac{1}{3} \times CI \times \Sigma_1 \times X$$

$$= \frac{1}{3} \times 1.0 \times 1087.7 \times \frac{100}{1.025} = 35\,372$$

Volume A = 35 372 cu. m

Volume B = +3 044 cu. m

Total volume = 38 416 cu. m

$$XY = \frac{\Sigma_2}{\Sigma_1} \times CI = \frac{\text{Moment}}{\text{Volume A}}$$

$$= \frac{3260}{1087.7} \times 1.0 = 3 \text{ m below 7 m waterline}$$

XY = 3 m

KX = 7 m

KX − XY = KY, so <u>KY = 4 m</u>

Volume	KG_{keel}	Moments about keel
35 372	4	141 488
+3 044	0.5	+1 522
38 416		143 010

Moments about the keel

$$\underline{KB} = \frac{\text{Total moment}}{\text{Total volume}} = \frac{143\ 010}{38\ 416} = \underline{3.72\ \text{metres}}$$

$$= 0.531 \times d$$

Summary

When using Simpson's Rules for ship calculations always use the following procedure:

1. Make a sketch using the given information.
2. Insert values into a *table* as shown in worked examples.
3. Use tabulated summations to finally *calculate* requested values.

Exercise 10

1 A ship's load water-plane is 60 m long. The lengths of the half-ordinates commencing from forward are as follows:

0.1, 3.5, 4.6, 5.1, 5.2, 5.1, 4.9, 4.3 and 0.1 m, respectively.

Calculate the area of the water-plane, the TPC in salt water, and the position of the centre of flotation, from amidships.

2 The half-ordinates of a ship's water-plane, which is 60 m long, commencing from forward, are as follows:

0, 3.8, 4.3, 4.6, 4.7, 4.7, 4.5, 4.3 and 1 m, respectively.

Find the area of the water-plane, the TPC, the coefficient of fineness of the water-plane area, and the position of the centre of flotation, from amidships.

3 The breadths at the load water-plane of a ship 90 metres long, measured at equal intervals from forward, are as follows:

0, 3.96, 8.53, 11.58, 12.19, 12.5, 11.58, 5.18, 3.44 and 0.30 m, respectively.

If the load draft is 5 metres, and the block coefficient is 0.6, find the FWA and the position of the centre of flotation, from amidships.

4 The areas of a ship's water-planes, commencing from the load draft of 24 metres, and taken at equal distances apart, are:

2000, 1950, 1800, 1400, 800, 400 and 100 sq. m, respectively.

The lower area is that of the ship's outer bottom. Find the displacement in salt water, the Fresh Water Allowance, and the height of the centre of buoyancy above the keel.

5 The areas of vertical transverse sections of a forward hold, spaced equidistantly between bulkheads, are as follows:

800, 960, 1100 and 1120 sq. m, respectively.

The length of the hold is 20 m. Find how many tonnes of coal (stowing at 4 cu. m per tonne) it will hold.

6 A ship 90 metres long is floating on an even keel at 6 m draft. The half-ordinates, commencing from forward, are as follows:

0, 4.88, 6.71, 7.31, 7.01, 6.40 and 0.9 m, respectively.

The half-ordinates 7.5 metres from bow and stern are 2.13 m and 3.35 m, respectively. Find the area of the water-plane and the change in draft if 153 tonnes of cargo is loaded with its centre of gravity vertically over the centre of flotation. Find also the position of the centre of flotation.

7 The areas of a ship's water-planes commencing from the load water-plane and spaced at equidistant intervals down to the inner bottom, are:

2500, 2000, 1850, 1550, 1250, 900 and 800 sq. m, respectively.

Below the inner bottom is an appendage 1 metre deep which has a mean area of 650 sq. m. The load draft is 7 metres. Find the load displacement in salt water, the Fresh Water Allowance, and the height of the centre of buoyancy above the keel.

8 A ship's water-plane is 80 metres long. The breadths commencing from forward are as follows:

0, 3.05, 7.1, 9.4, 10.2, 10.36, 10.3, 10.0, 8.84, 5.75 and 0 m, respectively.

The space between the first three and the last three ordinates is half of that between the other ordinates. Calculate the area of the water-plane, and the position of the centre of flotation.

9 Three consecutive ordinates in a ship's water-plane area are:

6.3, 3.35 and 0.75 m, respectively.

The common interval is 6 m. Find the area contained between the last two ordinates.

10 The transverse horizontal ordinates of a ship's amidships section commencing from the load waterline and spaced at 1 metre intervals are as follows:

16.30, 16.30, 16.30, 16.00, 15.50, 14.30 and 11.30 m, respectively.

Below the lowest ordinate there is an appendage of 8.5 sq m. Find the area of the transverse section.

11 The following table gives the area of a ship's water-plane at various drafts:

Draft (m)	6	7	8
Area (sq. m)	700	760	800

Find the volume of displacement and approximate mean TPC between the drafts of 7 and 8 m.

12 The areas of a ship's water-planes, commencing from the load water-plane and spaced 1 metre apart, are as follows:

800, 760, 700, 600, 450 and 10 sq. m, respectively.

Midway between the lowest two water-planes the area is 180 sq. m. Find the load displacement in salt water, and the height of the centre of buoyancy above the keel.

Chapter 11
Second moments of area – moments of inertia

The second moment of an element of an area about an axis is equal to the product of the area and the square of its distance from the axis. In some textbooks, this second moment of area is called the 'moment of inertia'. Let dA in Figure 11.1 represent an element of an area and let y be its distance from the axis AB.

The second moment of the element about AB is equal to $dA \times y^2$.

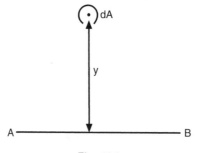

Fig. 11.1

To find the second moment of a rectangle about an axis parallel to one of its sides and passing through the centroid.

In Figure 11.2, l represents the length of the rectangle and b represents the breadth. Let G be the centroid and let AB, an axis parallel to one of the sides, pass through the centroid.

Consider the elementary strip which is shown shaded in the figure. The second moment (i) of the strip about the axis AB is given by the equation:

$$i = l dx \times x^2$$

Let I_{AB} be the second moment of the whole rectangle about the axis AB then:

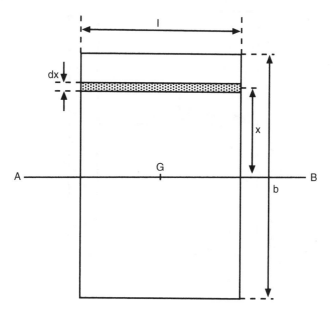

Fig. 11.2

$$I_{AB} = \int_{-b/2}^{+b/2} lx^2dx$$

$$I_{AB} = l\int_{-b/2}^{+b/2} x^2dx$$

$$= l\left[\frac{x^3}{3}\right]_{-b/2}^{+b/2}$$

$$I_{AB} = \frac{lb^3}{12}$$

To find the second moment of a rectangle about one of its sides.

Consider the second moment (i) of the elementary strip shown in Figure 11.3 about the axis AB:

$$i = ldx \times x^2$$

Let I_{AB} be the second moment of the rectangle about the axis AB, then:

$$I_{AB} = \int_{O}^{b} lx^2dx$$

$$= l\left[\frac{x^3}{3}\right]_{O}^{b}$$

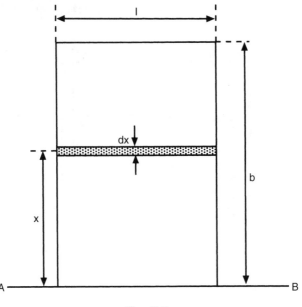

Fig. 11.3

or

$$I_{AB} = \frac{lb^3}{3}$$

The theorem of parallel axes

The second moment of an area about an axis through the centroid is equal to the second moment about any other axis parallel to the first reduced by the product of the area and the square of the perpendicular distance between the two axes. Thus, in Figure 11.4, if G represents the centroid of the area (A) and the axis OZ is parallel to AB, then:

$$I_{OZ} = I_{AB} - Ay^2 = \text{parallel axis theorem equation}$$

To find the second moment of a ship's waterplane area about the centreline.

In Figure 11.5:

$$\text{Area of elementary strip} = ydx$$
$$\text{Area of waterplane} = \int_O^L ydx$$

It has been shown in Chapter 10 that the area under the curve can be found by Simpson's Rules, using the values of y, the half-breadths, as ordinates.

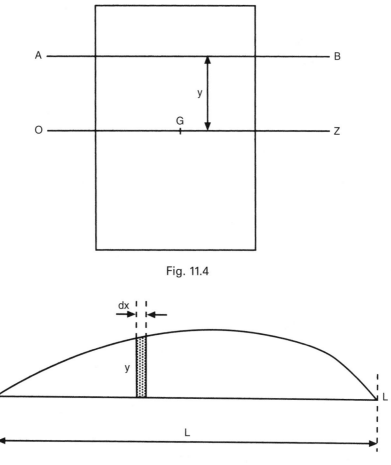

Fig. 11.4

Fig. 11.5

The second moment of a rectangle about one end is given by $\dfrac{lb^3}{3}$, and therefore the second moment of the elementary strip about the centreline is given by $\dfrac{y^3 dx}{3}$ and the second moment of the half waterplane about the centreline is given by:

$$\int_O^L \frac{y^3}{3}\,dx$$

Therefore, if I_{CL} is the second moment of the whole waterplane area about the centreline, then:

$$I_{CL} = \frac{2}{3}\int_O^L y^3 dx$$

The integral part of this expression can be evaluated by Simpson's Rules using the values of y^3 (i.e. the half-breadths cubed), as ordinates, and I_{CL} is found by multiplying the result by $\frac{2}{3}$. I_{CL} is also known as the 'moment of inertia about the centreline'.

Example 1

A ship's waterplane is 18 metres long. The half-ordinates at equal distances from forward are as follows:

$$0, 1.2, 1.5, 1.8, 1.8, 1.5 \text{ and } 1.2 \text{ metres, respectively}$$

Find the second moment of the waterplane area about the centreline.

$\frac{1}{2}$ ord.	$\frac{1}{2}$ ord.^{3}	SM	Products for I_{CL}
0	0	1	0
1.2	1.728	4	6.912
1.5	3.375	2	6.750
1.8	5.832	4	23.328
1.8	5.832	2	11.664
1.5	3.375	4	13.500
1.2	1.728	1	1.728
			$63.882 = \Sigma_1$

$$I_{CL} = \frac{2}{9} \times CI \times \Sigma_1$$

$$I_{CL} = \frac{2}{9} \times \frac{18}{6} \times 63.882$$

$$= \underline{42.588 \text{ m}^4}$$

To find the second moment of the waterplane area about a transverse axis through the centre of flotation.

$$\text{Area of elementary strip} = y \, dx$$
$$I_{AB} \text{ of the elementary strip} = x^2 y \, dx$$
$$I_{AB} \text{ of the waterplane area} = 2 \int_O^L x^2 y \, dx$$

Once again the integral part of this expression can be evaluated by Simpson's Rules using the values of $x^2 y$ as ordinates and the second moment about AB is found by multiplying the result by two.

Let OZ be a transverse axis through the centre of flotation. The second moment about OZ can then be found by the theorem of parallel axes shown in Figure 11.6, i.e.

$$I_{OZ} = I_{AB} - A\bar{X}^2$$

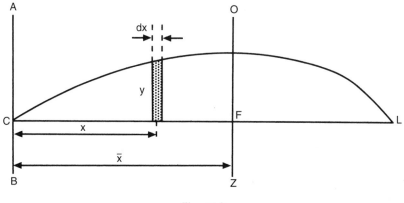

Fig. 11.6

Example 2

A ship's waterplane is 18 metres long. The half-ordinates at equal distances from forward are as follows:

0, 1.2, 1.5, 1.8, 1.8, 1.5 and 1.2 metres, respectively

Find the second moment of the waterplane area about a transverse axis through the centre of flotation.

$\frac{1}{2}$ord.	SM	Area func.	Lever	Moment func.	Lever	Inertia func.
0	1	0	0	0	0	0
1.2	4	4.8	1	4.8	1	4.8
1.5	2	3.0	2	6.0	2	12.0
1.8	4	7.2	3	21.6	3	64.8
1.8	2	3.6	4	14.4	4	57.6
1.5	4	6.0	5	30.0	5	150.0
1.2	1	1.2	6	7.2	6	43.2
		$25.8 = \Sigma_1$		$84.0 = \Sigma_2$		$332.4 = \Sigma_3$

$$\text{Area of waterplane} = \frac{1}{3} \times CI \times \Sigma_1 \times 2$$

$$= \frac{1}{3} \times \frac{18}{6} \times 25.8 \times 2$$

$$= 51.6 \text{ sq. m}$$

Distance of the centre of flotation from forward
$$= \frac{\Sigma_2}{\Sigma_1} \times CI$$

$$= \frac{84}{25.8} \times \frac{18}{6}$$

$$= 9.77 \text{ m}$$

$$= 0.77 \text{ m aft of amidships}$$

$$I_{AB} = \frac{1}{3} \times (CI)^3 \times \Sigma_3 \times 2$$

$$= \frac{1}{3} \times \left(\frac{18}{6}\right)^3 \times 332.4 \times 2 = 5983 \text{ m}^4$$

$$I_{OZ} = I_{AB} - A\bar{X}^2$$
$$= 5983 - 51.6 \times 9.77^2$$
$$= 5983 - 4925$$

Ans. $I_{OZ} = 1058 \text{ metres}^4$

There is a quicker and more efficient method of obtaining the solution to the above problem. Instead of using the foremost ordinate at the datum, use the midship ordinate. Proceed as follows:

$\frac{1}{2}$ord.	SM	Area func.	Lever$_\text{⦻}$	Moment func.	Lever$_\text{⦻}$	Inertia func.
0	1	0	−3	0	−3	0
1.2	4	4.8	−2	−9.6	−2	+19.2
1.5	2	3.0	−1	−3.0	−1	+3.0
1.8	4	7.2	0	0	0	0
1.8	2	3.6	+1	+3.6	+1	+3.6
1.5	4	6.0	+2	+12.0	+2	+24.0
1.2	1	1.2	+3	+3.6	+3	+10.8
		25.8 = Σ_1		+6.6 = Σ_2		60.6 = Σ_3

$$\text{Area of waterplane} = \frac{1}{3} \times \Sigma_1 \times h \times 2 \qquad h = \frac{18}{6} = 3 \text{ m}$$

$$= \frac{1}{3} \times 25.8 \times 3 \times 2$$

$$= 51.6 \text{ m}^2 \text{ (as before)}$$

$$\Sigma_2 = +6.6$$

The +ve sign shows centre of flotation is in aft body

$$\text{Centre of flotation from } \text{⦻} = \frac{\Sigma_2}{\Sigma_1} \times h$$

$$= +\frac{6.6}{25.8} \times 3$$

∴ Centre of flotation = +0.77 m or 0.77 m aft amidships (as before)

$$I_\text{⦻} = \tfrac{1}{3} \times \Sigma_3 \times h^3 \times 2 = \tfrac{1}{3} \times 60.6 \times 3^3 \times 2$$
$$\therefore I_\text{⦻} = 1090.8 \text{ m}^4$$

But

$$I_{LCF} = I_{\cancel{90}} - A\bar{y} = 1090.8 - (51.6 \times 0.77^2)$$
$$= 1090.8 - 30.6$$
$$= \underline{1060 \text{ m}^4}$$

i.e. very close to previous answer of 1058 m^4.

With this improved method the levers are much less in value. Consequently the error is decreased when predicting $LCF_{\cancel{90}}$ and I_{LCF}. $I_{\cancel{90}}$ is also known as the 'moment of inertia about amidships'. I_{LCF} is also known as the 'moment of inertia about the LCF'.

Summary

When using Simpson's Rules for second moments of area the procedure should be as follows:

1. Make a sketch from the given information.
2. Use a moment table and insert values.
3. Using summations obtained in the table proceed to calculate area, LCF, $I_{\cancel{90}}$, I_{LCF}, I_{CL}, etc.
4. Remember: sketch, table, calculation.

Exercise 11

1 A large square has a smaller square cut out of its centre such that the second moment of the smaller square about an axis parallel to one side and passing through the centroid is the same as that of the portion remaining about the same axis. Find what proportion of the area of the original square is cut out.

2 Find the second moment of a square of side 2a about its diagonals.

3 Compare the second moment of a rectangle $40 \text{ cm} \times 30 \text{ cm}$ about an axis through the centroid and parallel to the 40 cm side with the second moment about an axis passing through the centroid and parallel to the 30 cm side.

4 An H-girder is built from 5 cm thick steel plate. The central web is 25 cm high and the overall width of each of the horizontal flanges is 25 cm. Find the second moment of the end section about an axis through the centroid and parallel to the horizontal flanges.

5 A ship's waterplane is 36 m long. The half-ordinates, at equidistant intervals, commencing from forward, are as follows:

0, 4, 5, 6, 6, 5 and 4 m, respectively

Calculate the second moment of the waterplane area about the centreline and also about a transverse axis through the centre of flotation.

6   A ship's waterplane is 120 metres long. The half-ordinates at equidistant intervals from forward are as follows:

0, 3.7, 7.6, 7.6, 7.5, 4.6 and 0.1 m, respectively

Calculate the second moment of the waterplane area about the centreline and about a transverse axis through the centre of flotation.

7   A ship of 12 000 tonnes displacement is 150 metres long at the waterline. The half-ordinates of the waterplane at equidistant intervals from forward are as follows:

0, 4, 8.5, 11.6, 12.2, 12.5, 12.5, 11.6, 5.2, 2.4 and 0.3 m, respectively

Calculate the longitudinal and transverse BM'

8   The half-ordinates of a ship's waterplane at equidistant intervals from forward are as follows:

0, 1.3, 5.2, 8.3, 9.7, 9.8, 8.3, 5.3 and 1.9 m, respectively

If the common interval is 15.9 metres, find the second moment of the waterplane area about the centreline and a transverse axis through the centre of flotation.

9   A ship's waterplane is 120 metres long. The half-ordinates commencing from aft are as follows:

0, 1.3, 3.7, 7.6, 7.6, 7.5, 4.6, 1.8 and 0.1 m, respectively

The spacing between the first three and the last three half-ordinates is half of that between the other half-ordinates. Calculate the second moment of the waterplane area about the centreline and about a transverse axis through the centre of flotation.

10   A ship's waterplane is 90 metres long between perpendiculars. The half-ordinates at this waterplane are as follows:

Station	AP	$\frac{1}{2}$	1	2	3	4	5	$5\frac{1}{2}$	FP
$\frac{1}{2}$ Ords (m)	0	2	4.88	6.71	7.31	7.01	6.40	2	0.9

Calculate the second moment of the waterplane area about the centreline and also about a transverse axis through the centre of flotation.

Chapter 12
Calculating KB, BM and metacentric diagrams

The method used to determine the final position of the centre of gravity is examined in Chapter 13. To ascertain the GM for any condition of loading it is necessary also to calculate the KB and BM (i.e. KM) for any draft.

To find KB

The centre of buoyancy is the centre of gravity of the underwater volume.

For a box-shaped vessel on an even keel, the underwater volume is rectangular in shape and the centre of buoyancy will be at the half-length, on the centre line, and at half the draft as shown in Figure 12.1(a).

Therefore, for a box-shaped vessel on an even keel: KB = $\frac{1}{2}$ draft.

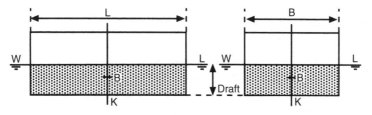

Fig. 12.1(a) Box-shaped vessel.

For a vessel which is in the form of a triangular prism as shown in Figure 12.1(b) the underwater section will also be in the form of a triangular prism. The centroid of a triangle is at 2/3 of the median from the apex. Therefore the centre of buoyancy will be at the half-length, on the centre line, but the KB = 2/3 draft.

For an ordinary ship the KB may be found fairly accurately by Simpson's Rules as explained in Chapter 10. The approximate depth of the centre of buoyancy of a ship *below* the waterline usually lies between 0.44 × draft

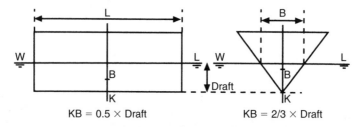

KB = 0.5 × Draft   KB = 2/3 × Draft

Fig. 12.1(b)   Triangular-shaped vessel.

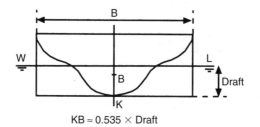

KB ≈ 0.535 × Draft

Fig. 12.1(c)   Ship-shaped vessel.

and 0.49 × draft. A closer approximation of this depth can be obtained by using Morrish's formula, which states:

$$\text{Depth of centre of buoyancy } below \text{ waterline} = \frac{1}{3}\left(\frac{d}{2} + \frac{V}{A}\right)$$

where
d = mean draft
V = volume of displacement
A = area of the water-plane

The derivation of this formula is as follows:

In Figure 12.2, let ABC be the curve of water-plane areas plotted against drafts to the load waterline. Let DE = V/A and draw EG parallel to the base cutting the diagonal FD in H.

It must first be shown that area DAHC is equal to area DABC.

$$\text{Rectangle AH} = \text{Rectangle HC}$$
$$\therefore \text{Triangle AGH} = \text{Triangle HEC}$$

and

$$\text{Area AHCD} = \text{Area AGED}$$
$$\text{Area AGED} = V/A \times A$$
$$= V$$

but

$$\text{Area DABC} = \text{V}$$

$$\therefore \text{Area DAHC} = \text{Area DABC}$$

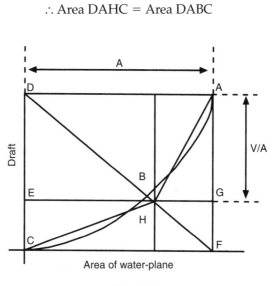

Fig. 12.2

The distance of the centroid of DABC below AD is the distance of the centre of buoyancy below the load waterline. It is now assumed that the centroid of the area DAHC is the same distance below the load waterline as the centroid of area DABC.

To find the distance of the centroid of area DAHC below AD.

$$\frac{\text{Area AGH}}{\text{Area AGED}} = \frac{\frac{1}{2}\text{AG} \times \text{GH}}{\text{AG} \times \text{AD}}$$

$$= \frac{1}{2}\frac{\text{GH}}{\text{AD}}$$

$$= \frac{1}{2}\frac{\text{GH}}{\text{AF}}$$

$$= \frac{1}{2}\frac{\text{AF} - \text{AG}}{\text{AF}}$$

$$= \frac{1}{2}\left(\frac{\text{d} - \text{AG}}{\text{d}}\right)$$

$$\therefore \text{Area AGH} = \frac{1}{2}\frac{(\text{d} - \text{V/A})}{\text{d}} \times \text{Area AGED}$$

The centroid of AGED is $\frac{1}{2}\frac{\text{V}}{\text{A}}$ from AD.

Now let triangle AGH be shifted to HEC.

The centroid of AGED will move parallel to the shift of the centroid of AGH and the vertical component of this shift (x) is given by:

$$x = \frac{AGH \times d/3}{AGED}$$

$$= \frac{\frac{1}{2}\left(\frac{d - V/A}{d}\right) \times \frac{d}{3} \times AGED}{AGED}$$

$$= \frac{1}{2}\left(\frac{d - V/A}{d}\right) \times \frac{d}{3}$$

$$= \frac{1}{6}(d - V/A)$$

The new vertical distance of the centroid *below* AD will now be given by:

$$\text{Distance } below \text{ AD} = \frac{1}{2}\frac{V}{A} + \frac{1}{6}\left(d - \frac{V}{A}\right)$$

$$= \frac{1}{3}\frac{V}{A} + \frac{1}{6}d$$

$$= \frac{1}{3}\left(\frac{d}{2} + \frac{V}{A}\right)$$

Therefore the distance of the centre of buoyancy *below* the load waterline is given by the formula:

$$\text{Distance } below \text{ LWL} = \frac{1}{3}\left(\frac{d}{2} + \frac{V}{A}\right)$$

This is known as *Morrish's* or *Normand's formula* and will give very good results for merchant ships.

To find transverse BM

The Transverse BM is the height of the transverse metacentre above the centre of buoyancy and is found by using the formula:

$$BM = \frac{1}{V}$$

where

1 = the second moment of the water-plane area about the centre line,
V = the ship's volume of displacement.

The derivation of this formula is as follows:

Consider a ship inclined to a small angle (θ) as shown in Figure 12.3(a). Let 'y' be the half-breadth.

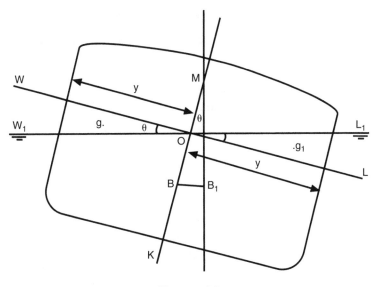

Fig. 12.3(a)

Since θ is a small angle then arc WW_1 = arc LL_1
$$= \theta y$$

Also:

Area of wedge WOW_1 = Area of wedge LOL_1
$$= \tfrac{1}{2}\theta y^2$$

Consider an elementary wedge of longitudinal length dx as in Figure 12.3(b).

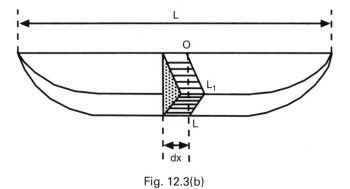

Fig. 12.3(b)

$$\text{The volume of this wedge} = \tfrac{1}{2}\theta y^2 \, dx$$

$$\text{The moment of the wedge about the centre line} = \tfrac{1}{2}\theta y^2 \, dx \times \tfrac{2}{3} y$$

$$= \tfrac{1}{3}\theta y^3 \, dx$$

$$\text{The total moment of both wedges about the centre line} = \tfrac{2}{3}\theta y^3 \, dx$$

$$\text{The sum of the moments of all such wedges} = \int_O^L \tfrac{2}{3}\theta y^3 \, dx$$

$$= \theta \int_O^L \tfrac{2}{3} y^3 \, dx$$

But

$$\int_O^L \tfrac{2}{3} y^3 \, dx = \left. \begin{array}{l} \text{The second moment} \\ \text{of the water-plane} \\ \text{area about the ship's centre line} \end{array} \right\} = I$$

$$\therefore \text{ The sum of the moments of the wedges} = I \times \theta$$

$$\text{But the sum of the moments} = v \times gg_1$$

where v is the volume of the immersed or emerged wedge.

$$\therefore \; I \times \theta = v \times gg_1$$

or

$$I = \frac{v \times gg_1}{\theta} \tag{I}$$

Now:

$$BB_1 = \frac{v \times gg_1}{V}$$

and

$$BB_1 = BM \times \theta$$

$$\therefore \; BM \times \theta = \frac{v \times gg_1}{V}$$

or

$$BM \times V = \frac{v \times gg_1}{\theta}$$

Substituting in (I) above

$$BM \times V = I$$

$$\therefore \; BM = \frac{I}{V}$$

For a rectangular water-plane area the second moment about the centre line is found by the formula:

$$I = \frac{LB^3}{12}$$

where
L = the length of the water-plane
B = the breadth of the water-plane

Thus, for a vessel having a rectangular water-plane area:

$$BM = \frac{LB^3}{12V}$$

For a box-shaped vessel:

$$BM = \frac{1}{V}$$

$$= \frac{LB^3}{12V}$$

$$= \frac{L \times B^3}{12 \times L \times B \times draft}$$

$$\therefore \ BM = \frac{B^2}{12d}$$

where
B = the beam of the vessel
d = any draft of the vessel
B = constant
d = variable

For a triangular-shaped prism:

$$BM = \frac{1}{V}$$

$$= \frac{LB^3}{12V}$$

$$= \frac{L \times B^3}{12(\frac{1}{2} \times L \times B \times draft)}$$

$$\therefore \ BM = \frac{B^2}{6d}$$

where
   B = the breadth *at the waterline*
   d = the corresponding draft
B, d = variables

Example 1

A box-shaped vessel is $24\,m \times 5\,m \times 5\,m$ and floats on an even keel at $2\,m$ draft. KG = 1.5 m. Calculate the initial metacentric height.

$$KB = \frac{1}{2}\,draft \qquad BM = \frac{B^2}{12d}$$

$$KB = 1\,m \qquad BM = \frac{5^2}{12 \times 2}$$

$$BM = 1.04\,m$$

$$
\begin{aligned}
KB &= 1.00\,m \\
BM &= +1.04\,m \\
KM &= 2.04\,m \\
KG &= -1.50\,m \\
GM &= 0.54\,m
\end{aligned}
$$

Ans. $\underline{GM = +0.54\,m}$

Example 2

A vessel is in the form of a triangular prism $32\,m$ long, $8\,m$ wide at the top and $5\,m$ deep. KG = 3.7 m. Find the initial metacentric height when floating on even keel at $4\,m$ draft F and A.

Let 'x' be the half-breadth *at the waterline*, as shown in Figure 12.4. Then

$$\frac{x}{4} = \frac{4}{5}$$

$$x = \frac{16}{5}$$

$$x = 3.2\,m$$

$$\therefore \ \text{The breadth at the waterline} = 6.4\,m$$

$$KB = 2/3\,draft \qquad BM = \frac{B^2}{6d}$$

$$= 2/3 \times 4$$

$$KB = 2.67\,m \qquad = \frac{6.4 \times 6.4}{6 \times 4}$$

$$BM = 1.71\,m$$

$$
\begin{aligned}
KB &= 2.67\,m \\
BM &= +1.71\,m \\
KM &= 4.38\,m \\
KG &= -3.70\,m \\
GM &= 0.68\,m
\end{aligned}
$$

Ans. $\underline{GM = +0.68\,m}$

Note how the breadth 'B' would decrease at the lower drafts. See also Figure 12.6(b).

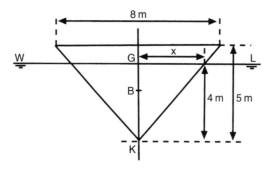

Fig. 12.4

Example 3

The second moment of a ship's water-plane area about the centre line is 20 000 m⁴ units. The displacement is 7000 tonnes whilst floating in dock water of density 1008 kg per cu. m. KB = 1.9 m and KG = 3.2 m. Calculate the initial metacentric height.

$$\text{Volume of water displaced} = \frac{7000 \times 1000}{1008} \text{ cu. m} = 6944 \text{ cu. m}$$

$$BM = \frac{1}{V}$$

$$\therefore BM = \frac{20\,000}{6944}$$

$$BM = 2.88 \text{ m}$$
$$KB = +1.90 \text{ m}$$
$$KM = 4.78 \text{ m}$$
$$KG = 3.20 \text{ m}$$

Ans. GM = +1.58 m

Metacentric diagrams

It has been mentioned in Chapter 6 that the officer responsible for loading a ship should aim to complete the loading with a GM which is neither too large nor too small. See table of typical GM values at the end of the Chapter 6 for merchant ships when fully loaded. A metacentric diagram is a figure in graph from which the KB, BM and thus the KM can be found for any draft by inspection. If the KG is known and the KM is found from the diagram, the difference will give the GM. Also, if a final GM be decided upon, the KM can be taken from the graph and the difference will give the required final KG.

The diagram is usually drawn for drafts between the light and loaded displacements, i.e. 3 m and 13 m, respectively overpage.

Figure 12.5 shows a metacentric diagram drawn for a ship having the following particulars:

Draft (m)	KB (m)	KM (m)
13	6.65	11.60
12	6.13	11.30
11	5.62	11.14
10	5.11	11.10
9	4.60	11.15
8	4.10	11.48
7	3.59	11.94
6	3.08	12.81
5	2.57	14.30
4	2.06	16.63
3	1.55	20.54
–	–	–

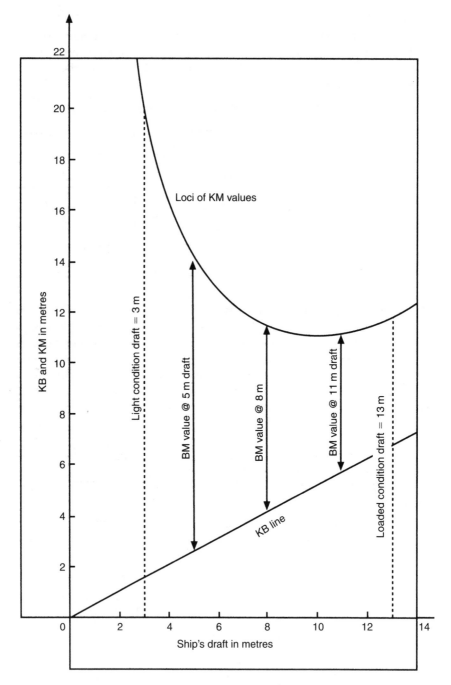

Fig. 12.5 Metacentric diagram for a ship-shaped vessel.

The following is a description of the method used in constructing this diagram. The scale on the left-hand side represents a scale of metres, and it is from this scale that all measurements are to be taken.

First the curve of the Centres of Buoyancy is plotted.

For each draft plot the corresponding KB. For example, plot 6.65 m @ 13 m, 6.13 m @ 12 m draft and so on to 1.55 m @ 3 m draft.

Join these points together to form the KB curve. In practice it will be very close to being a straight line because the curvature will be so small. See Figure 12.5.

Next the KM curve or Locus of Metacentres. For each draft plot the corresponding KM value given in the table.

At 13 m plot 11.60 m. At 12 m plot 11.30 m and so on down to plotting 20.54 m KM @ 3 m draft.

These points are then joined by a smooth curve as shown in Figure 12.5.

Note how it is possible for two different drafts to have the same value of KM in the range of drafts from 7 m to 13 m approximately.

For any draft being considered, the vertical distance between the KB line and the KM curve gives the BM value.

To find the KB's and KM's the vertical distances are measured from the base line to the curves.

Example 1

Construct the metacentric diagram for a box-shaped vessel 64 m long, 10 m beam and 6 m deep, for even keel drafts at 0.5 m intervals between the light draft 1 metre and the load draft 5 m. Also, from the diagram find:

(a) The minimum KM and the draft at which it occurs.
(b) The BM at 3.5 m.

Draft	$KB = \frac{1}{2} draft$	$BM = B^2/12d$	$KM = KB + BM$
1 m	0.5 m	8.33 m	8.83 m
1.5 m	0.75 m	5.56 m	6.31 m
2 m	1.0 m	4.17 m	5.17 m
2.5 m	1.25 m	3.33 m	4.58 m
3.0 m	1.5 m	2.78 m	4.28 m
3.5 m	1.75 m	2.38 m	4.13 m
4.0 m	2.00 m	2.08 m	4.08 m
4.5 m	2.25 m	1.85 m	4.10 m
5.0 m	2.5 m	1.67 m	4.17 m

See Figure 12.6(a) for KB and KM plotted against draft.

Explanation. To find the minimum KM, draw a horizontal tangent to the lowest point of the curve of metacentres, i.e. through A. The point where the tangent cuts the scale will give the minimum KM and the draft at which it occurs.

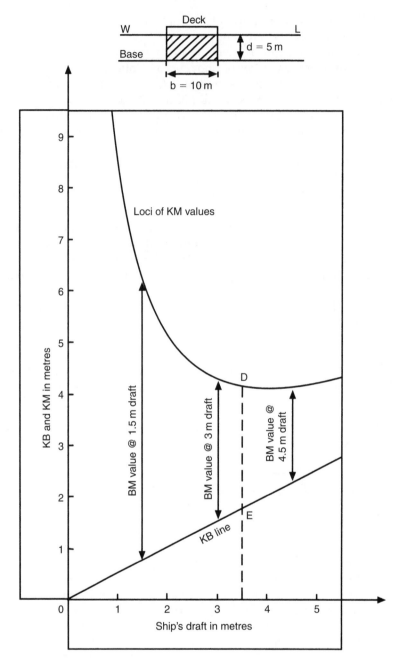

Fig. 12.6(a) Metacentric diagram for a box-shaped vessel.

Note. It is shown below that for a box-shaped vessel the minimum KM and the draft at which it occurs are both given by $B/\sqrt{6}$, where B is the beam.

Therefore, the answer to part (*a*) of the question is:

$$\text{Minimum KM} = 4.08 \text{ m occurring at } 4.08 \text{ m draft}$$

To find the BM at 3.5 m draft, measure the distance DE on the scale and it will give the BM (2.38 m).

Therefore, the answer to part (*b*) of the question is:

$$\text{BM at } 3.5 \text{ m draft} = 2.38 \text{ m}$$

To show that, for a box-shaped vessel, the minimum KM and the draft at which it occurs are both given by the expression $B/\sqrt{6}$, where B is equal to the vessel's beam.

$$\text{KM} = \text{KB} + \text{BM}$$

For a box-shaped vessel:

$$\text{KM} = \frac{d}{2} + \frac{B^2}{12d} \quad \text{------------------------------ (I)}$$

$$\frac{\partial \text{KM}}{\partial d} = \frac{1}{2} + \frac{B^2}{12d^2}$$

For minimum KM:

$$\frac{\partial \text{KM}}{\partial d} = O$$

$$\therefore O = \frac{1}{2} + \frac{B^2}{12d^2}$$

$$B^2 = 6d^2$$

and

$$d = B/\sqrt{6}$$

Substituting in Equation (I) above:

$$\text{Minimum KM} = \frac{B}{2\sqrt{6}} + \frac{B^2\sqrt{6}}{12B}$$

$$= \frac{6B + 6B}{12\sqrt{6}}$$

$$\text{Minimum KM} = B/\sqrt{6}$$

Figure 12.6(b) shows a metacentric diagram for a triangular-shaped underwater form with apex at the base. Note how the KM values have produced a straight line instead of the parabolic curve of the rectangular hull form. Note also how BM increases with *every increase* in *draft*.

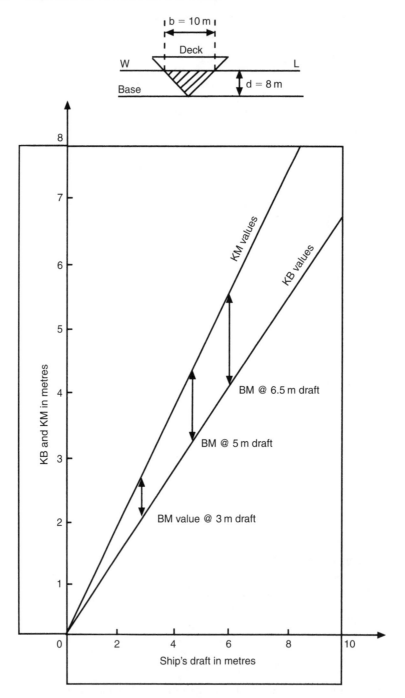

Fig. 12.6(b) Metacentric diagram for triangular-shaped vessel.

Exercise 12

1   A box-shaped vessel 75 m long, 12 m beam and 7 m deep, is floating on an even keel at 6 m draft. Calculate the KM.

2   Compare the initial metacentric heights of two barges, each 60 m long, 10 m beam at the waterline, 6 m deep, floating upright on an even keel at 3 m draft and having KG = 3 m. One barge is in the form of a rectangular prism and the other is in the form of a triangular prism, floating apex downwards.

3   Two box-shaped vessels are each 100 m long, 4 m deep, float at 3 m draft and have KG = 2.5 m. Compare their initial Metacentric Heights if one has 10 m beam and the other has 12 m beam.

4   Will a homogeneous log of square cross-section and relative density 0.7 have a positive initial Metacentric Height when floating in fresh water with one side parallel to the waterline? Verify your answer by means of a calculation.

5   A box-shaped vessel 60 m × 12 m × 5 m is floating on an even keel at a draft of 4 m. Construct a metacentric diagram for drafts between 1 m and 4 m. From the diagram find:
(a) the KM's at drafts of 2.4 m and 0.9 m,
(b) the draft at which the minimum KM occurs.

6   Construct a metacentric diagram for a box-shaped vessel 65 m × 12 m × 6 m for drafts between 1 m and 6 m. From the diagram find:
(a) the KM's at drafts of 1.2 m and 3.6 m,
(b) the minimum KM and the draft at which it occurs.

7   Construct a metacentric diagram for a box-shaped vessel 70 m long and 10 m beam, for drafts between 1 m and 6 m. From the diagram find:
(a) the KM's at drafts of 1.5 m and 4.5 m,
(b) the draft at which the minimum KM occurs.

8   A box-shaped vessel is 60 m long, 13.73 m wide and floats at 8 m even-keel draft in salt water.
(a) Calculate the KB, BM and KM values for drafts 3 m to 8 m at intervals of 1 m. From your results draw the Metacentric Diagram.
(b) At 3.65 m draft even keel, it is known that the VCG is 4.35 m above base. Using your diagram, estimate the transverse GM for this condition of loading.
(c) At 5.60 m draft even keel, the VCG is also 5.60 m above base. Using your diagram, estimate the GM for this condition of loading. What state of equilibrium is the ship in?

Draft (m)	3	4	5	6	7	8
KM (m)	6.75	5.94	5.64	5.62	5.75	5.96

Chapter 13
Final KG plus twenty reasons for a rise in G

When a ship is completed by the builders, certain written stability information must be handed over to the shipowner with the ship. Details of the information required are contained in the 1998 load line rules, parts of which are reproduced in Chapter 55. The information includes details of the ship's Lightweight, the Lightweight VCG and LCG, and also the positions of the centres of gravity of cargo and bunker spaces. This gives an initial condition from which the displacement and KG for any condition of loading may be calculated. The final KG is found by taking the moments of the weights loaded or discharged, about the keel. For convenience, when taking the moments, consider the ship to be on her beam ends.

In Figure 13.1(a), KG represents the original height of the centre of gravity above the keel, and W represents the original displacement. The original moment about the keel is therefore $W \times KG$.

Now load a weight w_1 with its centre of gravity at g_1 and discharge w_2 from g_2. This will produce moments about the keel of $w_1 \times Kg_1$ and $w_2 \times Kg_2$ in directions indicated in the figure. The final moment about the keel will be equal to the original moment plus the moment of the weight added minus the moment of the weight discharged. But the final moment must also be equal to the final displacement multiplied by the final KG as shown in Figure 13.1(b); i.e.

$$\text{Final moment} = \text{Final KG} \times \text{Final displacement}$$

or

$$\text{Final KG} = \frac{\text{Final moment}}{\text{Final displacement}}$$

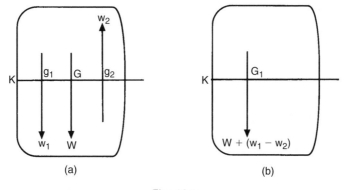

Fig. 13.1

Example 1

A ship of 6000 tonnes displacement has KG = 6 m and KM = 7.33 m. The following cargo is loaded:

1000 tonnes	KG 2.5 m
500 tonnes	KG 3.5 m
750 tonnes	KG 9.0 m

The following is then discharged:

450 tonnes of cargo KG 0.6 m

and

800 tonnes of cargo KG 3.0 m

Find the final GM.

Weight	KG	Moment about the keel
+6000	6.0	+36 000
+1000	2.5	+2500
+500	3.5	+1750
+750	9.0	+6750
+8250		+47 000
−450	0.6	−270
−800	3.0	−2400
+7000		+44 330

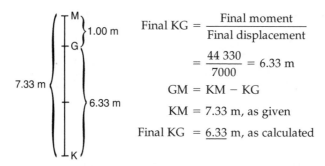

$$\text{Final KG} = \frac{\text{Final moment}}{\text{Final displacement}}$$

$$= \frac{44\,330}{7000} = 6.33 \text{ m}$$

$$\text{GM} = \text{KM} - \text{KG}$$

$$\text{KM} = 7.33 \text{ m, as given}$$

$$\text{Final KG} = \underline{6.33} \text{ m, as calculated}$$

Ans. Final GM = 1.00 m

Note. KM was assumed to be similar value at 6000 tonnes and 7000 tonnes displacement. This is feasible. As can be seen on Figure 6.2, it is possible to have the same KM at two different drafts.

Example 2

A ship of 5000 tonnes displacement has KG 4.5 m and KM 5.3 m. The following cargo is loaded:

2000 tonnes KG 3.7 m   and   1000 tonnes KG 7.5 m

Find how much deck cargo (KG 9 m) may now be loaded if the ship is to sail with a minimum GM of 0.3 m.

Let 'x' tonnes of deck cargo be loaded, so that the vessel sails with GM = 0.3 m.

Final KM = 5.3 m

Final GM = 0.3 m

Final KG = 5.0 m

$$\text{Final KG} = \frac{\text{Final moment}}{\text{Final displacement}} = 5.0 \text{ m}$$

$$\therefore 5 = \frac{37\,400 + 9x}{8000 + x} = \frac{\Sigma_v}{\Sigma_1}$$

$$40\,000 + 5x = 37\,400 + 9x$$

$$2600 = 4x$$

$$x = 650 \text{ tonnes}$$

Ans. Maximum to load = 650 tonnes

Weight	KG	Moment about the keel
5000	4.5	22 500
2000	3.7	7400
1000	7.5	7500
x	9.0	9x
$(8000 + x) = \Sigma_1$		$(37\,400 + 9x) = \Sigma_v$

Twenty reasons for a rise in G

When the vertical centre of gravity G rises, there will normally be a loss in the ship's stability. G may even rise above the transverse metacentre M to make the ship unstable. The master and mate onboard ship must be aware of changes in a ship that would cause such a rise in G. The following list gives reasons for such a rise:

1. Free-surface effects in partially filled tanks.
2. Collapse of a longitudinal division/bulkhead in a partially filled tank of liquid.
3. Icing up of superstructures.
4. Loading cargo in upper reaches of the vessel.
5. Water entering the ship through badly maintained hatches on upper deck and flooding the tween decks.
6. Hatches or bow doors inadvertently left open on the main deck.
7. Water landing on the deck from the sea in heavy weather conditions.
8. Raising of a weight from a deck using a mast and derrick.
9. Raising a weight low down in the ship to a higher position within the ship.
10. Timber deck cargo becoming saturated due to bad weather conditions.
11. Vessel making first contact with keel blocks in a dry dock at the stern.
12. A ship's first contact with a raised shelf or submerged wreck.
13. The raising of the sails on a yacht.
14. A bilging situation, causing free-surface effects.
15. A collapse of grain-boards or fish-boards.
16. A blockage of freeing ports or scuppers on the upper deck.
17. Passengers crowding on superstructure decks at time of departure or arrival.
18. Adding weight at a point *above* the ship's initial overall VCG.
19. Discharging a weight at a point *below* the ship's initial overall VCG.
20. Retrofits in accommodation decks and navigation spaces.

Exercise 13

1   A ship has a displacement of 1800 tonnes and KG = 3 m. She loads 3400 tonnes of cargo (KG = 2.5 m) and 400 tonnes of bunkers (KG = 5.0 m). Find the final KG.

2   A ship has a light displacement of 2000 tonnes and light KG = 3.7 m. She then loads 2500 tonnes of cargo (KG = 2.5 m) and 300 tonnes of bunkers (KG = 3 m). Find the new KG.

3   A ship sails with displacement 3420 tonnes and KG = 3.75 m. During the voyage bunkers were consumed as follows: 66 tonnes (KG = 0.45 m) and 64 tonnes (KG = 2 m). Find the KG at the end of the voyage.

4   A ship has displacement 2000 tonnes and KG = 4 m. She loads 1500 tonnes of cargo (KG = 6 m), 3500 tonnes of cargo (KG = 5 m) and 1520 tonnes of bunkers (KG = 1 m). She then discharges 2000 tonnes of cargo (KG = 2.5 m)

and consumes 900 tonnes of oil fuel (KG = 0.5 m) during the voyage. Find the final KG on arrival at the port of destination.

5   A ship has a light displacement of 2000 tonnes (KG = 3.6 m). She loads 2500 tonnes of cargo (KG = 5 m) and 300 tonnes of bunkers (KG = 3 m). The GM is then found to be 0.15 m. Find the GM with the bunkers empty.

6   A ship has a displacement of 3200 tonnes (KG = 3 m and KM = 5.5 m). She then loads 5200 tonnes of cargo (KG = 5.2 m). Find how much deck cargo having a KG = 10 m may now be loaded if the ship is to complete loading with a positive GM of 0.3 m.

7   A ship of 5500 tonnes displacement has KG 5 m, and she proceeds to load the following cargo:

> 1000 tonnes KG 6 m
> 700 tonnes KG 4 m
> 300 tonnes KG 5 m

She then discharges 200 tonnes of ballast KG 0.5 m. Find how much deck cargo (KG = 10 m) can be loaded so that the ship may sail with a positive GM of 0.3 metres. The load KM is 6.3 m.

8   A ship of 3500 tonnes light displacement and light KG 6.4 m has to load 9600 tonnes of cargo. The KG of the lower hold is 4.5 m, and that of the tween deck is 9 m. The load KM is 6.2 m and, when loading is completed, the righting moment at 6 degrees of heel is required to be 425 tonnes m. Calculate the amount of cargo to be loaded into the lower hold and tween deck, respectively. (Righting moment = W × GM × sin heel.)

9   A ship arrives in port with displacement 6000 tonnes and KG 6 m. She then discharges and loads the following quantities:

Discharge	1250 tonnes of cargo	KG 4.5 metres
	675 tonnes of cargo	KG 3.5 metres
	420 tonnes of cargo	KG 9.0 metres
Load	980 tonnes of cargo	KG 4.25 metres
	550 tonnes of cargo	KG 6.0 metres
	700 tonnes of bunkers	KG 1.0 metre
	70 tonnes of FW	KG 12.0 metres

During the stay in port 30 tonnes of oil (KG 1 m) are consumed. If the final KM is 6.8 m, find the GM on departure.

10   A ship has light displacement 2800 tonnes and light KM 6.7 m. She loads 400 tonnes of cargo (KG 6 m) and 700 tonnes (KG 4.5 m). The KG is then found to be 5.3 m. Find the light GM.

11   A ship's displacement is 4500 tonnes and KG 5 m. The following cargo is loaded:

> 450 tonnes KG 7.5 m
> 120 tonnes KG 6.0 m
> 650 tonnes KG 3.0 m

Find the amount of cargo to load in a tween deck (KG 6 m) so that the ship sails with a GM of 0.6 m. (The load KM is 5.6 m.)

12   A ship of 7350 tonnes displacement has KG 5.8 m and GM 0.5 m. Find how much deck cargo must be loaded (KG 9 m) if there is to be a metacentric height of not less than 0.38 m when loading is completed.

13   A ship is partly loaded and has a displacement of 9000 tonnes, KG 6 m, and KM 7.3 m. She is to make a 19-day passage consuming 26 tonnes of oil per day (KG 0.5 m). Find how much deck cargo she may load (KG 10 m) if the GM on arrival at the destination is to be not less than 0.3 m.

Chapter 14
Angle of list

Consider a ship floating upright as shown in Figure 14.1. The centres of gravity and buoyancy are on the centreline. The resultant force acting on the ship is zero, and the resultant moment about the centre of gravity is zero.

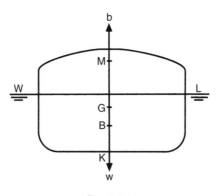

Fig. 14.1

Now let a weight already on board the ship be shifted transversely such that G moves to G_1 as in Figure 14.2(a). This will produce a listing moment of $W \times GG_1$, and the ship will list until G_1 and the centre of buoyancy are in the same vertical line as in Figure 14.2(b).

In this position G_1 will also lie vertically under M so long as the angle of list is small. Therefore, if the final positions of the metacentre and the centre of gravity are known, the final list can be found, using trigonometry, in the triangle GG_1M which is right-angled at G.

The final position of the centre of gravity is found by taking moments about the keel and about the centreline.

Note. It will be found more convenient in calculations, when taking moments, to consider the ship to be upright throughout the operation.

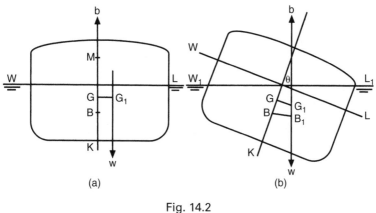

Fig. 14.2

Example 1

A ship of 6000 tonnes displacement has KM = 7.3 m and KG = 6.7 m, and is floating upright. A weight of 60 tonnes already on board is shifted 12 m transversely. Find the resultant list.

Figure 14.3(a) shows the initial position of G before the weight was shifted and Figure 14.3(b) shows the final position of G after the weight has been shifted.

When the weight is shifted transversely the ship's centre of gravity will also shift transversely, from G to G_1. The ship will then list θ degrees to bring G_1 vertically under M the metacentre:

$$GG_1 = \frac{w \times d}{W}$$

$$= \frac{60 \times 12}{6000}$$

$$GG_1 = 0.12\,m$$

$$GM = KM - KG = 7.3 - 6.7 = 0.6\,m$$

In triangle GG_1M:

$$\tan \theta = \frac{GG_1}{GM}$$

$$= \frac{0.12}{0.60} = 0.20$$

Ans. List = $11° 18\frac{1}{2}'$

Example 2

A ship of 8000 tonnes displacement has KM = 8.7 m and KG = 7.6 m. The following weights are then loaded and discharged.

Load 250 tonnes cargo KG 6.1 m and centre of gravity 7.6 m to starboard of the centreline.

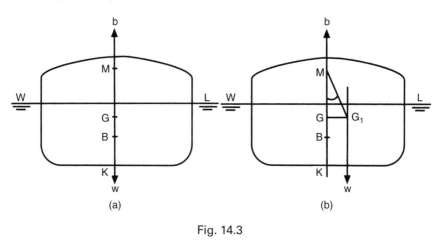

Fig. 14.3

Load 300 tonnes fuel oil KG 0.6 m and centre of gravity 6.1 m to port of the centreline.

Discharge 50 tonnes of ballast KG 1.2 m and centre of gravity 4.6 m to port of the centreline.

Find the final list.

Note. In this type of problem find the final KG by taking moments about the keel, and the final distance of the centre of gravity from the centreline by taking moments about the centreline.

Moments about the keel

Weight	KG	Moments about the keel
8000	7.6	60 800
250	6.1	1 525
300	0.6	180
8550		62 505
−50	1.2	−60
8500		62 445

$$\text{Final KG} = \frac{\text{Final moment}}{\text{Final displacement}}$$

$$= \frac{62\,445}{8500}$$

Final KG = 7.35 m

KM =	8.70 m
Final KG =	− 7.35 m
Final GM =	1.35 m

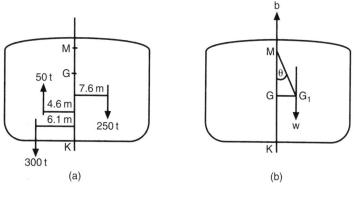

Fig. 14.4

Moments about the centreline (as in Figure 14.4(a))
For levers to port, use +ve sign.
For levers to starboard, use −ve sign.

w	d	Listing moment	
		To port +ve	To starboard −ve
+250	−7.6	−	−1900
−50	+4.6	−	−230
+300	+6.1	+1830	−
		+1830	−2130
			+1830
	Final moment		−300

Let the final position of the centre of gravity be as shown in Figure 14.4(b).

$$\therefore \text{ Final listing moment } = W \times GG_1$$

or

$$GG_1 = \frac{\text{Final moment}}{\text{Final displacement}}$$

$$= \frac{-300}{8500} = -0.035 \, \text{m}$$

$GG_1 = 0.035 \, \text{m}$ to starboard, because of the −ve sign used in table.

Since the final position of the centre of gravity must lie vertically under M, it follows that the ship will list θ degrees to starboard.

$$\tan \theta = \frac{GG_1}{GM}$$

$$= \frac{-0.035}{1.35} = -0.0259$$

$$\therefore \theta = 1° \ 29'$$

Ans. Final list = 1° 29' to starboard

Example 3

A ship of 8000 tonnes displacement has a GM = 0.5 m. A quantity of grain in the hold, estimated at 80 tonnes, shifts and, as a result, the centre of gravity of this grain moves 6.1 m horizontally and 1.5 m vertically. Find the resultant list.

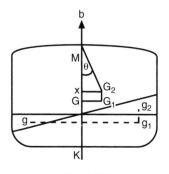

Fig. 14.5

Referring to Figure 14.5, let the centre of gravity of the grain shift from g to g_2. This will cause the ship's centre of gravity to shift from G to G_2 in a direction parallel to gg_2. The horizontal components of these shifts are g to g_1 and G to G_1, respectively, whilst the vertical components are g_1g_2 and G_1G_2.

$$GG_1 = \frac{w \times d}{W} \qquad\qquad G_1G_2 = \frac{w \times d}{W}$$

$$= \frac{80 \times 6.1}{8000} \qquad\qquad = \frac{80 \times 1.5}{8000}$$

$$GG_1 = 0.061 \, m \qquad\qquad G_1G_2 = 0.015 \, m$$

In Figure 14.5:

$$GX = G_1G_2 \qquad XG_2 = GG_1 \qquad \tan \theta = \frac{XG_2}{MX}$$

$$GX = 0.015 \, m \qquad XG_2 = 0.061 \, m$$

$$\underline{GM = 0.500 \, m} \qquad\qquad\qquad \tan \theta = \frac{0.061}{0.485} = 0.126$$

$$\underline{XM = 0.485 \, m} \qquad\qquad\qquad \tan \theta = 0.126$$

Ans. List = 7° 12'

Example 4

A ship of 13 750 tonnes displacement, GM = 0.75 m, is listed $2\frac{1}{2}$ degrees to starboard and has yet to load 250 tonnes of cargo. There is space available in each side of No. 3 between deck (centre of gravity, 6.1 m out from the centre-line). Find how much cargo to load on each side if the ship is to be upright on completion of loading.

Load 'w' tonnes to port and (250 − w) tonnes to starboard.

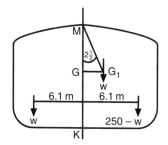

Fig. 14.6

In triangle GG₁M:

$$GG_1 = GM \tan \theta$$
$$= 0.75 \tan 2\tfrac{1}{2}°$$
$$= 0.75 \times 0.0437$$
$$GG_1 = 0.0328 \, m$$

Moments about the centreline

w	d	Listing moment	
		To port	To starboard
w	6.1	6.1 w	−
13 750	0.0328	−	451
250 − w	6.1	−	1525 − 6.1 w
		6.1 w	1976 − 6.1 w

If the ship is to complete loading upright, then:

Moment to port = Moment to starboard
6.1 w = 1976 − 6.1 w
w = 161.97 tonnes

Ans. Load 161.97 tonnes to port and 88.03 tonnes to starboard

Note. 161.97 + 88.03 = 250 tonnes of cargo (as given in question).

Example 5

A ship of 9900 tonnes displacement has KM = 7.3 m and KG = 6.4 m. She has yet to load two 50 tonne lifts with her own gear and the first lift is to be placed on deck on the inshore side (KG 9 m and centre of gravity 6 m out from the centreline). When the derrick plumbs the quay its head is 15 m above the keel and 12 m out from the centreline. Calculate the maximum list during the operation.

Note. The maximum list will obviously occur when the first lift is in place on the deck and the second weight is suspended over the quay as shown in Figure 14.7.

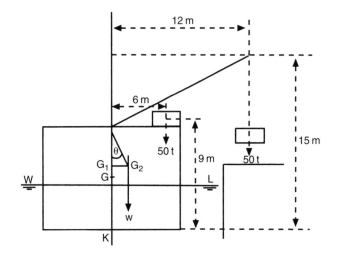

Fig. 14.7

Moments about the keel

Weight	KG	Moment
9900	6.4	63 360
50	9.0	450
50	15.0	750
10 000		64 560

$$\text{Final KG} = \frac{\text{Final moment}}{\text{Final displacement}}$$

$$= \frac{64\,560}{10\,000}$$

$$\text{Final KG} = 6.456\,\text{m} \ (KG_1)$$

That is, a rise of 0.056 m above the original KG of 6.4 m.

Moment about the centreline

w	d	Listing moment	
		To port	To starboard
50	12	–	600
50	6	–	300
			900

$$\text{Listing moment} = 900 \text{ tonnes m}$$
$$\text{But listing moment} = W \times G_1G_2$$
$$\therefore \ W \times G_1G_2 = 900$$
$$G_1G_2 = \frac{900}{10\,000}$$
$$G_1G_2 = 0.09\,\text{m}$$
$$GG_1 = KG_1 - KG = 6.456 - 6.400 = 0.056\,\text{m}$$
$$GM = KM - KG = 7.3 - 6.4$$
$$\therefore \ GM = 0.9\,\text{m}$$

In triangle G_1G_2M

$$G_1M = GM - GG_1$$
$$= 0.9\,\text{m} - 0.056\,\text{m}$$
$$G_1M = 0.844\,\text{m}$$
$$\tan \theta = \frac{G_1G_2}{G_1M}$$
$$= \frac{0.09}{0.844}$$
$$\tan \theta = 0.1066$$

Ans. Maximum List = 6° 6'

Summary
1. Always make a *sketch* from the given information.
2. Use a moment of weight table.
3. Use values from table to *calculate* the final requested data.

Exercise 14

1 A ship of 5000 tonnes displacement has KG 4.2 m and KM 4.5 m, and is listed 5 degrees to port. Assuming that the KM remains constant, find the final list if 80 tonnes of bunkers are loaded in No. 2 starboard tank whose centre of gravity is 1 metre above the keel and 4 metres out from the centreline.

2 A ship of 4515 tonnes displacement is upright, and has KG 5.4 m and KM 5.8 m. It is required to list the ship 2 degrees to starboard and a weight of 15 tonnes is to be shifted transversely for this purpose. Find the distance through which it must be shifted.

3 A ship of 7800 tonnes displacement has a mean draft of 6.8 m and is to be loaded to a mean draft of 7 metres. GM = 0.7 m and TPC = 20 tonnes. The ship is at present listed 4 degrees to starboard. How much more cargo can be shipped in the port and starboard tween deck, centres of gravity 6 m and 5 m, respectively, from the centreline, for the ship to complete loading and finish upright.

4 A ship of 1500 tonnes displacement has KB 2.1 m, KG 2.7 m and KM 3.1 m, and is floating upright in salt water. Find the list if a weight of 10 tonnes is shifted transversely across the deck through a distance of 10 metres.

5 A weight of 12 tonnes, when moved transversely across the deck through a distance of 12 m, causes a ship of 4000 tonnes displacement to list 3.8 degrees to starboard. KM = 6 m. Find the KG.

6 A quantity of grain, estimated at 100 tonnes, shifts 10 m horizontally and 1.5 m vertically in a ship of 9000 tonnes displacement. If the ship's original GM was 0.5 m, find the resulting list.

7 A ship of 7500 tonnes displacement has KM 8.6 m, KG 7.8 m and 20 m beam. A quantity of deck cargo is lost from the starboard side (KG 12 m, and centre of gravity 6 m in from the rail). If the resulting list is 3 degrees 20 minutes to port, find how much deck cargo was lost.

8 A ship of 12 500 tonnes displacement, KM 7 m and KG 6.4 m, has a 3 degree list to starboard and has yet to load 500 tonnes of cargo. There is space available in the tween decks, centres of gravity 6 m each side of the centreline. Find how much cargo to load on each side if the ship is to complete loading upright.

9 A ship is listed $2\frac{1}{2}$ degrees to port. The displacement is 8500 tonnes KM 5.5 m and KG 4.6 m. The ship has yet to load a locomotive of 90 tonnes mass on deck on the starboard side (centre of gravity 7.5 m from the centreline), and a tender of 40 tonnes. Find how far from the centreline the tender must be placed if the ship is to complete loading upright, and also find the final GM (KG of the deck cargo is 7 m).

10 A ship of 9500 tonnes displacement is listed $3\frac{1}{2}$ degrees to starboard, and has KM 9.5 m and KG 9.3 m. She loads 300 tonnes of bunkers in No. 3 double-bottom tank port side (KG 0.6 m and centre of gravity 6 m from the centreline), and discharges two parcels of cargo each of 50 tonnes from the

port side of No. 2 shelter deck (KG 11 m and centre of gravity 5 m from the centreline). Find the final list.

11   A ship of 6500 tonnes displacement is floating upright and has GM = 0.15 m. A weight of 50 tonnes, already on board, is moved 1.5 m vertically downwards and 5 m transversely to starboard. Find the list.

12   A ship of 5600 tonnes displacement is floating upright. A weight of 30 tonnes is lifted from the port side of No. 2 tween deck to the starboard side of No. 2 shelter deck (10 m horizontally). Find the weight of water to be transferred in No. 3 double-bottom tank from starboard to port to keep the ship upright. The distance between the centres of gravity of the tanks is 6 m.

13   A ship is just about to lift a weight from a jetty and place it on board. Using the data given below, calculate the angle of heel after the weight has just been lifted from this jetty. Weight to be lifted is 140 t with an outreach of 9.14 m. Displacement of ship prior to the lift is 10 060 tonnes. Prior to lift-off, the KB is 3.4 m. KG is 3.66 m, TPC_{SW} is 20, I_{NA} is 22 788 m^4, draft is 6.7 m in salt water. Height to derrick head is 18.29 m above the keel.

Chapter 15
Moments of statical stability

When a ship is inclined by an external force, such as wind and wave action, the centre of buoyancy moves out to the low side, parallel to the shift of the centre of gravity of the immersed and emerged wedges, to the new centre of gravity of the underwater volume. The force of buoyancy is considered to act vertically upwards through the centre of buoyancy, whilst the weight of the ship is considered to act vertically downwards through the centre of gravity. These two equal and opposite forces produce a moment or couple which may tend to right or capsize the ship. The moment is referred to as the *moment of statical stability* and may be defined as the moment to return the ship to the initial position when inclined by an external force.

A ship which has been inclined by an external force is shown in Figure 15.1.

The centre of buoyancy has moved from B to B_1 parallel to gg_1, and the force of buoyancy (W) acts vertically upwards through B_1. The weight of the ship (W) acts vertically downwards through the centre of gravity (G). The perpendicular distance between the lines of action of the forces (GZ) is called the *righting lever*. Taking moments about the centre of gravity, the

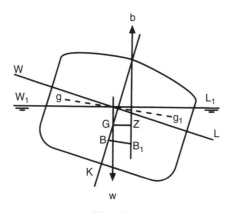

Fig. 15.1

moment of statical stability is equal to the product of the righting lever and the displacement, or:

$$\text{Moment of statical stability} = W \times GZ$$

The moment of statical stability at a small angle of heel

At small angles of heel the force of buoyancy may be considered to act vertically upwards through a fixed point called the initial metacentre (M). This is shown in Figure 15.2, in which the ship is inclined to a small angle (θ degrees).

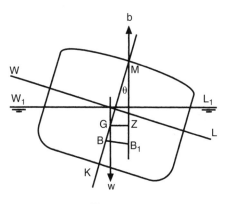

Fig. 15.2

$$\text{Moment of statical stability} = W \times GZ$$
$$\text{But in triangle GZM: GZ} = GM \sin \theta°$$
$$\therefore \text{Moment of statical stability} = W \times GM \times \sin \theta°$$

From this formula it can be seen that for any particular displacement at small angles of heel, the righting moments will vary directly as the initial metacentric height (GM). Hence, if the ship has a comparatively large GM she will tend to be 'stiff', whilst a small GM will tend to make her 'tender'. It should also be noticed, however, that the stability of a ship depends not only upon the size of the GM or GZ but also upon the displacement. Thus two similar ships may have identical GMs, but if one is at the light displacement and the other at the load displacement, their respective states of stability will be vastly different. The ship which is at the load displacement will be much more 'stiff' than the other.

Example 1

A ship of 4000 tonnes displacement has KG 5.5 m and KM 6.0 m. Calculate the moment of statical stability when heeled 5 degrees.

$$GM = KM - KG = 6.0 - 5.5 = 0.5 \text{ m}$$
$$\text{Moment of statical stability} = W \times GM \times \sin\theta°$$
$$= 4000 \times 0.5 \times \sin 5°$$
Ans. Moment of statical stability $= \underline{174.4 \text{ tonnes m}}$

Example 2

When a ship of 12 000 tonnes displacement is heeled $6\frac{1}{2}$ degrees the moment of statical stability is 600 tonnes m. Calculate the initial metacentric height.

$$\text{Moment of statical stability} = W \times GM \times \sin\theta°$$
$$\therefore GM = \frac{\text{Moment of statical stability}}{W \times \sin\theta°}$$
$$= \frac{600}{12\ 000\ \sin 6\frac{1}{2}°}$$

Ans. $\underline{GM = 0.44 \text{ m}}$

The moment of statical stability at a large angle of heel

At a large angle of heel the force of buoyancy can no longer be considered to act vertically upwards through the initial metacentre (M). This is shown in Figure 15.3, where the ship is heeled to an angle of more than 15 degrees. The centre of buoyancy has moved further out to the low side, and the vertical through B_1 no longer passes through (M), the initial metacentre. The righting lever (GZ) is once again the perpendicular distance between the vertical through G and the vertical through B_1, and the moment of statical stability is equal to $W \times GZ$.

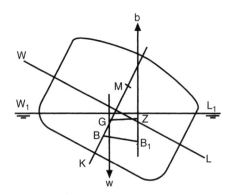

Fig. 15.3

But GZ is no longer equal to GM sin θ°. Up to the angle at which the deck edge is immersed, it may be found by using a formula known as the *wall-sided formula*; i.e.

$$GZ = (GM + \tfrac{1}{2}BM \tan^2 \theta) \sin \theta$$

The derivation of this formula is as follows:

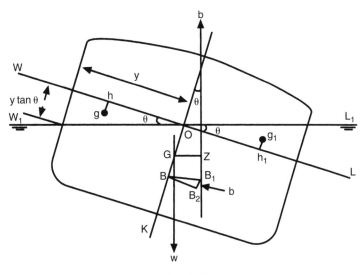

Fig. 15.4

Refer to the ship shown in Figure 15.4. When inclined the wedge WOW_1 is transferred to LOL_1 such that its centre of gravity shifts from g to g_1. This causes the centre of buoyancy to shift from B to B_1. The horizontal components of these shifts are hh_1 and BB_2, respectively, the vertical components being $(gh + g_1h_1)$ and B_1B_2, respectively.

Let BB_2 be 'a' units and let B_1B_2 be 'b' units.

Now consider the wedge LOL_1:

$$\text{Area} = \tfrac{1}{2} y^2 \tan \theta$$

Consider an elementary strip longitudinally of length dx as in Figure 15.5(b):

$$\text{Volume} = \tfrac{1}{2} y^2 \tan \theta \, dx$$

The total horizontal shift of the wedge (hh_1), is $2/3 \times 2y$ or $4/3 \times y$.

$$\therefore \quad \text{Moment of shifting this wedge} = \tfrac{4}{3} y \times \tfrac{1}{2} y^2 \tan \theta \, dx$$
$$= \tfrac{2}{3} y^3 \tan \theta \, dx$$

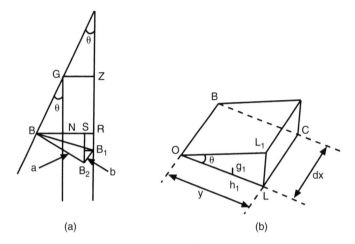

(a)   (b)

Fig. 15.5

The sum of the moment of all such wedges $= \int_O^L \frac{2}{3} y^3 \tan \theta \, dx$

$$= \tan \theta \int_O^L \frac{2}{3} y^3 \, dx$$

But the second moment of the water-plane area about the centre-line (I) $= \int_O^L \frac{2}{3} y^3 \, dx$

$\therefore$ Sum of the moment of all such wedges $= I \tan \theta$

$$BB_2 = \frac{v \times hh_1}{V}$$

or

$$V \times BB_2 = v \times hh_1$$

But, the sum of the moments of the wedges $= v \times hh_1$

$\therefore \ V \times BB_2 = I \tan \theta$

$$BB_2 = \frac{I}{V} \tan \theta$$

$$BB_2 = BM \tan \theta \longrightarrow \text{(a)}$$

The vertical shift of the wedge $= gh + g_1 h_1$

$$= 2gh$$

$\therefore$ The vertical moment of the shift $= v \times 2gh$

$$= 2vgh$$

In Figure 15.5(b):

$$OL = y \quad \text{and} \quad Oh_1 = \tfrac{2}{3} y$$

But

$$LL_1 = y \tan \theta$$
$$\therefore \quad g_1 h_1 = \tfrac{1}{3} y \tan \theta$$

$$\text{The volume of the wedge} = \tfrac{1}{2} y^2 \tan \theta \, dx$$
$$\text{The moment of the vertical shift} = \tfrac{1}{2} y^2 \tan \theta \, dx \times \tfrac{2}{3} y \tan \theta$$
$$= \tfrac{1}{3} y^3 \tan^2 \theta \, dx$$
$$\text{The vertical moment of all such wedges} = \int_O^L \tfrac{1}{3} y^3 \tan^2 \theta \, dx$$
$$= \tfrac{1}{2} I \tan^2 \theta$$
$$\therefore \quad \text{The moment of the vertical shift} = \tfrac{1}{2} I \tan^2 \theta$$

Also

$$B_1 B_2 = \frac{v \times 2gh}{V}$$

or

$$V \times b = 2vgh$$

but

$$2vgh = \text{The vertical moment of the shift}$$
$$\therefore \quad V \times b = \tfrac{1}{2} I \tan^2 \theta$$

or

$$b = \frac{I}{V} \times \frac{\tan^2 \theta}{2}$$
$$B_1 B_2 = \frac{BM \tan^2 \theta}{2} \quad\longrightarrow \quad \text{(b)}$$

Referring to Figure 15.5(a),

$$
\begin{aligned}
GZ &= NR \\
&= BR - BN \\
&= (BS + SR) - BN \\
&= a \cos \theta + b \sin \theta - BG \sin \theta \\
&= BM \tan \theta \cos \theta + \tfrac{1}{2} BM \tan^2 \theta \sin \theta - BG \sin \theta \quad \text{(from 'a' and 'b')} \\
&= BM \sin \theta + \tfrac{1}{2} BM \tan^2 \theta \sin \theta - BG \sin \theta \\
&= \sin \theta \, (BM + \tfrac{1}{2} BM \tan^2 \theta - BG) \\
GZ &= \sin \theta \, (GM + \tfrac{1}{2} BM \tan^2 \theta) \quad \text{(for } \theta \text{ up to } 25^\circ)
\end{aligned}
$$

This is the *wall-sided formula*.

Note. This formula may be used to obtain the GZ at any angle of heel so long as the ship's side at WW_1 is parallel to LL_1, but for small angles of heel (θ up to 5°), the term $\frac{1}{2}$ BM $\tan^2 \theta$ may be omitted.

Example 1

A ship of 6000 tonnes displacement has KB 3 m, KM 6 m and KG 5.5 m. Find the moment of statical stability at 25 degrees heel.

$$GZ = (GM + \tfrac{1}{2}BM \tan^2 \theta) \sin \theta$$
$$= (0.5 + \tfrac{1}{2} \times 3 \times \tan^2 25°)\,s$$
$$= 0.8262 \sin 25°$$
$$GZ = 0.35 \text{ m}$$
$$\text{Moment of statical stability} = W \times GZ$$
$$= 6000 \times 0.35$$

Ans. Moment of statical stability $= \underline{2100 \text{ tonnes m}}$

Example 2

A box-shaped vessel 65 m × 12 m × 8 m has KG 4 m, and is floating in salt water upright on an even keel at 4 m draft F and A. Calculate the moments of statical stability at (a) 5 degrees and (b) 25 degrees heel.

$W = L \times B \times \text{draft} \times 1.025$

$\quad = 65 \times 12 \times 4 \times 1.025$ tonnes

$W = 3198$ tonnes

$KB = \frac{1}{2}\,\text{draft}$

$KB = 2$ m

$BM = \dfrac{B^2}{12d}$

$\quad = \dfrac{12 \times 12}{12 \times 4}$

$BM = 3$ m

$$
\begin{array}{rl}
KB = & 2 \text{ m} \\
BM = & +3 \text{ m} \\
\hline
KM = & 5 \text{ m} \\
KG = & -4 \text{ m} \\
\hline
GM = & 1 \text{ m} \\
\hline
\end{array}
$$

At 5° heel

$$GZ = GM \sin \theta$$
$$= 1 \times \sin 5°$$
$$GZ = 0.0872$$
$$\text{Moment of statical stability} = W \times GZ$$
$$= 3198 \times 0.0872$$
$$= 278.9 \text{ tonnes m}$$

At 25° heel

$$GZ = (GM + \tfrac{1}{2} BM \tan^2 \theta) \sin \theta$$
$$= (1 + \tfrac{1}{2} \times 3 \times \tan^2 25°) \sin 25°$$
$$= (1 + 0.3262) \sin 25°$$
$$= 1.3262 \ \sin 25°$$
$$GZ = 0.56 \text{ metres}$$
$$\text{Moment of statical stability} = W \times GZ$$
$$= 3198 + 0.56$$
$$= 1790.9 \text{ tonnes m}$$

Ans. (a) 278.9 tonnes m and (b) 1790.9 tonnes m.

The moment of statical stability at a large angle of heel may also be calcu-lated using a formula known as *Attwood's formula*; i.e.

$$\text{Moment of statical stability} = W\left(\frac{v \times hh_1}{V} - BG \sin \theta\right)$$

The derivation of this formula is as follows:

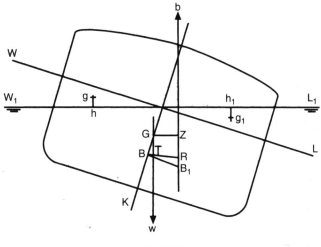

Fig. 15.6

$$\text{Moment of statical stability} = W \times GZ$$
$$= W(BR - BT)$$

Let v = the volume of the immersed or emerged wedge

hh_1 = the horizontal component of the shift of the centre of gravity of the wedge

V = the underwater volume of the ship

BR = the horizontal component of the shift of the centre of buoyancy.

$$BT = BG \sin \theta$$

also

$$BR = \frac{v \times hh_1}{V}$$

$$\therefore \text{Moment of statical stability} = W\left(\frac{v \times hh_1}{V} - BG \sin \theta\right)$$

Exercise 15

1   A ship of 10 000 tonnes displacement has GM 0.5 m. Calculate the moment of statical stability when the ship is heeled 7(3/4) degrees.

2   When a ship of 12 000 tonnes displacement is heeled 5(1/4) degrees the moment of statical stability is 300 tonnes m, KG 7.5 m. Find the height of the metacentre above the keel.

3   Find the moment of statical stability when a ship of 10 450 tonnes displacement is heeled 6 degrees if the GM is 0.5 m.

4   When a ship of 10 000 tonnes displacement is heeled 15 degrees, the righting lever is 0.2 m, KM 6.8 m. Find the KG and the moment of statical stability.

5   A box-shaped vessel 55 m × 7.5 m × 6 m has KG 2.7 m, and floats in salt water on an even keel at 4 m draft F and A. Calculate the moments of statical stability at (a) 6 degrees heel and (b) 24 degrees heel.

6   A ship of 10 000 tonnes displacement has KG 5.5 m, KB 2.8 m and BM 3 m. Calculate the moments of statical stability at (a) 5 degrees heel and (b) 25 degrees heel.

7   A box-shaped vessel of 3200 tonnes displacement has GM 0.5 m and beam 15 m, and is floating at 4 m draft. Find the moments of statical stability at 5 degrees and 25 degrees heel.

8   A ship of 11 000 tonnes displacement has a moment of statical stability of 500 tonnes m when heeled 5 degrees. Find the initial metacentric height.

9   (a) Write a brief description on the characteristics associated with an 'Angle of Loll'.

(b) For a box-shaped barge, the breadth is 6.4 m and the draft is 2.44 m even keel, with a KG of 2.67 m.

Using the given wall-sided formula, calculate the GZ ordinates up to an angle of heel of 20°, in 4° increments. From the results construct a statical stability curve up to 20° angle of heel. Label the important points on this constructed curve:

$$GZ = \sin \theta(GM + \tfrac{1}{2} BM \tan^2 \theta)$$

Chapter 16
Trim or longitudinal stability

Trim may be considered as the longitudinal equivalent of list. Trim is also known as 'longitudinal stability'. It is in effect transverse stability turned through 90°. Instead of trim being measured in degrees it is measured as the difference between the drafts forward and aft. If difference is zero then the ship is on even keel. If forward draft is greater than aft draft, the vessel is trimming *by the bow*. If aft draft is greater than the forward draft, the vessel is trimming *by the stern*.

Consider a ship to be floating at rest in still water and on an even keel as shown in Figure 16.1.

The centre of gravity (G) and the centre of buoyancy (B) will be in the same vertical line and the ship will be displacing her own weight of water. So W = b.

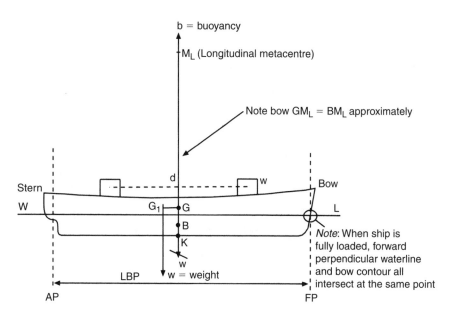

b = buoyancy

M_L (Longitudinal metacentre)

Note bow GM_L = BM_L approximately

Stern

W

d

w

Bow

G_1 G

L

B

K

w

w = weight

Note: When ship is fully loaded, forward perpendicular waterline and bow contour all intersect at the same point

LBP

AP

FP

Fig. 16.1

Now let a weight 'w', already on board, be shifted aft through a distance 'd', as shown in Figure 16.1. This causes the centre of gravity of the ship to shift from G to G_1, parallel to the shift of the centre of gravity of the weight shifted, so that:

$$GG_1 = \frac{w \times d}{W}$$

or

$$W \times GG_1 = w \times d$$

A trimming moment of $W \times GG_1$ is thereby produced.
  But

$$W \times GG_1 = w \times d$$

$$\therefore \text{ The trimming moment } = w \times d$$

The ship will now trim until the centres of gravity and buoyancy are again in the same vertical line, as shown in Figure 16.2. When trimmed, the wedge of buoyancy LFL_1 emerges and the wedge WFW_1 is immersed. Since the ship, when trimmed, must displace the same weight of water as when on an even keel, the volume of the immersed wedge must be equal to the volume of the emerged wedge and F, the point about which the ship trims, is the centre of gravity of the water-plane area. The point F is called the 'centre of flotation' or 'tipping centre'.

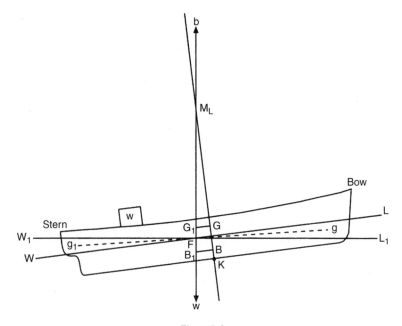

Fig. 16.2

A vessel with a rectangular water-plane has its centre of flotation on the centre line amidships but, on a ship, it may be a little forward or abaft amidships, depending on the shape of the water-plane. In trim problems, unless stated otherwise, it is to be assumed that the centre of flotation is situated amidships.

Trimming moments are taken about the centre of flotation since this is the point about which rotation takes place.

The longitudinal metacentre (M_L) is the point of intersection between the verticals through the longitudinal positions of the centres of buoyancy. The vertical distance between the centre of gravity and the longitudinal metacentre (GM_L) is called the longitudinal metacentric height.

BM_L is the height of the longitudinal metacentre above the centre of buoyancy and is found for any shape of vessel by the formula:

$$BM_L = \frac{I_L}{V}$$

where

I_L = the longitudinal second moment of the water-plane
        about the centre of flotation
V = the vessel's volume of displacement

The derivation of this formula is similar to that for finding the transverse BM.
   For a rectangular water-plane area:

$$I_L = \frac{BL^3}{12}$$

where

L = the length of the water-plane
B = the breadth of the water-plane

   Thus, for a vessel having a rectangular water-plane:

$$BM_L = \frac{BL^3}{12V}$$

For a box-shaped vessel:

$$BM_L = \frac{I_L}{V}$$

$$= \frac{BL^3}{12V}$$

$$= \frac{BL^3}{12 \times L \times B \times d}$$

$$BM_L = \frac{L^2}{12d}$$

where

L = the length of the vessel ⎫   Hence, BM_L is independent
d = the draft of the vessel ⎭   of ships Br. Mld

For a triangular prism:

$$BM_L = \frac{I_L}{V}$$

$$= \frac{BL^3}{12 \times \frac{1}{2} \times L \times B \times d}$$

$$BM_L = \frac{L^2}{6d} \text{, so again is independent of Br. Mld}$$

It should be noted that the distance BG is small when compared with BM_L or GM_L and, for this reason, BM_L may, without appreciable error, be substituted for GM_L in the formula for finding MCT 1 cm.

The moment to change trim one centimetre (denoted by MCT 1 cm or MCTC)

The MCT 1 cm, or MCTC, is the moment required to change trim by 1 cm, and may be calculated by using the formula:

$$MCT\,1\,cm = \frac{W \times GM_L}{100L}$$

where

  W = the vessel's displacement in tonnes
GM_L = the longitudinal metacentric height in metres
  L = the vessel's length in metres.

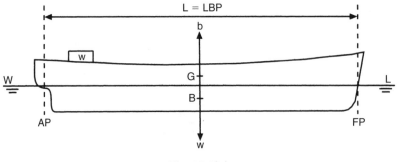

Fig. 16.3(a)

The derivation of this formula is as follows:

Consider a ship floating on an even keel as shown in Figure 16.3(a). The ship is in equilibrium.

Now shift the weight 'w' forward through a distance of 'd' metres. The ship's centre of gravity will shift from G to G_1, causing a trimming moment of $W \times GG_1$, as shown in Figure 16.3(b).

The ship will trim to bring the centres of buoyancy and gravity into the same vertical line as shown in Figure 16.3(c). The ship is again in equilibrium.

Let the ship's length be L metres and let the tipping centre (F) be l metres from aft.

The longitudinal metacentre (M_L) is the point of intersection between the verticals through the centre of buoyancy when on an even keel and when trimmed.

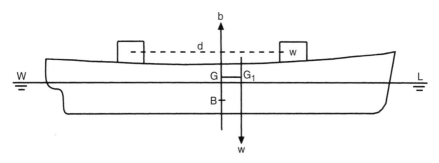

Fig. 16.3(b)

$$GG_1 = \frac{w \times d}{W} \text{ and } GG_1 = GM_L \tan \theta$$

$$\therefore \tan \theta = \frac{w \times d}{W \times GM_L}$$

but

$$\tan \theta = \frac{t}{L} \text{ (see Figure 16.4(b))}$$

Let the change of trim due to shifting the weight be 1 cm. Then $w \times d$ is the moment to change trim 1 cm.

$$\therefore \tan \theta = \frac{1}{100L}$$

but

$$\tan \theta = \frac{w \times d}{W \times GM_L}$$

$$\therefore \tan \theta = \frac{MCT \, 1 \, cm}{W \times GM_L}$$

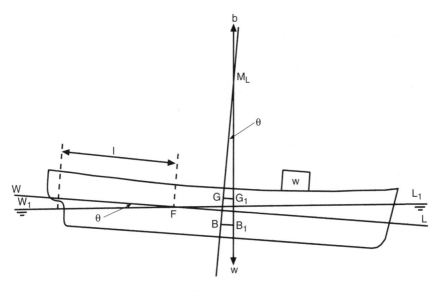

Fig. 16.3(c)

or

$$\frac{\text{MCT 1 cm}}{\text{W} \times \text{GM}_\text{L}} = \frac{1}{100\text{L}}$$

and

$$\text{MCT 1 cm} = \frac{\text{W} \times \text{GM}_\text{L}}{100\text{L}} \text{ tonnes m/cm}$$

To find the change of draft forward and aft due to change of trim

When a ship changes trim it will obviously cause a change in the drafts forward and aft. One of these will be increased and the other decreased. A formula must now be found which will give the change in drafts due to change of trim.

Consider a ship floating upright as shown in Figure 16.4(a). F_1 represents the position of the centre of flotation which is l metres from aft. The ship's length is L metres and a weight 'w' is on deck forward.

Let this weight now be shifted aft a distance of 'd' metres. The ship will trim about F_1 and change the trim 't' cms by the stern as shown in Figure 16.4(b).

W_1C is a line drawn parallel to the keel.

'A' represents the new draft aft and 'F' the new draft forward. The trim is therefore equal to A−F and, since the original trim was zero, this must also be equal to the change of trim.

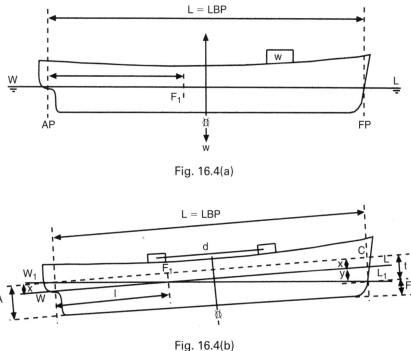

Fig. 16.4(a)

Fig. 16.4(b)

Let 'x' represent the change of draft aft due to the change of trim and let 'y' represent the change forward.

In the triangles WW_1F_1 and W_1L_1C, using the property of similar triangles:

$$\frac{x \text{ cm}}{1 \text{ m}} = \frac{t \text{ cm}}{L \text{ m}}$$

or

$$x \text{ cm} = \frac{1 \text{ m} \times t \text{ cm}}{L \text{ m}}$$

$$\therefore \text{ Change of draft aft in cm} = \frac{1}{L} \times \text{Change of trim in cm}$$

where

$1 = $ the distance of centre of flotation from aft in metres, and
$L = $ the ship's length in metres

It will also be noticed that $x + y = t$

$\therefore$ Change of draft F in cm = Change of trim − Change of draft A

The effect of shifting weights already on board

Example 1

A ship 126 m long is floating at drafts of 5.5 m F and 6.5 m A. The centre of flotation is 3 m aft of amidships. MCT 1 cm = 240 tonnes m. Displacement = 6000 tonnes. Find the new drafts if a weight of 120 tonnes already on board is shifted forward a distance of 45 metres.

$$\text{Trimming moment} = w \times d$$
$$= 120 \times 45$$
$$= 5400 \text{ tonnes m by the head}$$
$$\text{Change of trim} = \frac{\text{Trimming moment}}{\text{MCT 1 cm}}$$
$$= \frac{5400}{240}$$
$$= 22.5 \text{ cm by the head}$$
$$\text{Change of draft aft} = \frac{l}{L} \times \text{Change of trim}$$
$$= \frac{60}{126} \times 22.5$$
$$= 10.7 \text{ cm}$$
$$\text{Change of draft forward} = \frac{66}{126} \times 22.5$$
$$= 11.8 \text{ cm}$$

Original drafts	6.500 m A	5.500 m F
Change due to trim	−0.107 m	+0.118 m
Ans. New drafts	6.393 m A	5.618 m F

Example 2

A box-shaped vessel 90 m × 10 m × 6 m floats in salt water on an even keel at 3 m draft F and A. Find the new drafts if a weight of 64 tonnes already on board is shifted a distance of 40 metres aft.

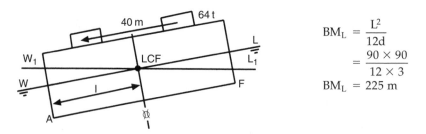

$$BM_L = \frac{L^2}{12d}$$
$$= \frac{90 \times 90}{12 \times 3}$$
$$BM_L = 225 \text{ m}$$

Fig. 16.5

$$W = L \times B \times d \times 1.025$$
$$= 90 \times 10 \times 3 \times 1.025$$
$$W = 2767.5 \text{ tonnes}$$
$$\text{MCT 1 cm} = \frac{W \times GM_L}{100L}$$

Since BG is small compared with GM_L:

$$\text{MCT 1 cm} \approx \frac{W \times BM_L}{100L}$$
$$= \frac{2767.5 \times 225}{100 \times 90}$$
$$\text{MCT 1 cm} = 69.19 \text{ tonnes m/cm}$$

$$\text{Change of trim} = \frac{w \times d}{\text{MCT 1 cm}}$$
$$= \frac{64 \times 40}{69.19}$$
$$\text{Change of trim} = 37 \text{ cm by the stern}$$
$$\text{Change of draft aft} = \frac{1}{L} \times \text{Change of trim}$$
$$= \frac{1}{2} \times 37 \text{ cm}$$
$$\text{Change of draft aft} = 18.5 \text{ cm}$$
$$\text{Change of draft forward} = 18.5 \text{ cm}$$

Original drafts	3.000 m A	3.000 m F
Change due to trim	+0.185 m	−0.185 m
Ans. New drafts	3.185 m A	2.815 m F

The effect of loading and/or discharging weights

When a weight is loaded at the centre of flotation it will produce no trimming moment, but the ship's drafts will increase uniformly so that the ship displaces an extra weight of water equal to the weight loaded. If the weight is now shifted forward or aft away from the centre of flotation, it will cause a change of trim. From this it can be seen that when a weight is loaded away from the centre of flotation, it will cause both a bodily sinkage and a change of trim.

Similarly, when a weight is being discharged, if the weight is first shifted to the centre of flotation it will produce a change of trim, and if it is then discharged from the centre of flotation the ship will rise bodily. Thus, both a change of trim and bodily rise must be considered when a weight is being discharged away from the centre of flotation.

Example 1

A ship 90 m long is floating at drafts 4.5 m F and 5.0 m A. The centre of flotation is 1.5 m aft of amidships. TPC 10 tonnes. MCT 1 cm. 120 tonnes m. Find the new drafts if a total weight of 450 tonnes is loaded in a position 14 m forward of amidships.

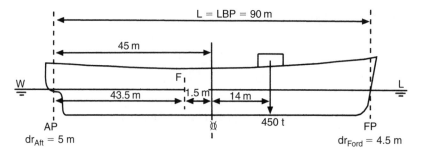

Fig. 16.6

$$\text{Bodily sinkage} = \frac{w}{\text{TPC}}$$

$$= \frac{450}{10}$$

$$\text{Bodily sinkage} = 45 \text{ cm}$$

$$\text{Change of trim} = \frac{\text{Trim moment}}{\text{MCT 1 cm}}$$

$$= \frac{450 \times 15.5}{120}$$

$$\text{Change of trim} = 58.12 \text{ cm by the head}$$

$$\text{Change of draft aft} = \frac{1}{L} \times \text{Change of trim}$$

$$= \frac{43.5}{90} \times 58.12$$

$$\text{Change of draft aft} = 28.09 \text{ cm}$$

$$\text{Change of draft forward} = \frac{46.5}{90} \times 58.12$$

$$\text{Change of draft forward} = 30.03 \text{ cm}$$

Original drafts	5.000 m A	4.500 m F
Bodily sinkage	+0.450 m	+0.450 m
	5.450 m	4.950 m
Change due trim	−0.281 m	+0.300 m
Ans. New drafts	5.169 m A	5.250 m F

Note. In the event of more than one weight being loaded or discharged, the net weight loaded or discharged is used to find the net bodily increase or decrease in draft, and the resultant trimming moment is used to find the change of trim.

Also, when the net weight loaded or discharged is large, it may be necessary to use the TPC and MCT 1 cm at the original draft to find the approximate new drafts, and then rework the problem using the TPC and MCT 1 cm for the mean of the old and the new drafts to find a more accurate result.

Example 2

A box-shaped vessel $40\,m \times 6\,m \times 3\,m$ is floating in salt water on an even keel at 2 m draft F and A. Find the new drafts if a weight of 35 tonnes is discharged from a position 6 m from forward. MCT 1 cm = 8.4 tonnes m.

$$\begin{aligned} TPC &= \frac{WPA}{97.56} \\ &= \frac{40 \times 6}{97.56} \\ TPC &= 2.46 \text{ tonnes} \end{aligned}$$

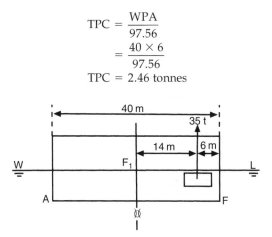

Fig. 16.7

$$\begin{aligned} \text{Bodily rise} &= \frac{w}{TPC} \\ &= \frac{35}{2.46} \\ \text{Bodily rise} &= 14.2 \text{ cm} \\ \text{Change of trim} &= \frac{w \times d}{MCT \ 1 \ cm} \\ &= \frac{35 \times 14}{8.4} \\ \text{Change of trim} &= 58.3 \text{ cm by the stern} \\ \text{Change of draft aft} &= \frac{1}{L} \times \text{Change of trim} \\ &= \frac{1}{2} \times 58.3 \text{ cm} \\ \text{Change of draft aft} &= 29.15 \text{ cm} \\ \text{Change of draft forward} &= \frac{1}{2} \times 58.3 \\ \text{Change of draft forward} &= 29.15 \text{ cm} \end{aligned}$$

Original drafts	2.000 m A	2.000 m F
Bodily rise	−0.140 m	−0.140 m
	1.860 m	1.860 m
Change due trim	+0.290 m	−0.290 m
Ans. <u>New drafts</u>	<u>2.150 m A</u>	<u>1.570 m F</u>

Example 3

A ship 100 m long arrives in port with drafts 3 m F and 4.3 m A. TPC 10 tonnes. MCT 1 cm 120 tonnes m. The centre of flotation is 3 m aft of amidships. If 80 tonnes of cargo is loaded in a position 24 m forward of amidships and 40 tonnes of cargo is discharged from 12 m aft of amidships, what are the new drafts?

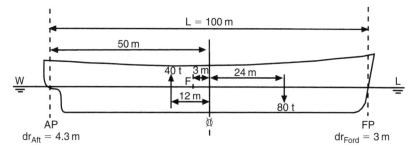

Fig. 16.8

$$\text{Bodily sinkage} = \frac{w}{\text{TPC}}$$

Cargo loaded	80 tonnes
Cargo discharged	<u>40 tonnes</u>
<u>Net loaded</u>	<u>40 tonnes</u>

$$= \frac{40}{10}$$

$$\text{Bodily sinkage} = 4 \text{ cm}$$

To find the change of trim take moments about the centre of flotation.

Weight	Distance from C.F.	Moment to change trim by	
		head	stern
+80	−27	2160	–
−40	+9	360	–
		2520	–

$$\text{Change of trim} = \frac{\text{Trim moment}}{\text{MCT 1 cm}}$$

$$= \frac{2520}{120}$$

$$\text{Change of trim} = 21 \text{ cm by the head}$$

$$\text{Change of draft aft} = \frac{1}{L} \times \text{Change of trim}$$

$$= \frac{47}{100} \times 21$$

$$\text{Change of draft aft} = 9.87 \text{ cm}$$

$$\text{Change of draft forward} = \frac{53}{100} \times 21$$

$$\text{Change of draft forward} = 11.13 \text{ cm}$$

Original drafts	4.300 m A	3.000 m F
Bodily sinkage	+0.040 m	+0.040 m
	4.340 m	3.040 m
Change due trim	−0.099 m	+0.111 m
Ans. New drafts	4.241 m A	3.151 m F

Example 4

A ship of 6000 tonnes displacement has drafts 7 m F and 8 m A. MCT 1 cm 100 tonnes m, TPC 20 tonnes, centre of flotation is amidships; 500 tonnes of cargo is discharged from each of the following four holds:

No. 1 hold, centre of gravity 40 m forward of amidships
No. 2 hold, centre of gravity 25 m forward of amidships
No. 3 hold, centre of gravity 20 m aft of amidships
No. 4 hold, centre of gravity 50 m aft of amidships

The following bunkers are also loaded:

150 tonnes at 12 m forward of amidships
50 tonnes at 15 m aft of amidships

Find the new drafts forward and aft.

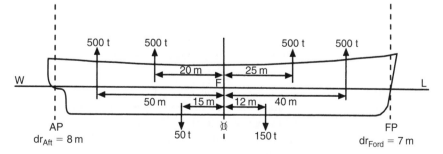

Fig. 16.9

Total cargo discharged   2000 tonnes

Total bunkers loaded   200 tonnes

Net weight discharged   1800 tonnes

$$\text{Bodily rise} = \frac{w}{\text{TPC}}$$

$$= \frac{1800}{20}$$

Bodily rise $= 90$ cm

Assume levers and moments aft of LCF are +ve.

Assume levers and moments forward of LCF are −ve.

Weight	Distance from C.F.	Moments
−500	−40	+20 000
−500	−25	+12 500
−500	+20	−10 000
−500	+50	−25 000
+150	−12	−1800
+50	+15	+750
		−3550

Resultant moment 3550 tonnes m by the head because of the −ve sign.

$$\text{Change of trim} = \frac{\text{Trim moment}}{\text{MCT 1 cm}}$$

$$= \frac{3550}{100}$$

Change of trim $= 35.5$ cm by the head

Since centre of flotation is amidships,

$$\text{Change of draft aft} = \text{Change of draft forward}$$

$$= \frac{1}{2} \text{ change of trim}$$

$$= 17.75 \text{ cm say } 0.18 \text{ m}$$

Original drafts	8.000 m A	7.000 m F
Bodily rise	−0.900 m	−0.900 m
	7.100 m	6.100 m
Change due trim	−0.180 m	+0.180 m
Ans. New drafts	6.920 m A	6.280 m F

Example 5

A ship arrives in port trimmed 25 cm by the stern. The centre of flotation is amidships. MCT 1 cm 100 tonnes m. A total of 3800 tonnes of cargo is to be discharged from 4 holds, and 360 tonnes of bunkers loaded in No. 4 double bottom tank; 1200 tonnes of the cargo is to be discharged from No. 2 hold and

600 tonnes from No. 3 hold. Find the amount to be discharged from Nos. 1 and 4 holds if the ship is to complete on an even keel.

Centre of gravity of No. 1 hold is 50 m forward of the centre of flotation
Centre of gravity of No. 2 hold is 30 m forward of the centre of flotation
Centre of gravity of No. 3 hold is 20 m abaft of the centre of flotation
Centre of gravity of No. 4 hold is 45 m abaft of the centre of flotation
Centre of gravity of No. 4 DB tank is 5 m abaft of the centre of flotation

Total cargo to be discharged from 4 holds	3800 tonnes
Total cargo to be discharged from Nos. 2 and 3	1800 tonnes
Total cargo to be discharged from Nos. 1 and 4	2000 tonnes

Let 'x' tonnes of cargo be discharged from No. 1 hold
Let (2000 − x) tonnes of cargo be discharged from No. 4 hold

Take moments about the centre of flotation, or as shown in Figure 16.10.

Original trim = 25 cm by the stern, i.e. +25 cm
Required trim = 0
Change of trim required = 25 cm by the head, i.e. −25 cm

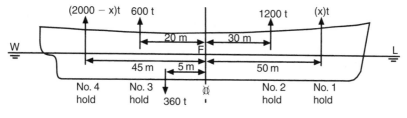

Fig. 16.10

Weight	Distance from C.F.	Moments	
		−ve	+ve
−x	−50	−	+50x
−1200	−30	−	+36 000
−600	+20	−12 000	−
−(2000 − x)	+45	−(90 000 − 45x)	−
+360	+5	−	+1800
		−102 000 + 45x	+37 800 + 50x

Trimming moment required = Change of trim × MCT 1 cm
$$= -25 \times 100 = -2500$$
Trimming moment required = 2500 tonnes m by the head
Resultant moment = Moment to change trim by head − MCT by stern
$$\therefore \quad -2500 = -102\,000 + 45x + 37\,800 + 50x$$
$$2500 = 64\,200 - 95x$$

or

$$95x = 61\,700$$

$$\underline{x = 649.5 \text{ tonnes}}$$

and

$$\underline{2000 - x = 1350.5 \text{ tonnes}}$$

Ans. Discharge 649.5 tonnes from No. 1 hold and 1350.5 tonnes from No. 4 hold.

Using trim to find the position of the centre of flotation

Example
A ship arrives in port floating at drafts of 4.50 m A and 3.80 m F. The following cargo is then loaded:

100 tonnes in a position 24 m aft of amidships
30 tonnes in a position 30 m forward of amidships
60 tonnes in a position 15 m forward of amidships

The drafts are then found to be 5.10 m A and 4.40 m F. Find the position of the longitudinal centre of flotation aft of amidships.
Original drafts 4.50 m A 3.80 m F give 0.70 m trim by the stern, i.e. +70 cm.
New drafts 5.10 m A 4.40 m F give 0.70 m trim by the stern, i.e. +70 cm.
Therefore there has been no change in trim, which means that

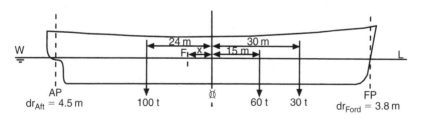

Fig. 16.11

The moment to change trim by the head = The moment to change trim by the stern.

Let the centre of flotation be 'x' metres aft of amidships. Taking moments, then,

$$100(24 - x) = 30(30 + x) + 60(15 + x)$$
$$2400 - 100x = 900 + 30x + 900 + 60x$$
$$190x = 600$$
$$x = 3.16 \, m$$

Ans. Centre of flotation is 3.16 metres aft of amidships.

Note. In this type of question it is usual to assume that the centre of flotation is aft of amidships, but this may not be the case. Had it been assumed that the centre of flotation was aft of amidships when in actual fact it was forward, then the answer obtained would have been minus.

Remember. Levers, moments and trim by the stern all have a +ve sign. Levers, moments and trim by the head all have a −ve sign.

Loading a weight to keep the after draft constant

When a ship is being loaded it is usually the aim of those in charge of the operation to complete loading with the ship trimmed by the stern. Should the ship's draft on sailing be restricted by the depth of water over a dock-sill or by the depth of water in a channel, then the ship will be loaded in such a manner as to produce this draft aft and be trimmed by the stern.

Assume now that a ship loaded in this way is ready to sail. It is then found that the ship has to load an extra weight. The weight must be loaded in such a position that the draft aft is not increased and also that the maximum trim is maintained.

If the weight is loaded at the centre of flotation, the ship's drafts will increase uniformly and the draft aft will increase by a number of centimetres equal to w/TPC. The draft aft must now be decreased by this amount.

Now let the weight be shifted through a distance of 'd' metres forward. The ship will change trim by the head, causing a reduction in the draft aft by a number of centimetres equal to $l/L \times$ Change of trim.

Therefore, if the same draft is to be maintained aft, the above two quantities must be equal. i.e.

$$\frac{l}{L} \times \text{Change of trim} = \frac{w}{TPC}$$

So

$$\text{Change of trim} = \frac{w}{TPC} \times \frac{L}{l} \text{ --------------- (I)}$$

But

$$\text{Change of trim} = \frac{w \times d}{\text{MCT 1 cm}} \text{ ----------------- (II)}$$

Equate (I) and (II)

$$\therefore \frac{w \times d}{\text{MCT 1 cm}} = \frac{w}{\text{TPC}} \times \frac{L}{l}$$

or

$$d = \frac{L \times \text{MCT 1 cm}}{l \times \text{TPC}}$$

where
d = the distance forward of the centre of flotation to load a weight to keep the draft aft constant
L = the ship's length, LBP
l = the distance of the centre of flotation to the stern

Example

A box-shaped vessel 60 m long, 10 m beam and 6 m deep is floating in salt water at drafts 4 m F and 4.4 m A. Find how far forward of amidships a weight of 30 tonnes must be loaded if the draft aft is to remain at 4.4 m.

$$\text{TPC}_{SW} = \frac{\text{WPA}}{97.56}$$

$$= \frac{60 \times 10}{97.56}$$

$$\text{TPC}_{SW} = 6.15 \text{ tonnes}$$

$$W = L \times B \times d \times \rho_{SW} \text{ tonnes}$$

$$= 60 \times 10 \times 4.2 \times 1.025$$

$$W = 2583 \text{ tonnes}$$

$$\text{BM}_L = \frac{L^2}{12d}$$

$$= \frac{60 \times 60}{12 \times 4.2}$$

$$\text{BM}_L = 71.42 \text{ metres}$$

$$\text{MCT 1 cm} \simeq \frac{W \times \text{BM}_L}{100L} \text{ because } \text{GM}_L \simeq \text{BM}_L$$

$$\text{MCT 1 cm} = \frac{2583 \times 71.42}{100 \times 60}$$

$$\text{MCT 1 cm} = 30.75 \text{ t m/cm}$$

$$d = \frac{L \times \text{MCT 1 cm}}{l \times \text{TPC}_{SW}}$$

$$= \frac{60}{30} \times 30.75 \times \frac{1}{6.15}$$

$$d = 10 \text{ metres from LCF}$$

LCF is at amidships.

Ans. Load the weight 10 metres forward of amidships.

Loading a weight to produce a required draft

Example

A ship 150 metres long arrives at the mouth of a river with drafts 5.5 m F and 6.3 m A. MCT 1 cm 200 tonnes m. TPC 15 tonnes. Centre of flotation is 1.5 m aft of amidships. The ship has then to proceed up the river where the maximum draft permissible is 6.2 m. It is decided that SW ballast will be run into the forepeak tank to reduce the draft aft to 6.2 m. If the centre of gravity of the forepeak tank is 60 metres forward of the centre of flotation, find the minimum amount of water which must be run in and also find the final draft forward.

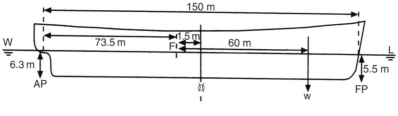

Fig. 16.12

(a) Load 'w' tonnes at the centre of flotation

$$\text{Bodily sinkage} = \frac{w}{\text{TPC}}$$

$$= \frac{w}{15} \text{ cm}$$

$$\text{New draft aft} = 6.3 \text{ m} + \frac{w}{15} \text{ cm} \ \text{-------------- (I)}$$

$$\text{Required draft aft} = 6.2 \text{ m} \text{---------- (II)}$$

$$\text{Equations (I)–(II)} = \underline{\text{Reduction required}} = 0.1 \text{ m} + \frac{w}{15} \text{ cm}$$

$$= 10 \text{ cm} + \frac{w}{15} \text{ cm}$$

$$= \left(10 + \frac{w}{15}\right) \text{cm ----- (III)}$$

(b) Shift 'w' tonnes from the centre of flotation to the forepeak tank

$$\text{Change of trim} = \frac{w \times d}{MCT\,1\,cm}$$

$$= \frac{60w}{200}$$

$$\text{Change of trim} = \frac{3w}{10}\,cm\text{ by the head}$$

$$\text{Change of draft aft due to trim} = \frac{l}{L} \times \text{Change of trim}$$

$$= \frac{73.5}{150} \times \frac{3w}{10}$$

$$\text{Change of draft aft due to trim} = 0.147w\,cm \quad\text{-------------------- (IV)}$$

$$\text{But change of draft required aft} = \left(10 + \frac{w}{15}\right)cm\text{ as per equation (III)}$$

$$0.147w = 10 + \frac{w}{15}\text{, i.e. equation (IV)} = \text{equation (III)}$$

$$2.205w = 150 + w$$

$$1.205w = 150$$

$$w = 124.5\text{ tonnes}$$

Therefore by loading 124.5 tonnes in the forepeak tank the draft aft will be reduced to 6.2 metres.

(c) To find the new draft forward

$$\text{Bodily sinkage} = \frac{w}{TPC}$$

$$= \frac{124.5}{15}$$

$$\text{Bodily sinkage} = 8.3\,cm$$

$$\text{Change of trim} = \frac{w \times d}{MCT\,1\,cm}$$

$$= \frac{124.5 \times 60}{200}$$

$$\text{Change of trim} = 37.35\,cm\text{ by the head}$$

$$\text{Change of draft aft due trim} = \frac{l}{L} \times \text{Change of trim}$$

$$\text{Change of draft aft due trim} = \frac{73.5}{150} \times 37.35$$

$$= 18.3\,cm$$

$$\text{Change of draft forward due trim} = \text{Change of trim} - \text{Change of draft aft}$$

$$= 37.35 - 18.3\,cm = 19.05\,cm\text{, or}$$

$$\text{Change of draft forward due trim} = \frac{76.5 \times 37.35}{150} = 19.05\,cm$$

Original drafts	6.300 m A	5.500 m F
Bodily sinkage	+ 0.080 m	+ 0.080 m
	6.380 m	5.580 m
Change due trim	− 0.180 m	+ 0.190 m
New drafts	6.200 m A	5.770 m F

Ans. Load 124.5 tonnes in forepeak tank. Final draft forward is 5.770 metres.

Using change of trim to find the longitudinal metacentric height (GM$_L$)

Earlier it was shown in this chapter that, when a weight is shifted longitudinally within a ship, it will cause a change of trim. It will now be shown how this effect may be used to determine the longitudinal metacentric height.

Consider Figure 16.13(a) which represents a ship of length 'L' at the waterline, floating upright on an even keel with a weight on deck forward. The centre of gravity is at G, the centre of buoyancy at B, and the longitudinal metacentre at M$_L$. The longitudinal metacentric height is therefore GM$_L$.

Now let the weight be shifted aft horizontally as shown in Figure 16.13(b). The ship's centre of gravity will also shift horizontally, from G to G$_1$, producing a trimming moment of W × GG$_1$ by the stern.

The ship will now trim to bring G$_1$ under M$_L$ as shown in Figure 16.13(c).

In Figure 16.13(c) W$_1$L$_1$ represents the new waterline, F the new draft forward and A the new draft aft. It was shown in Figure 16.4(b) and by the

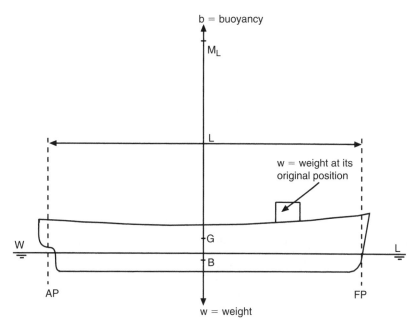

Fig. 16.13(a)

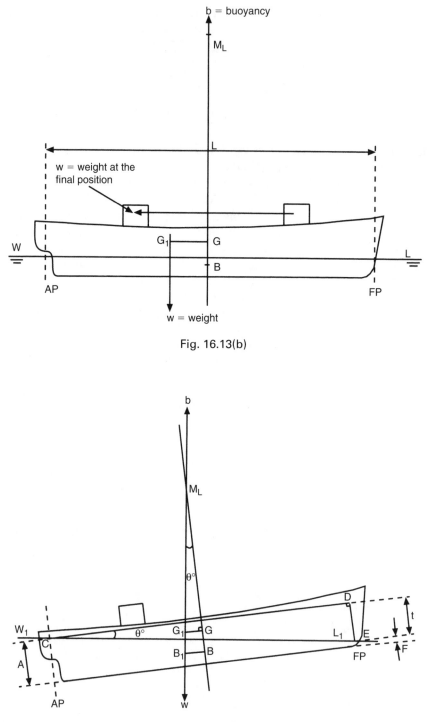

Fig. 16.13(b)

Fig. 16.13(c)

associated notes, that F−A is equal to the new trim (t) and since the ship was originally on an even keel, then 't' must also be equal to the change of trim.

If the angle between the new and old verticals is equal to θ, then the angle between the new and old horizontals must also be equal to θ (the angle between two straight lines being equal to the angle between their normals).

It will also be seen in Figure 16.13(c) that the triangles GG_1M_L and CDE are similar triangles.

$$\therefore \frac{GM_L}{GG_1} = \frac{L}{t}$$

or

$$GM_L = \frac{L}{t} \times GG_1$$

(All measurements are in metres.)

Example 1

When a weight is shifted aft in a ship 120 metres long, it causes the ship's centre of gravity to move 0.2 metres horizontally and the trim to change by 0.15 metres. Find the longitudinal metacentric height.

$$\therefore \frac{GM_L}{GG_1} = \frac{L}{t}$$

$$\therefore GM_L = \frac{L}{t} \times GG_1$$

$$= \frac{120 \times 0.2}{0.15}$$

Ans. $GM_L = 160$ metres

Example 2

A ship 150 metres long has a displacement of 7200 tonnes, and is floating upright on an even keel. When a weight of 60 tonnes, already on board, is shifted 24 metres forward, the trim is changed by 0.15 metres. Find the longitudinal metacentric height.

$$\frac{GM_L}{GG_1} = \frac{L}{t}$$

$$GM_L = GG_1 \times \frac{L}{t}$$

$$= \frac{w \times d}{W} \times \frac{L}{t}$$

$$= \frac{60 \times 24}{7200} \times \frac{150}{0.15}$$

Ans. $GM_L = 200$ metres

Which way will the ship trim?

As previously discussed, when a weight 'w' is placed onto a ship the LCG moves to a new position. This creates a temporary gap between the LCG and the LCB.

To obtain equilibrium, and for the ship to settle down at new end drafts, the vessel will have a change of trim. It will continue to do so until the LCB is once again vertically in line and below the new LCG position.

$$\text{Change of trim} = \frac{W \times \{LCB_{foap} - LCG_{foap}\}}{MCTC} \text{ cm}$$

$$\text{Change of trim also} = \frac{w \times d}{MCTC} \text{ cm}$$

$$foap = \text{forward of after perpendicular}$$

$$W \times \{LCB_{foap} - LCG_{foap}\} = w \times d \text{ also } = \text{Trimming moment in tm}$$

Observations:

Many naval architects measure LCG and LCB from amidships.
SQA/MCA advocate measuring LCG and LCB from the after perp.
Both procedures will end up giving the same change of trim.
LCG is the longitudinal overall centre of gravity for the lightweight plus all of the deadweight items on board the ship.
LCB is the 3-D longitudinal centroid of the underwater form of the vessel.
LCF is the 2-D centre of each waterplane.
W = ship's displacement in tonnes, that is lightweight + deadweight.
d = the horizontal lever from the LCF to the lcg of the added weight 'w'.
MCTC = moment to change trim one cm.

Always remember that:
If $\{LCB_{foap} - LCG_{foap}\}$ is *positive* in value ... ship will trim by the **STERN**.
If $\{LCB_{foap} - LCG_{foap}\}$ is *negative* in value ... ship will trim by the **BOW**.

Summary

1. Make a sketch from the given information.
2. Estimate the mean bodily sinkage.
3. Calculate the change of trim using levers measured from LCF.
4. Evaluate the trim ratio forward and aft at FP and AP, from the LCF position.
5. Collect the above calculated values to estimate the final end drafts.
6. In the solutions shown in the text, these final end drafts have been calculated to three decimal figures. In practice, naval architects and ship officers round off the drafts to two decimal places only. This gives acceptable accuracy.
7. Note how the formulae were written in *letters* first and then put the *figures* in. In the event of a mathematical error, marks will be given for a correct formula and for a correct sketch.

Exercise 16

Adding, discharging and moving weights

1 A ship of 8500 tonnes displacement has TPC 10 tonnes, MCT 1 cm = 100 tonnes m and the centre of flotation is amidships. She is completing loading under coal tips. Nos. 2 and 3 holds are full, but space is available in No. 1 hold (centre of gravity 50 m forward of amidships), and in No. 4 hold (centre of gravity 45 m aft of amidships). The present drafts are 6.5 m F and 7 m A, and the load draft is 7.1 m. Find how much cargo is to be loaded in each of the end holds so as to put the ship down to the load draft and complete loading on an even keel.

2 An oil tanker 150 m long, displacement 12 500 tonnes, MCT 1 cm 200 tonnes m, leaves port with drafts 7.2 m F and 7.4 m A. There is 550 tonnes of fuel oil in the foward deep tank (centre of gravity 70 m forward of the centre of flotation) and 600 tonnes in the after deep tank (centre of gravity 60 m aft of centre of flotation). The centre of flotation is 1 m aft of amidships. During the sea passage 450 tonnes of oil is consumed from aft. Find how much oil must be transferred from the forward tank to the after tank if the ship is to arrive on an even keel.

3 A ship 100 m long, and with a displacement of 2200 tonnes, has longitudinal metacentric height 150 m. The present drafts are 5.2 m F and 5.3 m A. Centre of flotation is 3 m aft of amidships. Find the new drafts if a weight of 5 tonnes already on board is shifted aft through a distance of 60 metres.

4 A ship is floating at drafts of 6.1 metres F and 6.7 metres A. The following cargo is then loaded:

>   20 tonnes in a position whose centre of gravity is 30 metres forward of amidships.
>   45 tonnes in a position whose centre of gravity is 25 metres forward of amidships.
>   60 tonnes in a position whose centre of gravity is 15 metres aft of amidships.
>   30 tonnes in a position whose centre of gravity is 3 metres aft of amidships.

The centre of flotation is amidships, MCT 1 cm = 200 tonnes m and TPC = 35 tonnes. Find the new drafts forward and aft.

5 A ship arrives in port trimmed 0.3 m by the stern and is to discharge 4600 tonnes of cargo from 4 holds; 1800 tonnes of the cargo is to be discharged from No. 2 and 800 tonnes from No. 3 hold. Centre of flotation is amidships, MCT 1 cm = 250 tonnes m.

>   The centre of gravity of No. 1 hold is 45 m forward of amidships.
>   The centre of gravity of No. 2 hold is 25 m forward of amidships.
>   The centre of gravity of No. 3 hold is 20 m aft of amidships.
>   The centre of gravity of No. 4 hold is 50 m aft of amidships.

Find the amount of cargo which must be discharged from Nos. 1 and 4 holds if the ship is to sail on an even keel.

6   A ship is 150 m long, displacement 12 000 tonnes, and is floating at drafts of 7 m F and 8 m A. The ship is to enter port from an anchorage with a maximum draft of 7.6 m. Find the minimum amount of cargo to discharge from a hold whose centre of gravity is 50 m aft of the centre of flotation (which is amidships), TPC 15 tonnes and MCT 1 cm = 300 tonnes m.

7   A ship 150 m × 20 m floats on an even keel at 10 m draft and has a block coefficient of fineness 0.8 and LGM of 200 metres. If 250 tonnes of cargo is discharged from a position 32 m from the centre of flotation, find the resulting change of trim.

8   A ship is floating in salt water at drafts of 6.7 m F and 7.3 m A. MCT 1 cm = 250 tonnes m. TPC 10 tonnes. Length of ship 120 metres. The centre of flotation is amidships; 220 tonnes of cargo is then discharged from a position 24 m forward of the centre of flotation. Find the weight of cargo which must now be shifted from 5 m aft of the centre of flotation to a position 20 m forward of the centre of flotation, to bring the draft aft to 7 metres. Find also the final draft forward.

9   A ship floats in salt water on an even keel displacing 6200 tonnes. KG = 5.5 m, KM = 6.3 m, and there is 500 tonnes of cargo yet to load. Space is available in No. 1 'tween deck (KG 7.6 m, centre of gravity 40 m forward of the centre of flotation) and in No. 4 lower hold (KG 5.5 m, centre of gravity 30 m aft of the centre of flotation). Find how much cargo to load in each space to complete loading trimmed 0.6 m by the stern, and find also the final GM. MCT 1 cm = 200 tonnes m.

10   A ship, floating at drafts of 7.7 m F and 7.9 m. A sustains damage in an end-on collision and has to lift the bow to reduce the draft forward to 6.7 m. The ship is about to enter a port in which the maximum permissible draft is 8.3 m. To do this it is decided to discharge cargo from No. 1 hold (centre of gravity 75 m forward of amidships) and No. 4 hold (centre of gravity 45 m aft of amidships). MCT 1 cm 200 tonnes m, TPC 15 tonnes. Centre of flotation is amidships. Find the minimum amount of cargo to discharge from each hold.

11   A ship 100 m long has centre of flotation 3 m aft of amidships and is floating at drafts 3.2 m F and 4.4 m A. TPC 10 tonnes. MCT 1 cm = 150 tonnes m; 30 tonnes of cargo is then discharged from 20 m forward of amidships and 40 tonnes is discharged from 12 m aft of amidships. Find the final drafts.

12   A ship 84 metres long is floating on an even keel at a draft of 5.5 metres; 45 tonnes of cargo is then loaded in a position 30 m aft of amidships. The centre of flotation is 1 m aft of amidships. TPC 15 tonnes. MCT 1 cm = 200 tonnes m. Find the final drafts.

13   A ship arrives in port with drafts 6.8 m. F and 7.2 m A; 500 tonnes of cargo is then discharged from each of 4 holds.

The centre of gravity of No. 1 hold is 40 m forward of amidships
The centre of gravity of No. 2 hold is 25 m forward of amidships
The centre of gravity of No. 3 hold is 20 m aft of amidships
The centre of gravity of No. 4 hold is 50 m aft of amidships

Also 50 tonnes of cargo is loaded in a position whose centre of gravity is 15 m aft of amidships, and 135 tonnes of cargo centre of gravity 40 m forward of amidships. TPC = 15 tonnes. MCT 1 cm = 400 tonnes m. The centre of flotation is amidships. Find the final drafts.

Using trim to find the position of the centre of flotation

14 A ship is floating at drafts 5.5 m F and 6.0 m A. The following cargo is then loaded:

97 tonnes centre of gravity 8 m forward of amidships
20 tonnes centre of gravity 40 m aft of amidships
28 tonnes centre of gravity 20 m aft of amidships

The draft is now 5.6 metres F and 6.1 metres A. Find the position of the centre of flotation relative to amidships.

15 Find the position of the centre of flotation of a ship if the trim remains unchanged after loading the following cargo:

100 tonnes centre of gravity 8 metres forward of amidships
20 tonnes centre of gravity 40 metres aft of amidships
28 tonnes centre of gravity 20 metres aft of amidships

16 A ship arrives in port with drafts 6.8 m F and 7.5 m A. The following cargo is discharged:

90 tonnes centre of gravity 30 m forward of amidships
40 tonnes centre of gravity 25 m aft of amidships
50 tonnes centre of gravity 50 m aft of amidships

The drafts are now 6.7 m F and 7.4 m A. Find the position of the centre of flotation relative to amidships.

Loading a weight to keep a constant draft aft

17 A ship is 150 m long, MCT 1 cm 400 tonnes m, TPC 15 tonnes. The centre of flotation is 3 m aft of amidships. Find the position in which to load a mass of 30 tonnes, with reference to the centre of flotation, so as to maintain a constant draft aft.

18 A ship 120 metres long, with maximum beam 15 m, is floating in salt water at drafts 6.6 m F and 7 m A. The block coefficient and coefficient of fineness of the water-plane is 0.75. Longitudinal metacentric height 120 m. Centre of flotation is amidships. Find how much more cargo can be loaded and in what position relative to amidships if the ship is to cross a bar with a maximum draft of 7 m F and A.

Loading a weight to produce a required draft

19 A ship 120 m long floats in salt water at drafts 5.8 m F and 6.6 m A. TPC 15 tonnes. MCT 1 cm = 300 tonnes m. Centre of flotation is amidships. What is the minimum amount of water ballast required to be taken into the forepeak tank (centre of gravity 60 m forward of the centre of flotation) to reduce the draft aft to 6.5 metres? Find also the final draft forward.

20 A ship leaves port with drafts 7.6 m F and 7.9 m A; 400 tonnes of bunkers are burned from a space whose centre of gravity is 15 m forward of the centre of flotation, which is amidships. TPC 20 tonnes. MCT 1 cm = 300 tonnes m. Find the minimum amount of water which must be run into the forepeak tank (centre of gravity 60 m forward of the centre of flotation) in order to bring the draft aft to the maximum of 7.7 m. Find also the final draft forward.

21 A ship 100 m long has MCT 1 cm 300 tonnes m requires 1200 tonnes of cargo to complete loading and is at present floating at drafts of 5.7 m F and 6.4 m A. She loads 600 tonnes of cargo in a space whose centre of gravity is 3 m forward of amidships. The drafts are then 6.03 m F and 6.67 m A. The remainder of the cargo is to be loaded in No. 1 hold (centre of gravity 43 m forward of amidships) and in No. 4 hold (centre of gravity 37 m aft of amidships). Find the amount which must be loaded in each hold to ensure that the draft aft will not exceed 6.8 metres. LCF is at amidships.

22 A ship 100 metres long is floating in salt water at drafts 5.7 m F and 8 m A. The centre of flotation is 2 m aft of amidships. Find the amount of water to run into the forepeak tank (centre of gravity 48 m forward of amidships) to bring the draft aft to 7.9 m. TPC 30 tonnes. MCT 1 cm = 300 tonnes m.

23 A ship 140 m long arrives off a port with drafts 5.7 m F and 6.3 m A. The centre of flotation is 3 m aft of amidships. TPC 30 tonnes. MCT 1 cm = 420 tonnes m. It is required to reduce the draft aft to 6.2 m by running water into the forepeak tank (centre of gravity 67 m forward of amidships). Find the minimum amount of water to load and also give the final draft forward.

Using change of trim to find GM$_L$

24 A ship 150 m long is floating upright on an even keel. When a weight already on board is shifted aft, it causes the ship's centre of gravity to shift 0.2 m horizontally and the trim to change by 0.15 m. Find the longitudinal metacentric height.

25 A ship of 5000 tonnes displacement is 120 m long and floats upright on an even keel. When a mass of 40 tonnes, already on board, is shifted 25 m forward horizontally, it causes the trim to change by 0.1 m. Find the longitudinal metacentric height.

26 A ship is 130 m LBP and is loaded ready for departure as shown in the table below. From her hydrostatic curves at 8 m even keel draft in salt water it is found that: MCTC is 150 tm/cm. LCF is 2.5 m forward ⦸, W is 12 795 tonnes and LCB is 2 m forward ⦸. Calculate the final end drafts for

this vessel. What is the final value for the trim? What is the Dwt value for this loaded condition?

Item	Weight in tonnes	LCG from amidships
Lightweight	3600	2.0 m aft
Cargo	8200	4.2 m forward
Oil fuel	780	7.1 m aft
Stores	20	13.3 m forward
Fresh water	100	20.0 m aft
Feed water	85	12.0 m aft
Crew and effects	10	At amidships

Chapter 17
Stability and hydrostatic curves

Cross curves of stability

GZ cross curves of stability

These are a set of curves from which the righting lever about an assumed centre of gravity for any angle of heel at any particular displacement may be found by inspection. The curves are plotted for an assumed KG and, if the actual KG of the ship differs from this, a correction must be applied to the righting levers taken from the curves.

Figure 17.1 shows a set of Stability Cross Curves plotted for an imaginary ship called M.V. 'Tanker', assuming the KG to be 9 metres. A scale of displacements is shown along the bottom margin and a scale of righting levers (GZs) in metres on the left-hand margin. The GZ scale extends from +4.5 m through 0 to −1 metre. The curves are plotted at 15 degree intervals of heel between 15 degrees and 90 degrees.

To find the GZs for any particular displacement locate the displacement concerned on the bottom scale and, through this point erect a perpendicular to cut all the curves. Translate the intersections with the curves horizontally to the left-hand scale and read off the GZs for each angle of heel.

Example 1

Using the Stability Cross Curves for M.V. 'Tanker' find the GZs at 15-degree intervals between 0 degrees and 90 degrees heel when the displacement is 35 000 tonnes, and KG = 9 metres.

Erect a perpendicular through 35 000 tonnes on the displacement scale and read off the GZs from the left-hand scale as follows:

Angle of heel	0°	15°	30°	45°	60°	75°	90°
GZ in metres	0	0.86	2.07	2.45	1.85	0.76	−0.5

Should the KG of the ship be other than 9 metres, a correction must be applied to the GZs taken from the curves to obtain the correct GZs. The corrections are tabulated in the block at the top right-hand side of Figure 17.1

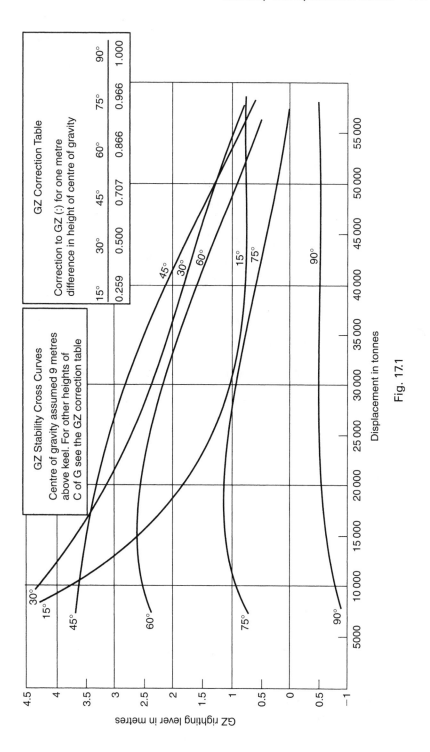

Fig. 17.1

and are given for each one metre difference between 9 metres and the ship's actual KG. To find the correction to the GZ, multiply the correction taken from the table for the angle of heel concerned, by the difference in KGs. To apply the correction: when the ship's KG is *greater* than 9 metres the ship is less stable and the correction must be *subtracted*, but when the KG is *less* than 9 metres the ship is more stable and the correction is to be *added*.

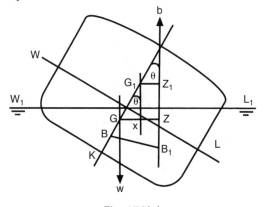

Fig. 17.2(a)

The derivation of the table is as follows:

In Figure 17.2(a), KG is 9 m, this being the KG for which this set of curves is plotted, and GZ represents the righting lever, as taken from the curves for this particular angle of heel.

Consider the case when the KG is greater than 9 m (KG_1 in Figure 17.2(a)). The righting lever is reduced to G_1Z_1. Let G_1X be perpendicular to GZ. Then:

$$G_1Z_1 = XZ$$
$$= GZ - GX$$

or

$$\text{Corrected GZ} = \text{Tabulated GZ} - \text{Correction}$$

Also, in triangle GXG_1:

$$GX = GG_1 \sin \theta°$$

or

$$\text{Correction} = GG_1 \sin \theta° \text{ where } \theta° \text{ is the angle of heel}$$

But GG_1 is the difference between 9 m and the ship's actual KG. Therefore, the corrections shown in the table on the Cross Curves for each one metre difference of KG are simply the Sines of the angle of heel.

Now consider the case where KG is less than 9 m (KG_2 in Figure 17.2(b)). The length of the righting lever will be increased to G_2Z_2.

Let GY be perpendicular to G_2Z_2 then:

$$G_2Z_2 = YZ_2 + G_2Y$$

but

$$YZ_2 = GZ$$

therefore

$$G_2Z_2 = GZ + G_2Y$$

or

$$\text{Corrected GZ} = \text{Tabulated GZ} + \text{Correction}$$

Also, in triangle GG_2Y:

$$G_2Y = GG_2 \sin \theta°$$

or

$$\text{Correction} = GG_2 \sin \text{heel}$$

It will be seen that this is similar to the previous result except that in this case the correction is to be *added* to the tabulated GZ.

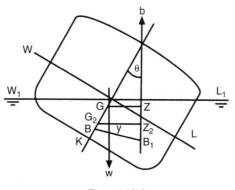

Fig. 17.2(b)

Example 2

Using the Stability Cross Curves for M.V. 'Tanker', find the GZs at 15-degree intervals between 0 degrees and 90 degrees when the displacement is 38 000 tonnes and the KG is 8.5 metres.

Heel	Tabulated GZ (KG 9 m)	Correction ($GG_1 \sin \theta°$)	Correct GZ (KG 8.5 m)
0°	0	$0.5 \times 0\quad\ = 0$	$0\ \ + 0\ \ = 0$
15°	0.81	$0.5 \times 0.259 = 0.129$	$0.81 + 0.13 = 0.94$
30°	1.90	$0.5 \times 0.5\quad\ = 0.250$	$1.90 + 0.25 = 2.15$
45°	2.24	$0.5 \times 0.707 = 0.353$	$2.24 + 0.35 = 2.59$
60°	1.70	$0.5 \times 0.866 = 0.433$	$1.70 + 0.43 = 2.13$
75°	0.68	$0.5 \times 0.966 = 0.483$	$0.68 + 0.48 = 1.16$
90°	−0.49	$0.5 \times 1.000 = 0.500$	$−0.49 + 0.50 = 0.01$

KN cross curves of stability

It has already been shown that the Stability Cross Curves for a ship are constructed by plotting the righting levers for an assumed height of the centre of gravity above the keel. In some cases the curves are constructed for an assumed KG of zero. The curves are then referred to as KN curves, KN being the righting lever measured from the keel. Figure 17.3(a) shows the KN curves for an imaginary ship called the M.V. 'Cargo-Carrier'.

To obtain the righting levers for a particular displacement and KG the values of KN are first obtained from the curves by inspection at the displacement concerned. The correct righting levers are then obtained by subtracting from the KN values a correction equal to the product of the KG and sin heel.

In Figure 17.3(b), let KN represent the ordinate obtained from the curves. Also, let the ship's centre of gravity be at G so that KG represents the actual height of the centre of gravity above the keel and GZ represents the length of the righting lever.

Now

$$GZ = XN$$
$$= KN - KX$$

or

$$GZ = KN - G \sin \theta$$

Thus, the righting lever is found by *always* subtracting from the KN ordinate a correction equal to KG sin heel.

Example 3

Find the righting levers for M.V. 'Cargo-Carrier' when the displacement is 40 000 tonnes and the KG is 10 metres.

Heel (θ)	KN	$\sin \theta$	KG $\sin \theta$	GZ = KN - KG $\sin \theta$
5°	0.90	0.087	0.87	0.03
10°	1.92	0.174	1.74	0.18
15°	3.11	0.259	2.59	0.52
20°	4.25	0.342	3.42	0.83
30°	6.30	0.500	5.00	1.30
45°	8.44	0.707	7.07	1.37
60°	9.39	0.866	8.66	0.73
75°	9.29	0.966	9.66	-0.37
90°	8.50	1.000	10.00	-1.50

Statical stability curves

The curve of statical stability for a ship in any particular condition of loading is obtained by plotting the righting levers against angle of heel as shown in Figures 17.4 and 17.5.

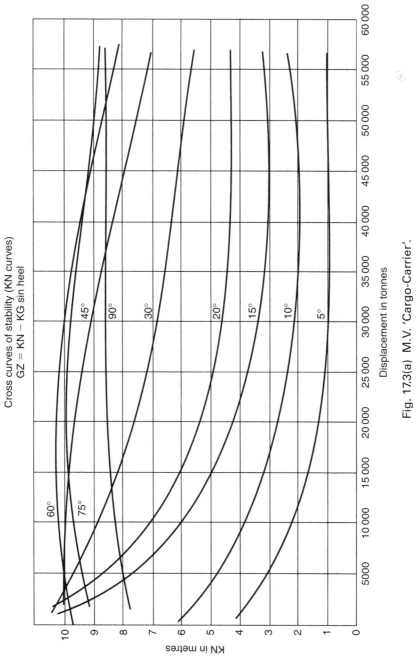

Cross curves of stability (KN curves)
GZ = KN – KG sin heel

KN in metres

Displacement in tonnes

Fig. 17.3(a) M.V. 'Cargo-Carrier'.

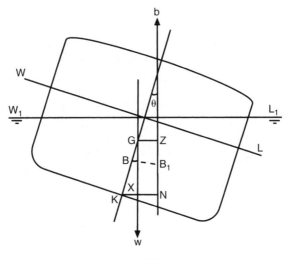

Fig. 17.3(b)

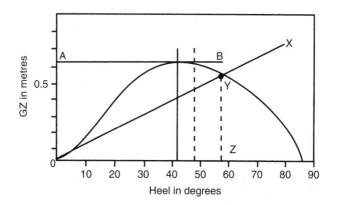

Fig. 17.4 Curve for a ship with positive initial metacentric height.

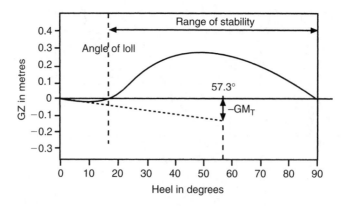

Fig. 17.5 Curve for a ship with negative initial metacentric height.

From this type of graph a considerable amount of stability information may be found by inspection:

The range of stability is the range over which the ship has positive righting levers. In Figure 17.4 the range is from 0 degrees to 86 degrees.

The angle of vanishing stability is the angle of heel at which the righting lever returns to zero, or is the angle of heel at which the sign of the righting levers changes from positive to negative. The angle of vanishing stability in Figure 17.4 is 86 degrees.

The maximum GZ is obtained by drawing a tangent to the highest point in the curve. In Figure 17.4, AB is the tangent and this indicates a maximum GZ of 0.63 metres. If a perpendicular is dropped from the point of tangency, it cuts the heel scale at the angle of heel at which the maximum GZ occurs. In the present case the maximum GZ occurs at 42 degrees heel.

The initial metacentric height (GM) is found by drawing a tangent to the curve through the origin (OX in Figure 17.4), and then erecting a perpendicular through an angle of heel of 57.3 degrees. Let the two lines intersect at Y. Then the height of the intersection above the base (YZ), when measured on the GZ scale, will give the initial metacentric height. In the present example the GM is 0.54 metres.

Figure 17.5 shows the stability curve for a ship having a negative initial metacentric height. At angles of heel of less than 18 degrees the righting levers are negative, whilst at angles of heel between 18 degrees and 90 degrees the levers are positive. The angle of loll in this case is 18 degrees, the range of stability is 18 degrees to 90 degrees, and the angle of vanishing stability is 90 degrees. (For an explanation of angle of loll see Chapter 6.) Note how the −ve GM is plotted at 57.3°.

Example 1

Using the Stability Cross Curves for M.V. 'Tanker', plot the curve of statical stability when the displacement is 33 500 tonnes and KG = 9.3 metres. From the curve find the following:

(a) The range of stability.
(b) The angle of vanishing stability.
(c) The maximum righting lever and the angle of heel at which it occurs.

Heel	Tabulated GZ (KG 9 m)	Correction to GZ (GG₁ sin heel)	Required GZ (KG 9.3 m)
0°	0	$0 \times 0 = 0$	$= 0$
15°	0.90	$0.3 \times 0.259 = 0.08$	$0.90 - 0.08 = 0.82$
30°	2.15	$0.3 \times 0.500 = 0.15$	$2.15 - 0.15 = 2.00$
45°	2.55	$0.3 \times 0.707 = 0.21$	$2.55 - 0.21 = 2.34$
60°	1.91	$0.3 \times 0.866 = 0.26$	$1.91 - 0.26 = 1.65$
75°	0.80	$0.3 \times 0.966 = 0.29$	$0.80 - 0.29 = 0.51$
90°	−0.50	$0.3 \times 1.000 = 0.30$	$-0.50 - 0.30 = -0.80$

(d) The initial metacentric height.
(e) The moment of statical stability at 25 degrees heel.

For the graph see Figure 17.6(a),

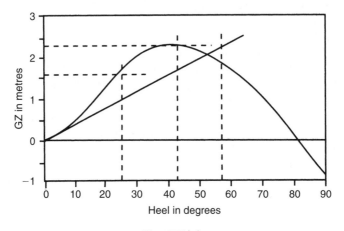

Fig. 17.6(a)

Answers from the curve:

(a) Range of Stability 0 degrees to 81 degrees.
(b) Angle of vanishing stability 81 degrees.
(c) Maximum GZ = 2.35 m occurring at 43 degrees heel.
(d) GM is 2.30 m.
(e) GZ at 25 degrees heel = 1.64 m.

$$\text{Moment of statical stability} = \text{W} \times \text{GZ}$$
$$= 33\,500 \times 1.64$$
$$= 54\,940 \text{ tonnes m}$$

Example 2
Construct the curve of statical stability for the M.V. 'Cargo-Carrier' when the displacement is 35 000 tonnes and KG is 9 metres. From the curve you have constructed find the following:

(a) The range of stability.
(b) The angle of vanishing stability.
(c) The maximum righting lever and the angle of the heel at which it occurs.
(d) The approximate initial metacentric height.

From the Stability Cross Curves:

Heel (θ)	KN	$\sin \theta$	$KG \sin \theta$	$GZ = KN - KG \sin \theta$
5°	0.9	0.087	0.783	0.12
10°	2.0	0.174	1.566	0.43
15°	3.2	0.259	2.331	0.87
20°	4.4	0.342	3.078	1.32
30°	6.5	0.500	4.500	2.00
45°	8.75	0.707	6.363	2.39
60°	9.7	0.866	7.794	1.91
75°	9.4	0.966	8.694	0.71
90°	8.4	1.000	9.000	−0.60

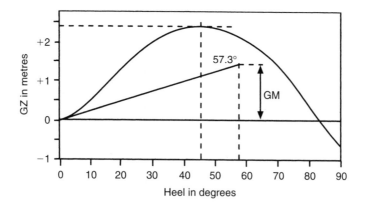

Fig. 17.6(b) This is the curve of statical stability.

Answers from the curve:

(a) Range of stability 0° to $83\frac{3}{4}°$.
(b) Angle of vanishing stability $83\frac{3}{4}°$.
(c) Maximum GZ = 2.39 m occurring at 45° heel.
(d) Approximate GM = 1.4 m.

Hydrostatic curves

Hydrostatic information is usually supplied to the ship's officer in the form of a table or a graph. Figure 17.7 shows the hydrostatic curves for the imaginary ship M.V. 'Tanker'. The various items of hydrostatic information are plotted against draft.

When information is required for a specific draft, first locate the draft on the scale on the left-hand margin of the figure. Then draw a horizontal line through the draft to cut all of the curves on the figure. Next draw a perpendicular through the intersections of this line with each of the curves in turn and read off the information from the appropriate scale.

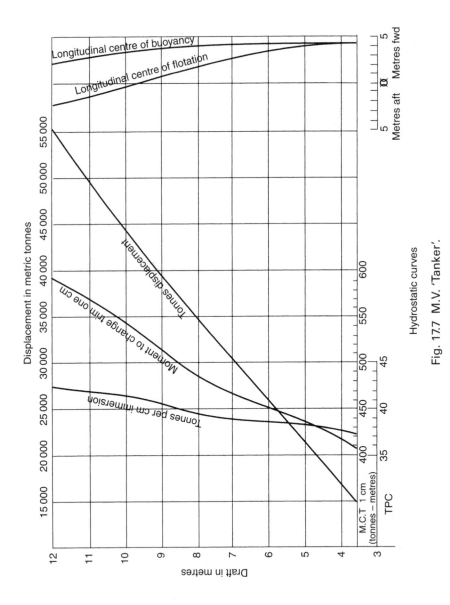

Hydrostatic curves

Fig. 17.7 M.V. 'Tanker'.

Example 1

Using the hydrostatic curves for M.V. 'Tanker', take off all of the information possible for the ship when the mean draft is 7.6 metres.

1. TPC = 39.3 tonnes.
2. MCT 1 cm = 475 tonnes-metres.
3. Displacement = 33 000 tonnes.
4. Longitudinal centre of flotation is 2.2 m forward of amidships.
5. Longitudinal centre of buoyancy is 4.0 m forward of amidships.

When information is required for a specific displacement, locate the displacement on the scale along the top margin of the figure and drop a perpendicular to cut the curve marked 'Displacement'. Through the intersection draw a horizontal line to cut all of the other curves and the draft scale. The various quantities can then be obtained as before.

Example 2

From the hydrostatic curves take off the information for M.V. 'Tanker' when the displacement is 37 500 tonnes.

1. Draft = 8.55 m.
2. TPC = 40 tonnes.
3. MCT 1 cm = 500 tonnes m.
4. Longitudinal centre of flotation is 1.2 m forward of amidships.
5. Longitudinal centre of buoyancy is 3.7 m forward of amidships.

The curves themselves are produced from calculations involving Simpson's Rules. These involve half-ordinates, areas, moments and moments of inertia for each water line under consideration.

Using the hydrostatic curves

After the end drafts have been taken it is necessary to interpolate to find the 'mean draft'. This is the draft immediately *below the LCF* which may be aft, forward or even at amidships. This draft can be labelled d_H.

If d_H is taken as being simply the average of the two end drafts then in large full-form vessels (supertankers) and fine-form vessels (container ships) an appreciable error in the displacement can occur. (See Fig. 17.8.)

Let us assume the true mean draft 'd_H' is 6 m. The Naval Architect or mate on board ship draws a horizontal line parallel to the SLWL at 6 m on the vertical axis right across all of the hydrostatic curves.

At each intersection with a curve and this 6 m line, he or she projects downwards and reads off on the appropriate scale on the 'x' axis.

For our hydrostatic curves, at a mean draft of 6 m, for example, we would obtain the following:

TPC = 19.70 t   Displace^{MT} = 10 293 t

MCTC = 152.5 tm/cm   LCF$_⦶$ = 0.05 m forward ⦶

LCB$_⦶$ = 0.80 m forward ⦶   KML = 207.4 m

KM$_T$ = 7.46 m

These values can then be used to calculate *the new end drafts* and *transverse stability*, if weights are *added* to the ship, *discharged* from the ship or simply *moved* longitudinally or transversely within the ship.

LCF$_{\not{X}}$ and LCB$_{\not{X}}$ are distances measured from amidships ($\not{X}$), or forward of the Aft Perp.

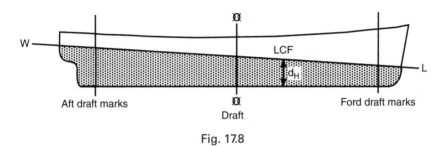

Fig. 17.8

Nowadays these values can be put on a spreadsheet in a computer package. When the hydrostatic draft d$_H$ is keyed, the hydrostatic values appertaining to this draft are then displayed, ready for use.

A set of hydrostatic values has been calculated for a 135.5 m general cargo ship of about 10 000 tonnes deadweight. These are shown on next page. From those values a set of hydrostatic curves were drawn. These are shown in Figure 17.9.

Hydrostatic values for Fig. 17.9   These are for a 135.5 m LBP general cargo ship.

Draft d$_H$ (see below) m	TPC tonnes	KB m	Displace^{MT} tonnes	KM$_L$ m	MCTC tm/cm	KM$_T$ m	LCF$_⋈$ m	LCB$_⋈$ m
9 m	20.64	4.80	16 276	146.5	167.4	7.71	2.10 aft	0.45 aft
8 m	20.36	4.27	14 253	161.8	162.9	7.54	1.20 aft	⋈
7 m	20.06	3.74	12 258	181.5	158.1	7.44	0.50 aft	0.45 forward
6 m	19.70	3.21	10 293	207.4	152.5	7.46	0.05 forward	0.80 forward
5 m	19.27	2.68	8361	243.2	145.9	7.70	0.42 forward	1.05 forward
4 m	18.76	2.15	6486	296.0	138.3	8.28	0.70 forward	1.25 forward
3 m	18.12	1.61	4674	382.3	129.1	9.56	0.83 forward	1.30 forward
2.5 m	17.69	1.35	3785	449.0	123.0	10.75	0.85 forward	1.30 forward

Note. LCF and LCB are now measured ford of Aft Perp (foap).

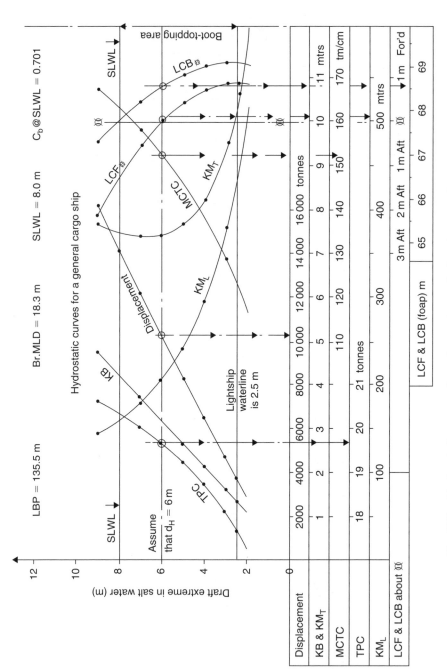

Fig. 17.9   Hydrostatic curves: based on values tabulated on previous page.

Exercise 17

1   Plot the curve of stability for M.V. 'Tanker' when the displacement is 34 500 tonnes and KG = 9 m. From this curve find the approximate GM, the range of stability, the maximum GZ and the angle of heel at which it occurs.

2   Plot the curve of statical stability for M.V. 'Tanker' when the displacement is 23 400 tonnes and KG = 9.4 m. From this curve find the approximate GM, the maximum moment of statical stability and the angle of heel at which it occurs. Find also the range of stability.

3   The displacement of M.V. 'Tanker' is 24 700 tonnes and KG = 10 m. Construct a curve of statical stability and state what information may be derived from it. Find also the moments of statical stability at 10 degrees and 40 degrees heel.

4   Using the cross curves of stability for M.V. 'Tanker':

(a) Draw a curve of statical stability when the displacement is 35 000 tonnes, KG = 9.2 metres, and KM = 11.2 m.

(b) From this curve determine the moment of statical stability at 10 degrees heel and state what other information may be obtained from the curve.

5   Plot the stability curve for M.V. 'Tanker' when the displacement is 46 800 tonnes and KG = 8.5 metres. From this curve find the approximate GM, the range of stability, the maximum GZ and the angle of heel at which it occurs.

6   From the hydrostatic curves for M.V. 'Tanker' find the mean draft when entering a dock where the density of the water is 1009 kg per cu. m, if the mean draft in salt water is 6.5 m.

7   Find the mean draft of M.V. 'Tanker' in dock water of density 1009 kg per cu. m when the displacement is 35 400 tonnes.

8   M.V. 'Tanker' is 200 m long and is floating at the maximum permissible draft aft. There is one more parcel of heavy cargo to load on deck to put her down to her marks at the load displacement of 51 300 tonnes. Find the position, relative to amidships, to load the parcel so as to maintain the maximum possible trim by the stern.

9   Using the hydrostatic curves for M.V. 'Tanker' find the displacement, MCT 1 cm and the TPC, when the ship is floating in salt water at a mean draft of 9 metres.

   Also find the new mean draft if she now enters a dock where the density of the water is 1010 kg per cu. m.

10   M.V. 'Tanker' is 200 m long and has a light displacement of 16 000 tonnes. She has on board 100 tonnes of cargo, 300 tonnes of bunkers, and 100 tonnes of fresh water and stores. The ship is trimmed 1.5 m by the stern. Find the new drafts if the 100 tonnes of cargo already on board is now shifted 50 m forward.

11   Construct the curve of statical stability for M.V. 'Cargo-Carrier' when the displacement is 35 000 tonnes and KG is 8 metres. From this curve find

(a) the range of stability,

   (b) the angle of vanishing stability and
   (c) the maximum GZ and the heel at which it occurs.
12 Construct the curve of statical stability for M.V. 'Cargo-Carrier' when the
    displacement is 28 000 tonnes and the KG is 10 metres. From the curve find
    (a) the range of stability,
    (b) the angle of vanishing stability and
    (c) the maximum GZ and the heel at which it occurs.
13 A vessel is loaded up ready for departure. KM is 11.9 m. KG is 9.52 m with
    a displacement of 20 550 tonnes. From the ship's Cross Curves of Stability,
    the GZ ordinates for a displacement of 20 550 tonnes and a VCG of 8 m
    above base are as follows

Angle of heel (θ)	0	15	30	45	60	75	90
GZ ordinate (m)	0	1.10	2.22	2.60	2.21	1.25	0.36

   Using this information, construct the ship's statical stability curve for this
   condition of loading and determine the following:
   (a) Maximum righting lever GZ.
   (b) Angle of heel at which this maximum GZ occurs.
   (c) Angle of heel at which the deck edge just becomes immersed.
   (d) Range of stability.
14 Using the table of KN ordinates below calculate the righting levers for a
    ship when her displacement is 40 000 tonnes and her actual KG is 10 m.
    Draw the resulting Statical Stability curve and from it determine:
    (a) Maximum GZ value.
    (b) Approximate GM value.
    (c) Righting moment at the angle of heel of 25°.
    (d) Range of Stability.

Angle of heel (θ)	0	5	10	15	20	30	45	60
KN ordinates (m)	0	0.90	1.92	3.11	4.25	6.30	8.44	9.39

Angle of heel (θ)	75	90
KN ordinates (m)	9.29	8.50

Chapter 18
Increase in draft due to list

Box-shaped vessels

The draft of a vessel is the depth of the lowest point below the waterline. When a box-shaped vessel is floating upright on an even keel as in Figure 18.1(a) the draft is the same to port as to starboard. Let the draft be 'd' and let the vessel's beam be 'b'.

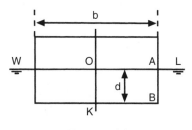

Fig. 18.1(a)

Now consider the same vessel when listed θ degrees as shown in Figure 18.1(b). The depth of the lowest point or draft is now increased to 'D' (XY).

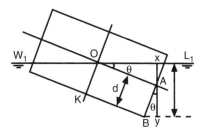

Fig. 18.1(b)

In triangle OXA:

$$OA = \tfrac{1}{2}b \text{ and angle } AXO = 90° \qquad \therefore AX = \tfrac{1}{2}b \sin \theta°$$

In triangle ABY:

$$AB = d \text{ and angle } AYB = 90° \qquad \therefore AY = d \cos \theta°$$

$$D = XY$$
$$= AX + AY$$
$$D = \tfrac{1}{2}b \sin \theta + d \cos \theta$$

or

$$\text{New draft} = \tfrac{1}{2} \text{ beam sin list} + \text{Old draft cos list}$$

Note. It will usually be found convenient to calculate the draft when listed by using the triangles AOX and ABY.

Example 1
A box-shaped ship with 12 m beam is floating upright at a draft of 6.7 m. Find the increase in draft if the vessel is now listed 18 degrees.

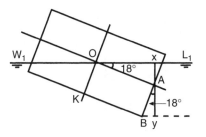

Fig. 18.2(a)

In triangle OAX:

$$\text{Angle AOX} = 18°$$

Therefore use COS 18°

$$OA \text{ (Dist.)} = 6.0 \text{ m}$$
$$AX \text{ (Dep.)} = 1.85 \text{ m}$$

In triangle ABY:

$$\text{Angle BAY} = 18°$$

Therefore use COS 18°

$$
\begin{aligned}
\text{AB (Dist.)} &= \quad 6.7 \text{ m}° \\
\text{AY (D. Lat.)} &= \quad 6.37 \text{ m} \\
\text{AX} &= \quad 1.85 \text{ m} \\
\text{AY} &= +6.37 \text{ m} \\
\text{New draft} &= \quad 8.22 \text{ m} \\
\text{Old draft} &= +6.70 \text{ m} \\
& \quad \overline{1.52 \text{ m}}
\end{aligned}
$$

Ans. Increase = 1.52 m

Example 2

Using the dimensions of the ship given in Example 1, proceed to calculate the increase in draft at 3° intervals from 0° to 18°. Plot a graph of draft increase $\propto \theta$:

$$\tfrac{1}{2} \text{ beam } \sin \theta = \tfrac{1}{2} \times 12 \times \sin \theta = 6 \sin \theta \tag{I}$$

$$\text{Old draft } \cos \theta = 6.7 \cos \theta \tag{II}$$

Increase in draft = (I) + (II) − Old draft in metres.

Angle of list	$6 \sin \theta$	$6.7 \cos \theta$	Old draft	Increase in draft (m)
0°	0	6.70	6.70	0
3°	0.31	6.69	6.70	0.30
6°	0.63	6.66	6.70	0.59
9°	0.94	6.62	6.70	0.86
12°	1.25	6.55	6.70	1.10
15°	1.55	6.47	6.70	1.32
18°	1.85	6.37	6.70	1.52

The above results clearly show the increase in draft or loss of underkeel clearance when a vessel lists.

Ships in the late 1990s and early 2000s are being designed with shorter lengths and wider breadths mainly to reduce first cost and hogging/sagging characteristics.

These wider ships are creating problems sometimes when initial underkeel clearance is only 10 per cent of the ship's static draft. It only requires a small angle of list for them to go aground in way of the bilge strakes.

One ship, for example, is the supertanker *Esso Japan*. She has an LBP of 350 m and a width of 70 m at amidships. Consequently extra care is required should possibilities of list occur.

Figure 18.2(b) shows the requested graph of draft increase $\propto \theta$.

Vessels having a rise of floor

Example

A ship has 20 m beam at the waterline and is floating upright at 6 m draft. If the rise of floor is 0.25 m, calculate the new draft if the ship is now listed 15 degrees (see Figure 18.3).

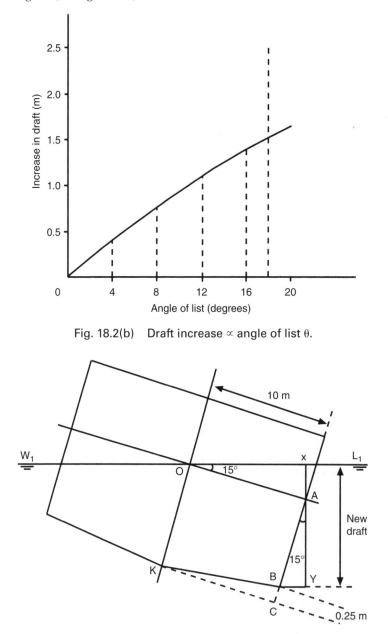

Fig. 18.2(b)   Draft increase ∝ angle of list θ.

Fig. 18.3

In triangle OAX:

$$\text{Angle O} = 15°$$

Therefore use COS 15°

$$\text{AO (Dist.)} = 10 \text{ m}$$
$$\text{AX (Dep.)} = 2.59 \text{ m}$$

In triangle ABY:

$$\text{Angle A} = 15°$$

Therefore use COS 15°

$$\text{AB (Dist.)} = \text{AC} - \text{BC}$$
$$= \text{Old draft} - \text{Rise of floor}$$
$$= 6 \text{ m} - 0.25 \text{ m}$$
$$\text{AB (Dist.)} = 5.75 \text{ m}$$
$$\text{AY (D. Lat)} = 5.55 \text{ m}$$
$$\text{AX} = 2.59 \text{ m}$$
$$\text{AY} = +5.55 \text{ m}$$
$$\text{XY} = \underline{8.14 \text{ m}}$$

Ans. New draft = 8.14 metres

If the formula is to be used to find the new draft it must now be amended to allow for the rise of floor as follows:

$$\text{New draft} = \tfrac{1}{2} \text{ beam sin list} + (\text{Old draft} - \text{Rise}) \cos \text{list}$$

Note. In practise, the shipbuilder's naval architect usually calculates the increase in draft for the ship at various angles of list and suppliers the information to the ship's officer by means of a table on the plans of the ship.

The rise of floor, similar to tumblehome on merchant ships, has almost become obsolete on merchant ships of today.

Exercise 18

1 A box-shaped vessel with 10 m beam, and 7.5 m deep, floats upright at a draft of 3 m. Find the increase of draft when listed 15 degrees to starboard.

2 A ship 90 m long, 15 m beam at the waterline, is floating upright at a draft of 6 m. Find the increase of draft when the ship is listed 10 degrees, allowing 0.15 m rise of floor.

3 A box-shaped vessel 60 m × 6 m × 4 m draft is floating upright. Find the increase in the draft if the vessel is now listed 15 degrees to port.

4 A box-shaped vessel increases her draft by 0.61 m when listed 12 degrees to starboard. Find the vessel's beam if the draft, when upright, was 5.5 m.

5 A box-shaped vessel with 10 m beam is listed 10 degrees to starboard and the maximum draft is 5 m. A weight already on board is shifted transversely across the deck, causing the vessel to list to port. If the final list is 5 degrees to port, find the new draft.

Chapter 19
Water pressure

The water pressure at any depth is due to the weight of water above the point in question and increases uniformly with depth below the surface. From this it can be seen that the pressure at any depth will vary with the depth below the surface and the density of the water.

Pressure at any depth

Consider an area of 1 sq m at the water level as shown by ABCD in Figure 19.1. The water pressure on this area is zero as there is no water above it.

Now consider the area EFGH which is submerged to a depth of 1 m below the surface. There is 1 cu. m of water above it. Let the density of the water be 'w' tonnes per cu. m. Then the pressure on this area is 'w' tonnes

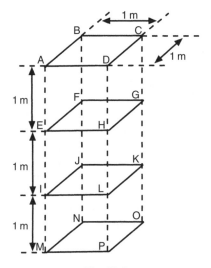

Fig. 19.1

per sq. m, and the total pressure or thrust on an area of 'A' sq. m whose centre of gravity is 1 m below the surface, will be wA tonnes.

Next consider the area IJKL whose centre of gravity is 2 m below the surface. The pressure is now due to 2 cu. m of water and will be 2 w tonnes per sq. m. Also, the total pressure or thrust on any area whose centre of gravity is 2 m below the surface, is 2 wA tonnes.

Similarly, the pressure on the area MNOP will be 3 w tonnes per sq. metre, and the total pressure or thrust on any area whose centre of gravity is 3 m below the surface will be 3 wA tonnes.

The following formulae may now be deduced:

$$\text{Total pressure or thrust} = w \times A \times h \text{ tonnes}$$

where
h = the depth of the centre of gravity below the surface
w = the density of the water in tonnes/m^3
A = the area in sq. m

Example 1

Find the water pressure and the total thrust on the flat plate keel of a ship which is floating on an even keel at 4 metres mean draft in water of density 1.024 t/m^3 per cu. m. The keel plate is 35 m long and 1.5 m wide:

$$\text{Pressure} = 4 \times 1.024 = 4.1 \text{ tonnes/m}^2$$
$$\text{Total thrust} = w \times A \times h$$
$$= 1.024 \times 35 \times 1.5 \times 4$$
$$\text{Total thrust} = 215.04 \text{ tonnes}$$

Example 2

A lock gate which is 15 m wide has salt water on one side to a depth of 8 m and fresh water on the other side to a depth of 9 m. Find the resultant thrust on the lock gate and state on which side of the gate it acts.

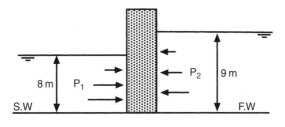

Fig. 19.2

Let P_1 = Thrust on the SW side, and P_2 = Thrust on the FW side

$$\therefore P_1 = \text{Pressure on the SW side}$$
$$P_1 = \rho_{SW} \times A \times h_1$$
$$= 1.025 \times 15 \times 8 \times \tfrac{8}{2}$$
$$P_1 = 492 \text{ tonnes}$$
$$\therefore P_2 = \text{Pressure on the FW side}$$
$$P_2 = \rho_{FW} \times A \times h_2$$
$$= 1.000 \times 15 \times 9 \times \tfrac{9}{2}$$
$$P_2 = 607.5 \text{ tonnes}$$
$$\text{Resultant thrust} = P_2 - P_1$$
$$= 607.5 - 492$$

Ans. Resultant thrust = 115.5 tonnes on the FW side

Example 3

Find the pressure and total thrust on two deep tank lids, each 4 m × 2.5 m, resting on coamings 30 cm high, when the tank is filled with salt water to a head of 9.3 m above the crown of the tank.

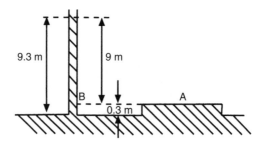

Fig. 19.3

$$\text{Pressure at A} = \text{Pressure at B (see Figure 19.3)}$$
$$= w \times A \times h$$
$$= 1.025 \times 1.0 \text{ m}^2 \times 9.0$$
$$\text{Pressure on lids} = 9.225 \text{ tonnes per sq. m}$$
$$\text{Total thrust} = 2 \times \text{thrust on one lid}$$
$$= 2 \times (P \times A)$$
$$= 2 \times 9.225 \times 4 \times 2.5 \text{ tonnes}$$
$$\text{Total thrust} = 184.5 \text{ tonnes}$$

Ans. Pressure = 9.225 tonnes per sq. m and total thrust = 184.5 tonnes

Exercise 19

1 A sealed box is made of metal and is capable of withstanding a pressure of 15.54 kN per sq. m. Find to what depth the box may be submerged in salt water before it collapses.

2 Find the water pressure, in kilo-newtons per sq. m, on the keel of a ship which is floating upright on an even keel in salt water at a draft of 6 metres.

3 A ship is floating on an even keel in salt water at a draft of 7 metres. She has a flat plate keel which is 100 m long and 2 m wide. Find the water pressure and the total thrust on the keel due to the water.

4 Find the water pressure and thrust on a flat plate keel 6 m long and 1.5 m wide when the ship is floating on an even keel at 6 m draft in salt water.

5 Find the water pressure and thrust on the inner bottom plating of a double bottom tank (10 m × 10 m) when the tank is filled with fresh water to a head of 9 m above the crown of the tank.

6 A deep tank has two lids, each 4 m × 3 m, resting on coamings which are 30 cm high. Find the water pressure on each of the two lids and the total thrust on both lids when the tank is filled with salt water to a head of 6.3 m above the crown of the tank.

7 A rectangular plate 2 m × 1 m is held vertically in fresh water with the 2 m edge on the waterline. Find the thrust on one side of the plate.

8 A lock gate 12 m wide has water of density 1010 kg per cu. m to a depth of 10 m on one side and water of density 1015 kg per cu. m to a depth of 6 m on the other side. Find the resultant thrust on the lock gate and state on which side it acts.

9 A lock gate is 20 m wide and has water of density 1016 kg per cu. m on one side to a depth of 10 m and salt water on the other side to a depth of 7.5 m. Find the resultant thrust and state on which side of the gate it will act.

10 Find the total water pressure acting on a box-shaped barge keel plating which is 60 m long 10 m wide, and is floating upright in salt water on an even keel at 5 m draft.

11 A box-shaped barge has a rectangular collision bulkhead 10 m wide and 6 m deep. Find the total water pressure on this bulkhead if the forepeak is flooded to a depth of 5 m with salt water.

Chapter 20
Combined list and trim

When a problem involves a change of both list and trim, the two parts must be treated quite separately. It is usually more convenient to tackle the trim part of the problem first and then the list, but no hard and fast rule can be made on this point.

Example 1

A ship of 6000 tonnes displacement has KM = 7 m, KG = 6.4 m and MCT 1 cm = 120 tonnes m. The ship is listed 5 degrees to starboard and trimmed 0.15 m by the head. The ship is to be brought upright and trimmed 0.3 m by the stern by transferring oil from No. 2 double bottom tank to No. 5 double bottom tank. Both tanks are divided at the centre line and their centres of gravity are 6 m out from the centre line. No. 2 holds 200 tonnes of oil on each side and is full. No. 5 holds 120 tonnes on each side and is empty. The centre of gravity of No. 2 is 23.5 m forward of amidships and No. 5 is 21.5 m aft of amidships. Find what transfer of oil must take place and give the final distribution of the oil. (Neglect the effect of free surface on the GM.) Assume that LCF is at amidships:

(a) To bring the ship to the required trim

$$\text{Present trim} = 0.15 \text{ m by the head } \supset$$
$$\underline{\text{Required trim} = 0.30 \text{ m by the stern } \subset}$$
$$\text{Change of trim} = 0.45 \text{ m by the stern } \subset$$
$$= 45 \text{ cm by the stern } \subset$$
$$\text{Trim moment} = \text{Change of trim} \times \text{MCT 1 cm}$$
$$= 45 \times 120$$
$$\text{Trim moment} = 5400 \text{ tonnes m by the stern } \subset$$

Let 'w' tonnes of oil be transferred aft to produce the required trim.

$$\therefore \ \text{Trim moment} = w \times d$$
$$= 45w \text{ tonnes m}$$
$$\therefore \ 45w = 5400$$
$$w = 120 \text{ tonnes}$$

From this it will be seen that, if 120 tonnes of oil is transferred aft, the ship will then be trimmed 0.30 m by the stern.

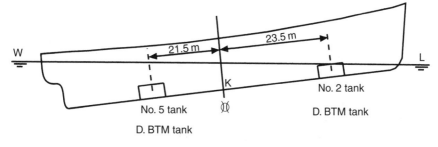

Fig. 20.1(a)

(b) To bring the ship upright

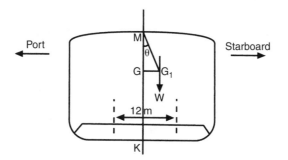

Fig. 20.1(b)   Looking forward.

$$KM = 7.0 \text{ m}$$
$$KG = \underline{-6.4 \text{ m}}$$
$$GM = 0.6 \text{ m}$$

In triangle GG_1M:

$$GG_1 = GM \times \tan\theta$$
$$= 0.6 \times \tan 5°$$
$$GG_1 = 0.0525 \text{ m}$$

Let 'x' tonnes of oil be transferred from starboard to port.

$$\text{Moment to port} = x \times d$$
$$= 12x \text{ tonnes m}$$

$$\text{Initial moment to starboard} = W \times GG_1$$
$$= 6000 \times 0.0525$$
$$= 315 \text{ tonnes m}$$

But if the ship is to complete the operation upright:

$$\text{Moment to starboard} = \text{Moment to port}$$

or

$$315 = 12x$$
$$x = 26.25 \text{ tonnes}$$

The ship will, therefore, be brought upright by transferring 26.25 tonnes from starboard to port.

From this it can be seen that, to bring the ship to the required trim and upright, 120 tonnes of oil must be transferred from forward to aft and 26.25 tonnes from starboard to port. This result can be obtained by taking 120 tonnes from No. 2 starboard and by putting 93.75 tonnes of this oil in No. 5 starboard and the remaining 26.25 tonnes in No. 5 port tank.

The distributions would then be as follows:

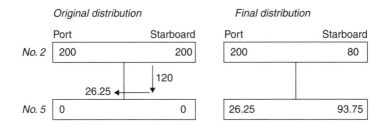

Note. There are, of course, alternative methods by which this result could have been obtained, but in each case a total of *120 tonnes* of oil must be transferred aft and *26.25 tonnes* must be transferred from starboard to port.

Exercise 20

1 A tanker has displacement of 10 000 tonnes, KM = 7 m, KG = 6.4 m and MCT 1 cm = 150 tonnes m. There is a centre line bulkhead in both No. 3 and No. 8 tanks. Centre of gravity of No. 3 tank is 20 m forward of the centre of flotation and the centre of gravity of No. 8 tank is 30 m aft of the centre of flotation. The centre of gravity of all tanks is 5 m out from the centre line. At present the ship is listed 4 degrees to starboard and trimmed 0.15 m by the head. Find what transfer of oil must take place if the ship is to complete upright and trimmed 0.3 m by the stern.

2 A ship of 10 000 tonnes displacement is listed 5 degrees to port and trimmed 0.2 m by the head. KM = 7.5 m, KG = 6.8 m and MCT 1 cm = 150 tonnes m. Centre of flotation is amidships. No. 1 double bottom tank is divided at the centre line, each side holds 200 tonnes of oil and the tank is full. No. 4 double bottom tank is similarly divided each side having a

capacity of 150 tonnes, but the tank is empty. The centre of gravity of No. 1 tank is 45 m forward of amidships and the centre of gravity of No. 4 tank is 15 m aft of amidships. The centre of gravity of all tanks is 5 m out from the centre line. It is desired to bring the ship upright and trimmed 0.3 m by the stern by transferring oil. If the free surface effect on GM be neglected, find what transfer of oil must take place and also the final distribution of the oil.

3   A ship of 6000 tonnes displacement, KG = 6.8 m, is floating upright in salt water, and the draft is 4 m F and 4.3 m A. KM = 7.7 m TPC = 10 tonnes. MCT 1 cm = 150 tonnes m. There is a locomotive to discharge from No. 2 lower hold (KG = 3 m, and centre of gravity 30 m forward of the centre of flotation which is amidships). If the weight of the locomotive is 60 tonnes and the height of the derrick head is 18 m above the keel and 20 m out from the centre line when plumbing overside, find the maximum list during the operation and the drafts after the locomotive has been discharged. Assume KM is constant.

4   A ship displaces 12 500 tonnes, is trimmed 0.6 m by the stern and listed 6 degrees to starboard. MCT 1 cm = 120 tonnes m, KG = 7.2 m, KM = 7.3 m. No. 2 and No. 5 double bottom tanks are divided at the centre line. Centre of gravity of No. 2 is 15 m forward of the centre of flotation and centre of gravity of No. 5 is 12 m aft of the centre of flotation. Centre of gravity of all tanks is 4 m out from the centre line. The ship is to be brought upright and on to an even keel by transferring oil from aft to forward, taking equal quantities from each side of No. 5. Find the amounts of oil to transfer.

Chapter 21
Calculating the effect of free surface of liquids (FSE)

The effect of free surface of liquids on stability was discussed in general terms in Chapter 7, but the problem will now be studied more closely and the calculations involved will be explained.

When a tank is partially filled with a liquid, the ship suffers a virtual loss in metacentric height which can be calculated by using the formula:

$$\text{FSE} = \text{Virtual loss of GM} = \frac{i}{W} \times \rho \times \frac{1}{n^2} \text{ metres} \qquad \text{(I)}$$

where

      i = the second moment of the free surface about the centre line, in m^4

      w = the ship's displacement, in tonnes

      ρ = the density of the liquid in the tank, in tonnes/cu. m

      n = the number of the longitudinal compartments into which the tank is equally subdivided

$i \times p$ = free surface moment, in tonnes m

The derivation of this formula is as follows:

The ship shown in Figure 21.1(a) has an undivided tank which is partially filled with a liquid.

When the ship is inclined, a wedge of the liquid in the tank will shift from the high side to the low side such that its centre of gravity shifts from g to g_1. This will cause the centre of gravity of the ship to shift from G to G_1, where

$$GG_1 = \frac{w \times gg_1}{W}$$

Let

$$\rho_1 = \text{density of the liquid in the tank}$$

and

$$\rho_2 = \text{density of the water in which the ship floats}$$

then

$$GG_1 = \frac{v \times \rho_1 \times gg_1}{V \times \rho_2}$$

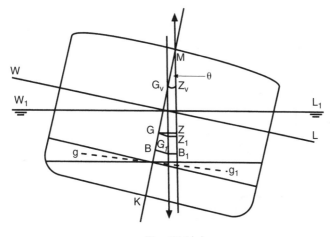

Fig. 21.1(a)

Had there been no free surface when the ship inclined, the righting lever would have been GZ. But, due to the liquid shifting, the righting lever is reduced to G_1Z_1 or G_vZ_v. The virtual reduction of GM is therefore GG_v.

For a small angle of heel:

$$GG_1 = GG_v \times \theta$$

$$\therefore \ GG_v \times \theta = \frac{v \times gg_1 \times \rho_1}{V \times \rho_2}$$

or

$$GG_V = \frac{v \times gg_1 \times \rho_1}{V \times \theta \times \rho_2}$$

From the proof of BM = I/V, I × θ = v × gg₁

Let i = the second moment of the free surface about the centre line. Then

$$GG_v = \frac{i \times \rho_1}{V \times \rho_2}$$

This is the formula to find the virtual loss of GM due to the free surface effect in an undivided tank.

Now assume that the tank is subdivided longitudinally into 'n' compartments of equal width as shown in Figure 21.1(b).

Let

$$l = \text{length of the tank}$$
$$b = \text{breadth of the tank}$$

The breadth of the free surface in each compartment is thus b/n, and the second moment of each free surface is given by $\dfrac{l \times (b/n)^3}{12}$

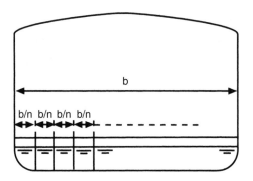

Fig. 21.1(b)

$$GG_v = \text{virtual loss of GM for one compartment}$$
$$\text{multiplied by the number of compartments}$$
$$= \frac{i \times \rho_1}{V \times \rho_2} \times n$$

where

$$i = \text{the second moment of the water-plane area}$$
$$\text{in one compartment about the centre line}$$
$$= \frac{l \times (b/n)^3 \times \rho_1}{12 \times V \times \rho_2} \times n$$
$$= \frac{l \times b^3 \times \rho_1}{12 \times V \times n^3 \times \rho_2} \times n$$

or

$$GG_v = \frac{i}{V} \times \frac{\rho_1}{\rho_2} \times \frac{1}{n^2}$$

where

i = the second moment of the water-plane area of
the whole tank about the centre line

But

$$W = V \times \rho_2$$

so

$$GG_v = \frac{i}{W} \times \rho_1 \times \frac{1}{n^2}$$

as shown in equation (I) at the beginning of this chapter.

This is the formula to find the virtual loss of GM due to the free surface effect in a tank which is subdivided longitudinally.

From this formula it can be seen that, when a tank is subdivided longitudinally, the virtual loss of GM for the undivided tank is divided by the square of the number of compartments into which the tank is divided. Also note that the actual weight of the liquid in the tank will have no effect whatsoever on the virtual loss of GM due to the free surface.

For a rectangular area of free surface, the second moment to be used in the above formula can be found as follows:

$$i = \frac{LB^3}{12}$$

where

L = the length of the free surface
B = the total breadth of the free surface, ignoring divisions

Note. Transverse subdivisions in partially filled tanks (slack tanks) *do not* have any influence on reducing free surface effects.

However, fitting longitudinal bulkheads *do* have a very effective influence in reducing this virtual loss in GM.

Example 1

A ship of 8153.75 tonnes displacement has KM = 8 m, KG = 7. 5 m, and has a double bottom tank 15 m × 10 m × 2 m which is full of salt water ballast. Find the new GM if this tank is now pumped out till half empty.

Note. The mass of the water pumped out will cause an actual rise in the position of the ship's centre of gravity and the free surface created will cause a virtual loss in GM. There are therefore two shifts in the position of the centre of gravity to consider.

In Figure 21.2 the shaded portion represents the water to be pumped out with its centre of gravity at position g. The original position of the ship's centre of gravity is at G.

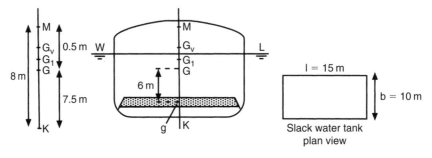

Fig. 21.2

Let GG_1 represent the actual rise of G due to the mass discharged.

$$\text{The mass of water discharged (w)} = 15 \times 10 \times 1 \times 1.025 \text{ tonnes}$$
$$w = 153.75 \text{ tonnes}$$
$$W_2 = W_1 - w = 8153.75 - 153.75$$
$$= 8000 \text{ tonnes}$$
$$GG_1 = \frac{w \times d}{W_2}$$
$$= \frac{153.75 \times 6}{8000}$$
$$GG_1 = 0.115 \text{ m}$$

Let G_1G_v represent the virtual loss of GM due to free surface or rise in G_1. Then:

$$G_1G_v = \frac{i}{W} \times \rho_1 \times \frac{1}{n^2}$$

as per equation (I) at the beginning of this chapter:

$$n = 1$$
$$\therefore \ G_1G_v = \frac{i}{W_2} \times \rho_{SW}$$

or

$$G_1G_v = \frac{1b^3}{12} \times \frac{\rho_{SW}}{W_2} \times \frac{1}{n^2}$$
$$\text{Loss in GM} = \text{FSE}$$

or

$$G_1 G_v = \frac{15 \times 10^3 \times 1.025}{12 \times 8000}$$

$$\underline{G_1 G_v = 0.160 \text{ m } \uparrow}$$

$$\text{Old KM} = \underline{8.000 \text{ m}}$$

$$\text{Old KG} = \underline{7.500 \text{ m}}$$

$$\text{Old GM} = 0.500 \text{ m}$$

$$\text{Actual rise of G} = 0.115 \text{ m}$$

$$0.385 \text{ m} = \text{GM}_{\text{solid}}$$

$$G_1 G_v = \text{Virtual rise of G} = \underline{0.160 \text{ m } \uparrow}$$

$$= \underline{0.225 \text{ m}}$$

Ans. New GM = $0.225\,\text{m}$ = GM_{fluid}

Hence G_1 has risen due to the discharge of the ballast water (loading change) and has also risen due to free surface effects.

Be aware that in some cases these two rises of G do not take G above M thereby making the ship unstable.

Example 2

A ship of 6000 tonnes displacement, floating in salt water, has a double bottom tank 20 m × 12 m × 2 m which is divided at the centre line and is partially filled with oil of relative density 0.82. Find the virtual loss of GM due to the free surface of the oil.

$$\text{Virtual loss of GM} = \frac{i}{W} \times \rho_{\text{oil}} \times \frac{1}{n^2}$$

$$= \frac{lb^3}{12} \times \rho_{\text{oil}} \times \frac{1}{W} \times \frac{1}{n^2}$$

$$= \frac{20 \times 12^3}{12 \times 6000} \times 0.820 \times \frac{1}{2^2}$$

Ans. Virtual loss of GM = 0.098 metres

Example 3

A ship of 8000 tonnes displacement has KM 7.5 m and KG 7.0 m. A double bottom tank is 12 m long, 15 m wide and 1 m deep. The tank is divided longitudinally at the centre line and both sides are full of salt water. Calculate the list if one side is pumped out until it is half empty. (See Fig. 21.3.)

$$\text{Mass of water discharge} = l_{\text{DB}} \times b_{\text{DB}} \times \frac{d_{\text{DB}}}{2} \times \rho_{\text{SW}}$$

$$\text{Mass of water discharge} = 12 \times 7.5 \times 0.5 \times 1.025$$

$$w = 46.125 \text{ tonnes}$$

$$\text{Vertical shift of G(GG}_1) = \frac{w \times d}{W_2}$$

$$= \frac{46.125 \times 6.25}{8000 - 46.125} = \frac{46.125 \times 6.25}{7953.875}$$

$$= 0.036 \text{ metres}$$

$$\text{Horizontal shift of G(G}_v\text{G}_2) = \frac{w \times d}{W - w}$$

$$= \frac{46.125 \times 3.75}{7953.875}$$

$$= 0.022 \text{ metres}$$

$$\text{Virtual loss of GM(G}_1\text{G}_v) = \frac{1 \times b_2^3}{12} \times \frac{\rho_{SW}}{W_2} \times \frac{1}{n^2}$$

$$= \frac{12 \times 7.5^3}{12} \times \frac{1.025}{7953.875} \times \frac{1}{1^2}$$

$$= 0.054 \text{ metres}$$

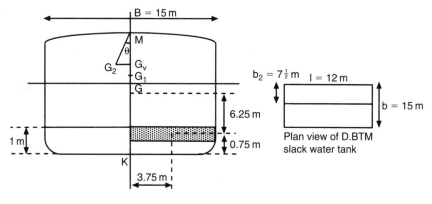

Fig. 21.3

$$\text{KM} = 7.500 \text{ metres}$$

$$\text{Original KG} = 7.000 \text{ metres}$$

$$\text{Original GM} = 0.500 \text{ metres} = \text{GM}_{\text{solid}}$$

$$\text{Vertical rise of G(GG}_1) = 0.036 \text{ metres} \quad \uparrow$$

$$\text{G}_1\text{M} = 0.464 \text{ metres} = \text{G}_1\text{M}_{\text{solid}}$$

$$\text{Virtual loss of GM(G}_1\text{G}_v) = 0.054 \text{ metres}$$

$$\text{New GM(G}_v\text{M}) = 0.410 \text{ metres} = \text{GM}_{\text{fluid}}$$

In triangle G_vG_2M:

$$\tan\theta = \frac{G_2G_v}{G_vM}$$

$$= \frac{0.022}{0.410} = 0.0536$$

Ans. List = 3°04′

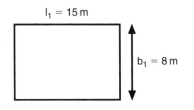

Stability factors
Not to scale

So again G has risen due to discharging water ballast. It has also risen due to free-surface effects. A further movement has caused G to move off the centreline and has produced this angle of list of 3°04′.

The following worked example shows the effect of subdivisions in slack tanks in relation to free surface effects (FSE):

Question: A ship has a displacement of 3000 tonnes. On the vessel is a rectangular double-bottom tank 15 m long and 8 m wide. This tank is partially filled with ballast water having a density of 1.025 t/m³.

If the GM_T without free surface effects is 0.18 m calculate the virtual loss in GM_T and the final GM_T when the double bottom tank has:

(a) no divisional bulkheads fitted,
(b) one transverse bulkhead fitted at mid-length,
(c) one longitudinal bulkhead fitted on ₵ of the tank,
(d) two longitudinal bulkheads fitted giving three equal divisions.

Answer

$l_1 = 15\,m$

$b_1 = 8\,m$

Fig. 21.4(a)

$$FSE = \text{virtual loss in } GM_T \text{ or rise in } G = \frac{I \times \rho_{SW}}{W}$$

$$= \frac{1 \times b_1^3 \times \rho_{SW}}{12 \times W} \quad (\text{see Fig. 21.4(a)})$$

$$\therefore \text{ virtual loss in } GM_T = \frac{15 \times 8^3 \times 1.025}{3000 \times 12}$$

$$= \underline{0.2187 \text{ m}} \uparrow$$

$$\therefore GM_T \text{ finally} = 0.1800 - 0.2187$$

$$= \underline{-0.0387 \text{ m}} \uparrow$$

i.e. unstable ship!!

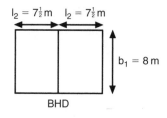

Fig. 21.4(b)

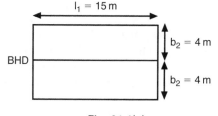

Fig. 21.4(c)

$$\text{FSE} = \text{virtual loss in GM}_T \text{ or rise in G} = \frac{2 @ l_2 \times b_1^3}{12 \times W} \times \rho_{SW} \text{ (see Fig. 21.4(b))}$$

$$\therefore \text{ virtual loss} = \frac{2 \times 7.5 \times 8^3 \times 1.025}{12 \times 3000}$$

$$= 0.2187 \text{ m} \uparrow$$

This is same answer as for part (a). Consequently it can be concluded that fitting transverse divisional bulkheads in tanks does not reduce the free surface effects. Ship is still unstable!!

$$\text{FSE} = \text{vertical loss in GM}_T \text{ or rise in G} = \frac{2 @ l_1 b_2^3}{12 \times W} \rho_{SW}$$

$$\therefore \text{ virtual loss in GM}_T = \frac{2 \times 15 \times 4^3 \times 1.025}{12 \times 3000} \text{ (see Fig. 21.4(c))}$$

$$= 0.0547 \text{ m} \uparrow \text{ i.e. } \tfrac{1}{4} \text{ of answer to part (a)}$$

Hence

$$\text{final GM}_T = 0.1800 - 0.0547 \text{ m} = +0.1253 \text{ m ship is stable.}$$

GM_T is now +ve, but below the minimum GM_T of 0.15 m that is allowable by DfT regulations.

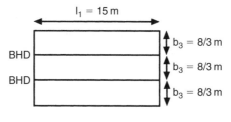

Fig. 21.4(d)

$$\text{FSE} = \text{virtual loss in } GM_T \text{ or rise in } G = \frac{3 @ l_1 \times b_3^3}{12 \times W} \times \rho_{SW} \quad \text{(see Fig. 21.4(d))}$$

$$\therefore \text{ Virtual loss in } GM_T = \frac{3 \times 15 \times \left(\dfrac{8}{3}\right)^3 \times 1.025}{12 \times W}$$

$$= 0.0243 \text{ m} \uparrow \text{ i.e. } \tfrac{1}{9} \text{ of answer to part (a)}$$

Hence

$$\text{Final } GM_T = 0.1800 - 0.0243 = +0.1557 \text{ m ship is stable}$$

Ship is stable and above DfT minimum GM_T value of 0.15 m.

So longitudinal divisional bulkheads (watertight or wash-bulkheads) are effective. They cut down rapidly the loss in GM_T. Note the $1/n^2$ law where n is the number of equal divisions made by the longitudinal bulkheads.

Free surface effects therefore depend on:

(i) density of slack liquid in the tank,
(ii) ship's displacement in tonnes,
(iii) dimensions and shape of the slack tanks,
(iv) bulkhead subdivision within the slack tanks.

The longitudinal divisional bulkheads referred to in examples in this chapter need not be absolutely watertight; they could have openings in them. Examples on board ship are the centreline wash bulkhead in the Fore Peak tank and in Aft Peak tank.

Exercise 21

1   A ship of 10 000 tonnes displacement is floating in dock water of density 1024 kg per cu. m, and is carrying oil of relative density 0.84 in a double-bottom tank. The tank is 25 m long, 15 m wide, and is divided at the centre line. Find the virtual loss of GM due to this tank being slack.

2   A ship of 6000 tonnes displacement is floating in fresh water and has a deep tank (10 m × 15 m × 6 m) which is undivided and is partly filled with nut oil of relative density 0.92. Find the virtual loss of GM due to the free surface.

3   A ship of 8000 tonnes displacement has KG = 3.75 m, and KM = 5.5 m. A double-bottom tank 16 m × 16 m × 1 m is subdivided at the centre line and is full of salt water ballast. Find the new GM if this tank is pumped out until it is half empty.

4   A ship of 10 000 tonnes displacement, KM 6 m, KG 5.5 m, is floating upright in dock water of density 1024 kg per cu. m. She has a double bottom tank 20 m × 15 m which is subdivided at the centre line and is partially filled with oil of relative density 0.96. Find the list if a mass of 30 tonnes is now shifted 15 m transversely across the deck.

5   A ship is at the light displacement of 3000 tonnes and has KG 5.5 m, and KM 7.0 m. The following weights are then loaded:

> 5000 tonnes of cargo KG 5 m
> 2000 tonnes of cargo KG 10 m
> 700 tonnes of fuel oil of relative density 0.96.

The fuel oil is taken into Nos. 2, 3 and 5 double bottom tanks, filling Nos. 3 and 5, and leaving No. 2 slack.

The ship then sails on a 20-day passage consuming 30 tonnes of fuel oil per day. On arrival at her destination Nos. 2 and 3 tanks are empty, and the remaining fuel oil is in No. 5 tank. Find the ship's GM's for the departure and arrival conditions.

Dimensions of the tanks:

> No. 2   15 × 15 m × 1 m
> No. 3   22 m × 15 m × 1 m
> No. 5   12 m × 15 m × 1 m

Assume that the KM is constant and that the KG of the fuel oil in every case is half of the depth of the tank.

6   A ship's displacement is 5100 tonnes, KG = 4 m, and KM = 4.8 m. A double-bottom tank on the starboard side is 20 m long, 6 m wide and 1 m deep and is full of fresh water. Calculate the list after 60 tonnes of this water has been consumed.

7   A ship of 6000 tonnes displacement has KG 4 m and KM 4.5 m. A double-bottom tank in the ship 20 m long and 10 m wide is partly full of salt-water ballast. Find the moment of statical stability when the ship is heeled 5 degrees.

8   A box-shaped vessel has the following data:

Length is 80 m, breadth is 12 m, draft even keel is 6 m, KG is 4.62 m.

A double bottom tank 10 m long, of full width and 2.4 m depth is then half-filled with water ballast having a density of $1.025 \, t/m^3$. The tank is located at amidships.

Calculate the new even keel draft and the new transverse GM after this water ballast has been put in the double bottom tank.

Chapter 22
Bilging and permeability

Bilging amidships compartments

When a vessel floats in still water it displaces its own weight of water. Figure 22.1(a) shows a box-shaped vessel floating at the waterline, WL. The weight of the vessel (W) is considered to act downwards through G, the centre of gravity. The force of buoyancy is also equal to W and acts upwards through B, the centre of buoyancy; b = W.

Now let an empty compartment amidships be holed below the waterline to such an extent that the water may flow freely into and out of the compartment. A vessel holed in this way is said to be 'bilged'.

Figure 22.1(b) shows the vessel in the bilged condition. The buoyancy provided by the bilged compartment is lost. The draft has increased and the vessel now floats at the waterline W_1L_1, where it is again displacing its own

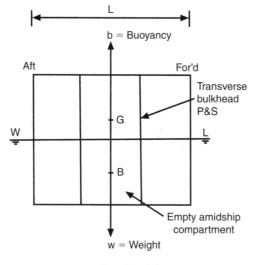

Fig. 22.1(a)

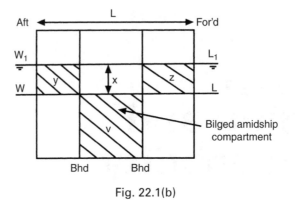

Fig. 22.1(b)

weight of water. 'X' represents the increase in draft due to bilging. The volume of lost buoyancy (v) is made good by the volumes 'y' and 'z'.

$$\therefore \ v = y + z$$

Let 'A' be the area of the water-plane before bilging, and let 'a' be the area of the bilged compartment. Then:

$$y + z = Ax - ax$$

or

$$v = x(A - a)$$
$$\text{Increase in draft} = x = \frac{v}{A - a}$$

i.e.

$$\text{Increase in draft} = \frac{\text{Volume of lost buoyancy}}{\text{Area of intact waterplane}}$$

Note. Since the distribution of weight within the vessel has not been altered the KG after bilging will be the same as the KG before bilging.

Example 1

A box-shaped vessel is 50 metres long and is floating on an even keel at 4 metres draft. A amidships compartment is 10 metres long and is empty. Find the increase in draft if this compartment is bilged. See Figure 22.1(c).

$$x = \frac{v}{A - a} = \frac{l \times B \times B}{(L - l)B}$$

let

$$B = \text{Breadth of the vessel}$$

then

$$x = \frac{10 \times B \times 4}{50 \times B - 10 \times B}$$

$$= \frac{40B}{40B}$$

$$\underline{\text{Increase in draft} = 1\,\text{metre}}$$

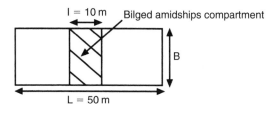

Fig. 22.1(c)

Example 2

A box-shaped vessel is 150 metres long × 24 metres wide × 12 metres deep, and is floating on an even keel at 5 metres draft. GM = 0.9 metres. A compartment amidships is 20 metres long and is empty. Find the new GM if this compartment is bilged.

$$\text{Old KB} = \tfrac{1}{2}\ \text{Old draft}$$

$$= 2.5\ \text{m}$$

$$\text{Old BM} = B^2/12d$$

$$= \frac{24 \times 24}{12 \times 5}$$

$$= 9.6\ \text{m}$$

$$\text{Old KB} = \underline{+2.5\ \text{m}}$$

$$\text{Old KM} = 12.1\ \text{m}$$

$$\text{Old GM} = \underline{-0.9\ \text{m}}$$

$$\text{KG} = \underline{11.2\ \text{m}}$$

This KG will not change after bilging has taken place:

$$x = \frac{v}{A - a}$$

$$= \frac{20 \times 24 \times 5}{(150 \times 24) - (20 \times 24)}$$

$$= \frac{2400}{130 \times 24}$$

$$\text{Increase in draft } = 0.77 \text{ m}$$
$$\text{Old draft } = \underline{5.00 \text{ m}}$$
$$\text{New draft } = \underline{5.77 \text{ m}} = \text{Say draft } d_2$$

$$\text{New KB} = \frac{1}{2} \text{ New draft} = \frac{d_2}{2}$$
$$= 2.89 \text{ m}$$
$$\text{New BM} = B^2/12d_2$$
$$= \frac{24 \times 24}{12 \times 5.77}$$
$$= 8.32 \text{ m}$$
$$+$$
$$\text{New KB} = 2.89 \text{ m}$$
$$\text{New KM} = 11.21 \text{ m}$$
$$-$$
$$\text{As before, KG} = 11.20 \text{ m}$$

Ans. New GM = 0.01 m

This is +ve but dangerously low in value!!

Permeability, μ

Permeability is the amount of water that can enter a compartment or tank after it has been bilged. When an empty compartment is bilged, the whole of the buoyancy provided by that compartment is lost. Typical values for permeability, μ, are as follows:

Empty compartment	$\mu = 100\%$
Engine room	$\mu = 80\%$ to 85%
Grain-filled cargo hold	$\mu = 60\%$ to 65%
Coal-filled compartment	$\mu = 36\%$ approximately
Filled water ballast tank	$\mu = 0\%$
(when ship is in salt water)	

Consequently, the higher the value of the permeability for a bilged compartment, the greater will be a ship's loss of buoyancy when the ship is bilged.

The permeability of a compartment can be found from the formula:

$$\mu = \text{Permeability} = \frac{\text{Broken stowage}}{\text{Stowage factor}} \times 100 \text{ per cent}$$

The broken stowage to be used in this formula is the broken stowage per tonne of stow.

When a bilged compartment contains cargo, the formula for finding the increase in draft must be amended to allow for the permeability. If 'μ' represents the permeability, expressed as a fraction, then the volume of lost buoyancy will be 'μv' and the area of the intact water-plane will be '$A - \mu v$' square metres. The formula then reads:

$$x = \frac{\mu v}{A - \mu a}$$

Example 1

A box-shaped vessel is 64 metres long and is floating on an even keel at 3 metres draft. A compartment amidships is 12 m long and contains cargo having a permeability of 25 per cent. Calculate the increase in the draft if this compartment be bilged.

$$x = \frac{\mu v}{A - \mu a}$$

$$= \frac{\frac{1}{4} \times 12 \times B \times 3}{(64 \times B) - (\frac{1}{4} \times 12 \times B)}$$

$$= \frac{9B}{61B}$$

Ans. Increase in draft = 0.15 m

Example 2

A box-shaped vessel 150 m × 20 m × 12 m is floating on an even keel at 5 metres draft. A compartment amidships is 15 metres long and contains timber of relative density 0.8, and stowage factor 1.5 cubic metres per tonne. Calculate the new draft if this compartment is now bilged.

The permeability 'μ' must first be found by using the formula given above; i.e.

$$\text{Permeability} = \frac{BS}{SF} \times 100 \text{ per cent} = '\mu'$$

The stowage factor is given in the question. The broken stowage per tonne of stow is now found by subtracting the space which would be occupied by one tonne of solid timber from that actually occupied by one tonne of timber in the hold. One tonne of fresh water occupies one cubic metre and the relative density of the timber is 0.8.

$$\therefore \text{ Space occupied by one tonne of solid timber} = \frac{1}{0.8}$$

$$= 1.25 \text{ cubic metres}$$

$$\text{Stowage factor} = \underline{1.50} \text{ cubic metres}$$

$$\therefore \text{ Broken stowage} = \underline{0.25} \text{ cubic metres}$$

$$\text{Permeability } `\mu' = \frac{BS}{SF} \times 100 \text{ per cent}$$

$$= \frac{0.25}{1.50} \times 100 \text{ per cent}$$

$$= 100/6 \text{ per cent}$$

$$\therefore `\mu' = 1/6 \text{ or } 16.67 \text{ per cent}$$

$$\text{Increase in draft } = x = \frac{\mu v}{A - \mu a}$$

$$= \frac{1/6 \times 15 \times 20 \times 5}{(150 \times 20) - ((1/6) \times 15 \times 20)}$$

$$= 250/2950 = 0.085 \text{ m}$$

$$\text{Increase in draft} = 0.085 \text{ metres}$$

$$\text{Old draft} = \underline{5.000 \text{ metres}} = \text{Draft } d_1$$

Ans. <u>New draft = 5.085 metres = Draft d_2</u>

When the bilged compartment does not extend above the waterline, the area of the intact water-plane remains constant as shown in Figure 22.2.

In this figure:

$$\mu v = Ax$$

Let

$$d = \text{Density of the water}$$

then

$$\mu v \times d = Ax \times d$$

but

$$\mu v \times d = \text{mass of water entering the bilged compartment}$$

$$Ax \times d = \text{mass of the extra layer of water displaced}$$

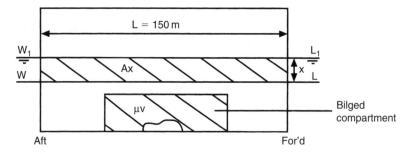

Fig. 22.2

Therefore, when the compartment is bilged, the extra mass of water displaced is equal to the buoyancy lost in the bilged compartment. It should be carefully noted, however, that although the effect on draft is similar to that of loading a mass in the bilged compartment equal to the lost buoyancy, no mass has in fact been loaded. The *displacement* after bilging is *the same* as the displacement before bilging and there is *no alteration* in the position of the vessel's *centre of gravity*. The increase in the draft is due solely to lost buoyancy.

Example 3

A ship is floating in salt water on an even keel at 6 metres draft. TPC is 20 tonnes. A rectangular-shaped compartment amidships is 20 metres long, 10 metres wide and 4 metres deep. The compartment contains cargo with permeability 25 per cent. Find the new draft if this compartment is bilged.

$$\text{Buoyancy lost} = \frac{25}{100} \times 20 \times 10 \times 4 \times 1.025 \text{ tonnes}$$

$$= 205 \text{ tonnes}$$

$$\text{Extra mass of water displaced} = \text{TPC} \times X \text{ tonnes}$$

$$\therefore X = w/\text{TPC}$$

$$= 205/20$$

$$\text{Increase in draft} = 10.25 \text{ cm}$$

$$= 0.1025 \text{ m}$$

$$\text{Plus the old draft} = \underline{6.0000 \text{ m}}$$

Ans. New draft = 6.1025 m

Note. The *lower* the permeability is the *less* will be the changes in end drafts after bilging has taken place.

Bilging end compartments

When the bilged compartment is situated in a position away from amidships, the vessel's mean draft will increase to make good the lost buoyancy but the trim will also change.

Consider the box-shaped vessel shown in Figure 22.3(a). The vessel is floating upright on an even keel, WL representing the waterline. The centre of buoyancy (B) is at the centre of the displaced water and the vessel's centre of gravity (G) is vertically above B. There is no trimming moment.

Now let the forward compartment which is X metres long be bilged. To make good the loss in buoyancy, the vessel's mean draft will increase as shown in Figure 22.3(b), where W_1L_1 represents the new waterline. Since there has been no change in the distribution of mass within the vessel, the centre of gravity will remain at G. It has already been shown that the effect on mean draft will be similar to that of loading a mass in the compartment equal to the mass of water entering the bilged space to the original waterline.

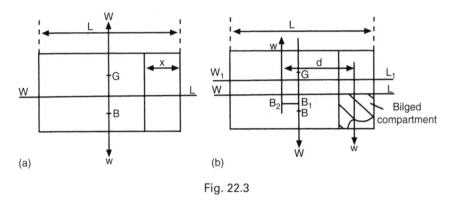

Fig. 22.3

The vertical component of the shift of the centre of buoyancy (B to B_1) is due to the increase in the mean draft. KB_1 is equal to half of the new draft. The horizontal component of the shift of the centre of buoyancy (B_1B_2) is equal to $X/2$.

A trimming moment of $W \times B_1B_2$ by the head is produced and the vessel will trim about the centre of flotation (F), which is the centre of gravity of the new water-plane area:

$$B_1B_2 = \frac{w \times d}{W}$$

or

$$W \times B_1B_2 = w \times d$$

but

$$W \times B_1B_2 = \text{Trimming moment}$$
$$\therefore \ w \times d = \text{Trimming moment}$$

It can therefore be seen that the effect on trim is similar to that which would be produced if a mass equal to the lost buoyancy were loaded in the bilged compartment.

Note. When calculating the TPC, MCTC, etc., it must be remembered that the information is required for the vessel in the bilged condition, using draft d_2 and intact length l_2.

Example 1
A box-shaped vessel 75 metres long × 10 metres wide × 6 metres deep is floating in salt water on an even keel at a draft of 4.5 metres. Find the new drafts if a forward compartment 5 metres long is bilged.

(a) First let the vessel sink bodily

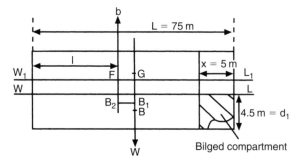

Fig. 22.4

$w = x \times B \times d_1 \times 1.025$ tonnes

$= 5 \times 10 \times 4.5 \times 1.025$

$w = 230.63$ tonnes

$$TPC = \frac{WPA}{97.56} = \frac{L_2 \times B}{97.56}$$

$$= \frac{70 \times 10}{97.56}$$

$$TPC = 7.175$$

Increase in draft $= w/TPC$

$$= 230.63/7.175$$

$$= 32.14 \text{ cm}$$

$$= 0.321 \text{ m}$$

$$+$$

Old draft $= \underline{4.500 \text{ m}} = $ draft d_1

New mean draft $= \underline{4.821 \text{ m}} = $ draft d_2

(b) Now consider the trim

$W = L \times B \times d_1 \times 1.025$ tonnes

$= 75 \times 10 \times 4.5 \times 1.025$

$W = 3459$ tonnes

$$BM_L = I_L/V$$

$$= \frac{BL_2^3}{12V}$$

$$= \frac{10 \times 70^3}{12 \times 75 \times 10 \times 4.5}$$

$$BM_L = 84.7 \text{ metres}$$

$$MCTC \simeq \frac{W \times BM_L}{100L}$$

$$= \frac{3459 \times 84.7}{100 \times 75}$$

$$= 39.05 \text{ tonnes m per cm}$$

$$\text{Change of trim} = \frac{\text{Moment changing trim}}{MCTC}$$

where

$$d = \frac{LBP}{2} = \frac{75}{2} = 37.5 \, m = \text{Lever from new LCF}$$

$$= \frac{230.6 \times 37.5}{39.05}$$

$$= 221.4 \text{ cm by the head}$$

After bilging, LCF has moved to F, i.e. $(L - x)/2$ from the stern

$$\text{Change of draft aft} = \frac{1}{L} \times \text{Change of trim}$$

$$= \frac{35}{75} \times 221.4$$

$$= 103.3 \text{ cm} = 1.033 \, m$$

$$\text{Change of draft forward} = \frac{40}{75} \times 221.4$$

$$= 118.1 \text{ cm} = 1.181 \, m$$

(c) Now find new drafts

Drafts before trimming	A 4.821 m	F 4.821 m
Change due to trim	− 1.033 m	+ 1.181 m
Ans. New drafts	A 3.788 m	F 6.002 m

Example 2

A box-shaped vessel 100 metres long × 20 metres wide × 12 metres deep is floating in salt water on an even keel at 6 metres draft. A forward compartment is 10 metres long, 12 metres wide and extends from the outer bottom to a watertight flat, 4 metres above the keel. The compartment contains cargo of permeability 25 per cent. Find the new drafts if this compartment is bilged.

$$\left.\begin{array}{l}\text{Mass of water entering}\\\text{the bilged compartment}\end{array}\right\} = \frac{25}{100} \times 10 \times 12 \times 4 \times 1.025$$

$$= 123 \text{ tonnes}$$

$$TPC_{sw} = \frac{WPA}{97.56}$$

$$\text{Increase in draft} = w/TPC$$

$$= \frac{100 \times 20}{97.56}$$

$$= 123/20.5$$

$$= 6 \text{ cm}$$

$$TPC = 20.5 \text{ tonnes}$$

$$\text{Increase in draft} = 0.06 \, m$$

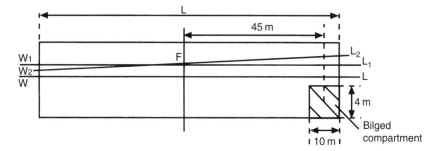

Fig. 22.5

$$W = L \times B \times d_1 \times 1.025$$
$$= 100 \times 20 \times 6 \times 1.025$$
$$W = \underline{12\ 300\ tonnes}$$

$$BM_L = \frac{I_L}{V} = \frac{BL^3}{12V} = \frac{B \times L^3}{12 \times L \times B \times d} = \frac{L^2}{12 \times d_1}$$

$$BM_L = \frac{100 \times 100}{12 \times 6}$$

$$\underline{BM_L} = 139\ \text{metres}$$

$$MCTC \simeq \frac{W \times BM_L}{100 \times L} = \frac{12\ 300 \times 139}{100 \times 100}$$

$$MCTC = 171\ \text{tonnes m per cm}$$

$$\text{Trimming moment} = W \times B_1B_2$$

$$= w \times d$$

$$\text{Trimming moment} = 123 \times 45\ \text{tonnes m}$$

$$\text{Change of trim} = \frac{\text{Trimming moment}}{MCTC} = \frac{123 \times 45}{171}$$

$$\text{Change of trim} = 32.4\ \text{cm by the head}$$

That is, 0.32 m by the head

Note. The centre of flotation, being the centroid of the water-plane area, remains amidships.

Old drafts	A	6.00 m	F	6.00 m
Bodily increase		+ 0.06 m		+ 0.06 m
		6.06 m		6.06 m
Change due to trim		− 0.16 m		+ 0.16 m
Ans. New drafts	A	5.90 m	F	6.22 m

Effect of bilging on stability

It has already been shown that when a compartment in a ship is bilged the mean draft is increased. The change in mean draft causes a change in the positions of the centre of buoyancy and the initial metacentre. Hence KM is changed and, since KG is constant, the GM will be changed.

Example 1

A box-shaped vessel 40 metres long, 8 metres wide and 6 metres deep, floats in salt water on an even keel at 3 metres draft. GM = 1 metre. Find the new GM if an empty compartment 4 metres long and situated amidships is bilged.

(a) Original condition before bilging

Find the KG

$$KB = \frac{d_1}{2} \qquad\qquad BM = \frac{I}{V}$$

$$= 1.5 \text{ metres} \qquad = \frac{LB^3}{12V} = \frac{B^2}{12 \times d_1}$$

$$= \frac{8 \times 8}{12 \times 3}$$

$$
\begin{array}{ll}
BM = & 1.78 \text{ m} \\
KB = & +1.50 \text{ m} \\
\hline
KM = & 3.28 \text{ m} \\
GM = & -1.00 \text{ m} \\
\hline
KG = & 2.28 \text{ m} \\
\end{array}
$$

(b) Vessel's condition after bilging

Find the new draft

$$\text{Lost buoyancy} = 4 \times 8 \times 3 \times 1.025 \text{ tonnes}$$

$$TPC_{sw} = \frac{WPA}{97.56} = \frac{36 \times 8}{97.56}$$

$$\text{Increase in draft} = \frac{\text{Lost buoyancy}}{TPC}$$

$$= 4 \times 8 \times 3 \times 1.025 \times \frac{100}{36 \times 8 \times 1.025} \text{ cm}$$

$$= 33.3 \text{ cm or } 0.33 \text{ m}$$

It should be noted that the increase in draft can also be found as follows:

$$\text{Increase in draft} = \frac{\text{Volume of lost buoyancy}}{\text{Area of intact water-place}} = \frac{4 \times 8 \times 3}{36 \times 8}$$

$$= 1/3 \text{ metres.}$$

$$\text{Original draft} = 3.000 \text{ m} = \text{Draft } d_1$$

$$\text{New draft} = 3.333 \text{ m} = \text{Draft } d_2$$

(c) Find the new GM

$$KB = \frac{d_2}{2} = 1.67 \text{ m}$$

BM = I/V   (*Note.* 'I' represents the second moment
of the intact water-place about the centreline)

$$= \frac{(L-1)B^3}{12 \times V} = \frac{36 \times 8^3}{12 \times 40 \times 8 \times 3}$$

$$\begin{array}{rl}
BM_2 = & 1.60 \text{ m} \\
& + \\
KB_2 = & 1.67 \text{ m} \\
\hline
KM_2 = & 3.27 \text{ m} \\
& - \\
KG = & 2.28 \text{ m} \text{ as before bilging occurred} \\
\hline
\text{Final } GM_2 = & 0.99 \text{ m}
\end{array}$$

GM_2 is +ve so vessel is in stable equilibrium.

Summary

When solving problems involving bilging and permeability it is suggested
that:

1. Make a sketch from given information.
2. Calculate mean bodily sinkage using w and TPC.
3. Calculate change of trim using GM_L or BM_L.
4. Collect calculated data to evaluate the final requested end drafts.

Exercise 22

Bilging amidships compartments

1 (a) Define permeability, 'μ'.
   (b) A box-shaped vessel 100 m long, 15 m beam floating in salt water, at a
       mean draft of 5 m, has an amidships compartment 10 m long which is
       loaded with a general cargo. Find the new mean draft if this compart-
       ment is bilged, assuming the permeability to be 25 per cent.
2 A box-shaped vessel 30 m long, 6 m beam and 5 m deep, has a mean draft of
   2.5 m. An amidships compartment 8 m long is filled with coal stowing at
   1.2 cu. m per tonne. 1 cu. m of solid coal weighs 1.2 tonnes. Find the increase
   in the draft if the compartment is holed below the waterline.
3 A box-shaped vessel 60 m long, 15 m beam, floats on an even keel at 3 m
   draft. An amidships compartment is 12 m long and contains coal (SF =
   1.2 cu.m per tonne and relative density = 1.28). Find the increase in the
   draft if this compartment is bilged.

4 A box-shaped vessel 40 m long, 6 m beam, is floating at a draft of 2 m F and A. She has an amidships compartment 10 m long which is empty. If the original GM was 0.6 m, find the new GM if this compartment is bilged.

5 If the vessel in Question 4 had cargo stowed in the central compartment such that the permeability was 20 per cent, find the new increase in the draft when the vessel is bilged.

6 A box-shaped vessel 60 m × 10 m × 6 m floats on an even keel at a draft of 5 m F and A. An amidships compartment 12 m long contains timber of relative density 0.8 and stowage factor 1.4 cu. m per tonne. Find the increase in the draft if this compartment is holed below the waterline.

7 A box-shaped vessel 80 m × 10 m × 6 m is floating upright in salt water on an even keel at 4 m draft. She has an amidships compartment 15 m long which is filled with timber (SF = 1.5 cu. m per tonne). 1 tonne of solid timber would occupy 1.25 cu. m of space. What would be the increase in the draft if this compartment is now bilged?

Bilging end compartments

8 A box-shaped vessel 75 m × 12 m is floating upright in salt water on an even keel at 2.5 m draft F and A. The forepeak tank which is 6 m long is empty. Find the final drafts if the vessel is now holed forward of the collision bulkhead.

9 A box-shaped vessel 150 m long, 20 m beam, is floating upright in salt water at drafts of 6 m F and A. The collision bulkhead is situated 8 m from forward. Find the new drafts if the vessel is now bilged forward of the collision bulkhead.

10 A box-shaped vessel 60 m long, 10 m beam, is floating upright in salt water on even keel at 4 m draft F and A. The collision bulkhead is 6 m from forward. Find the new drafts if she is now bilged forward of the collision bulkhead.

11 A box-shaped vessel 65 m × 10 m × 6 m is floating on an even keel in salt water at 4 m draft F and A. She has a forepeak compartment 5 m long which is empty. Find the new drafts if this compartment is now bilged.

12 A box-shaped vessel 64 m × 10 m × 6 m floats in salt water on an even keel at 5 m draft. A forward compartment 6 metres long and 10 metres wide, extends from the outer bottom to a height of 3.5 m, and is full of cargo of permeability 25 per cent. Find the new drafts if this compartment is now bilged.

Chapter 23
Dynamical stability

Dynamical stability is defined as the work done in inclining a ship.

Consider the ship shown in Figure 23.1. When the ship is upright the force 'W' acts upwards through B and downwards through G. These forces act throughout the inclination; b = w.

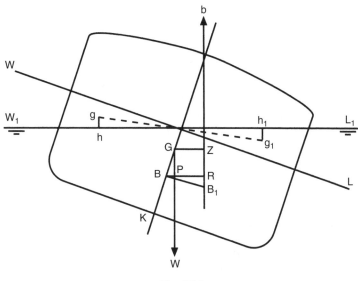

Fig. 23.1

$$\text{Work done} = \text{Weight} \times \text{Vertical separation of G and B}$$

or

$$\text{Dynamical stability} = W \times (B_1 Z - BG)$$
$$= W \times (B_1 R + RZ - BG)$$

$$= W \times \left[\frac{v(gh + g_1h_1)}{V} + PG - BG \right]$$

$$= W \times \left[\frac{v(gh + g_1h_1)}{V} + BG \cos \theta - BG \right]$$

$$\text{Dynamical stability} = W \left[\frac{v(gh + g_1h_1)}{V} - BG(1 - \cos \theta) \right]$$

This is known as *Moseley's formula* for dynamical stability.

If the curve of statical stability for a ship has been constructed the dynamical stability to any angle of heel may be found by multiplying the area under the curve to the angle concerned by the vessel's displacement. i.e.

$$\text{Dynamical stability} = W \times \text{Area under the stability curve}$$

The derivation of this formula is as follows:

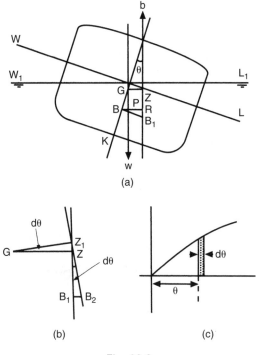

Fig. 23.2

Consider Figure 23.2(a) which shows a ship heeled to an angle θ. Now let the ship be heeled through a further very small angle $d\theta$. The centre of buoyancy B_1 will move parallel to W_1L_1 to the new position B_2 as shown in Figure 23.2(b).

B_2Z_1 is the new vertical through the centre of buoyancy and GZ_1 is the new righting arm. The vertical separation of Z and Z_1 is therefore $GZ \times d\theta$. But this is also the vertical separation of B and G. Therefore the dynamical stability from θ to $(\theta + d\theta)$ is $W \times (GZ \times d\theta)$.

Refer now to Figure 23.2(c) which is the curve of statical stability for the ship. At θ the ordinate is GZ. The area of the strip is $GZ \times d\theta$. But $W \times (GZ \times d\theta)$ gives the dynamical stability from θ to $(\theta + d\theta)$, and this must be true for all small additions of inclination:

$$\therefore \text{ Dynamical stability} = \int_O^\theta W \times GZ \times d\theta$$
$$= W \int_O^\theta GZ\, d\theta$$

Therefore the dynamical stability to any angle of heel is found by multiplying the area under the stability curve to that angle by the displacement.

It should be noted that in finding the area under the stability curve by the use of Simpson's Rules, the common interval must be expressed in *radians*:

$$57.3° = 1 \text{ radian}$$
$$1° = \frac{1}{57.3} \text{ radians}$$

or

$$x° = \frac{x}{57.3} \text{ radians}$$

Therefore to convert degrees to radians simply divide the number of degrees by 57.3.

Example 1

A ship of 5000 tonnes displacement has righting levers as follows:

Angle of heel	10°	20°	30°	40°
GZ (metres)	0.21	0.33	0.40	0.43

Calculate the dynamical stability to 40 degrees heel.

GZ	SM	Functions of area
0	1	0
0.21	4	0.84
0.33	2	0.66
0.40	4	1.60
0.43	1	0.43
		$3.53 = \Sigma_1$

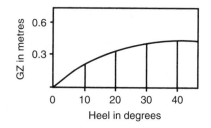

Fig. 23.3

$$h = 10°$$

$$h = \frac{10}{57.3} \text{ radians} = \text{Common interval CI}$$

$$\text{The area under the stability curve} = \frac{1}{3} \times \text{CI} \times \Sigma_1$$

$$= \frac{1}{3} \times \frac{10}{57.3} \times 3.53$$

$$= 0.2053 \text{ metre-radians}$$

$$\text{Dynamical stability} = \text{W} \times \text{Area under the stability curve}$$

$$= 5000 \times 0.2053$$

Ans. Dynamical stability = 1026.5 metre tonnes

Example 2

A box-shaped vessel 45 m × 10 m × 6 m is floating in salt water at a draft of 4 m F and A. GM = 0.6 m. Calculate the dynamical stability to 20 degrees heel.

$$BM = \frac{B^2}{12d} \qquad \text{Displacement} = 45 \times 10 \times 4 \times 1.025 \text{ tonnes}$$

$$= \frac{10 \times 10}{12 \times 4} \qquad \text{Displacement} = 1845 \text{ tonnes}$$

$$BM = 2.08 \text{ m}$$

Note. When calculating the GZ's 10 degrees may be considered a small angle of heel, but 20 degrees is a large angle of heel, and therefore, the wall-sided formula must be used to find the GZ.

GZ	SM	Products for area
0	1	0
0.104	4	0.416
0.252	1	0.252
		$0.668 = \Sigma_1$

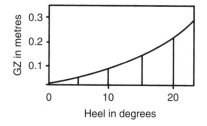

Fig. 23.4

At 10° heel:

$$GZ = GM \times \sin \theta$$
$$= 0.6 \times \sin 10°$$
$$GZ = 0.104 \text{ m}$$

At 20° heel:

$$GZ = (GM + \tfrac{1}{2} BM \tan^2 \theta) \sin \theta$$
$$= (0.6 + \tfrac{1}{2} \times 2.08 \times \tan^2 20°) \sin 20°$$
$$= (0.6 + 0.138) \sin 20°$$
$$= 0.738 \sin 20°$$
$$GZ = 0.252 \text{ m}$$

$$\text{Area under the curve} = \frac{1}{3} \times CI \times \Sigma_1$$
$$= \frac{1}{3} \times \frac{10}{57.3} \times 0.668$$
$$\text{Area under the curve} = 0.0389 \text{ metre-radians}$$
$$\text{Dynamical stability} = W \times \text{Area under the curve}$$
$$= 1845 \times 0.0389$$

Ans. Dynamical stability = 71.77 m tonnes

Exercise 23

1 A ship of 10 000 tonnes displacement has righting levers as follows:

Heel	10°	20°	30°	40°
GZ (m)	0.09	0.21	0.30	0.33

Calculate the dynamical stability to 40 degrees heel.

2 When inclined, a ship of 8000 tonnes displacement has the following righting levers:

Heel	15°	30°	45°	60°
GZ (m)	0.20	0.30	0.32	0.24

   Calculate the dynamical stability to 60 degrees heel.

3 A ship of 10 000 tonnes displacement has the following righting levers
   when inclined:

Heel	0°	10°	20°	30°	40°	50°
GZ (m)	0.0	0.02	0.12	0.21	0.30	0.33

   Calculate the dynamical stability to 50 degrees heel.

4 A box-shaped vessel 42 m × 6 m × 5 m, is floating in salt water on an even
   keel at 3 m draft and has KG = 2 m. Assuming that the KM is constant, calculate the dynamical stability to 15 degrees heel.

5 A box-shaped vessel 65 m × 10 m × 6 m is floating upright on an even keel
   at 4 m draft in salt water. GM = 0.6 m. Calculate the dynamical stability to
   20 degrees heel.

Chapter 24
Effect of beam and freeboard on stability

To investigate the effect of beam and freeboard on stability, it will be necessary to assume the stability curve for a particular vessel in a particular condition of loading. Let curve A in Figure 24.1 represent the curve of stability for a certain box-shaped vessel whose deck edge becomes immersed at about 17 degrees heel.

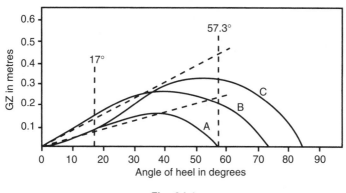

Fig. 24.1

The effect of increasing the beam

Let the draft, freeboard and KG remain unchanged, but increase the beam and consider the effect this will have on the stability curve.

For a ship-shaped vessel BM = I/V, and for a box-shaped vessel BM = $B^2/12d$. Therefore an increase in beam will produce an increase in BM. Hence the GM will also be increased, as will the righting levers at all angles of heel. The range of stability is also increased. The new curve of stability would appear as curve B in Figure 24.1.

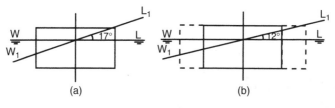

Fig. 24.2

It will be noticed that the curve, at small angles of heel, is much steeper than the original curve, indicating the increase in GM. Also, the maximum GZ and the range of stability have been increased whilst the angle of heel at which the deck edge becomes immersed, has been reduced. The reason for the latter change is shown in Figure 24.2. Angle θ reduces from 17° to 12°.

Figure 24.2(a) represents the vessel in her original condition with the deck edge becoming immersed at about 17 degrees. The increase in the beam, as shown in Figure 24.2(b), will result in the deck edge becoming immersed at a smaller angle of heel. When the deck edge becomes immersed, the breadth of the water-plane will decrease and this will manifest itself in the curve by a reduction in the rate of increase of the GZs with increase in heel.

The effect of increasing the freeboard

Now return to the original vessel. Let the draft, KG, and the beam, remain unchanged, but let the freeboard be increased from f_1 to f_2. The effect of this is shown by curve C in Figure 24.1.

There will be no effect on the stability curve from the origin up to the angle of heel at which the original deck edge was immersed. When the vessel is now inclined beyond this angle of heel, the increase in the freeboard will cause an increase in the water-plane area and, thus, the righting levers will also be increased. This is shown in Figure 24.2(c), where WL represents

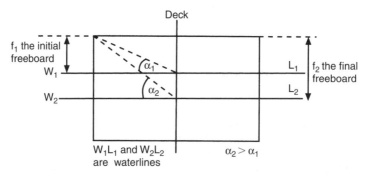

Fig. 24.2(c)

the original breadth of the water-plane when heeled x degrees, and WL_1 represents the breadth of the water-plane area for the same angle of heel but with the increased freeboard. Thus, the vessel can heel further over before her deck edge is immersed, because $\alpha_2 > \alpha_1$.

From the above it may be concluded that an increase in freeboard has no effect on the stability of the vessel up to the angle of heel at which the original deck edge became immersed, but beyond this angle of heel all of the righting levers will be increased in length. The maximum GZ and the angle at which it occurs will be increased as also will be the range of stability.

Summary

With increased beam

   GM_T and GZ increase.

   Range of stability increases.

   Deck edge immerses earlier.

   KB remains similar.

With increased freeboard

   GM_T and GZ increase.

   Range of stability increases.

   Deck edge immerses later at greater θ.

   KB decreases.

Chapter 25
Effects of side winds on stability

When wind forces are on the side of a ship, it is possible to determine the heeling arm. Consider Figure 25.1 and assume that:

P   = wind force = $2 \times 10^{-5} \times A_m \times (V_K)^2$ tonnes
V_K = the velocity in kts of the wind on side of the ship
$\check{y}$ = lever from ship's VCB to the centre of area exposed to the wind
Θ = angle of heel produced by the beam wind
W   = ship's displacement in tonnes
A_m = area at side of ship on which the wind acts in m^2
GZ   = righting lever necessary to return ship back to being upright.
$W \times GZ$ = ship's righting moment

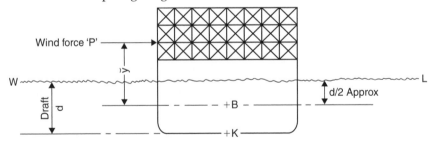

Fig. 25.1   Wind force on stowed containers.

$$\text{The wind heeling arm } (y_2) = \frac{(P \times \check{y} \times \cos^2\Theta)}{W} \text{ metres}$$

This wind heeling arm can be estimated for all angles of heel between 0° and 90°. These values can then be superimposed onto a statical stability curve (see Figure 25.2).

Figure 25.2 shows that:

Θ_1 = Where the two curves intersect to give the angle of steady heel. It is the angle the ship would list to if the wind was steady and there were no waves.

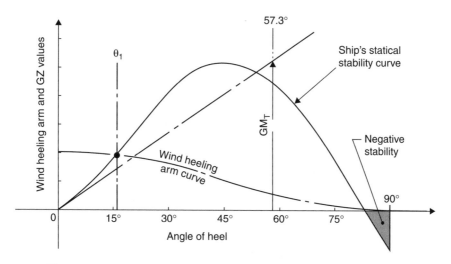

Fig. 25.2   GZ Values and wind heeling arm against angle of heel.

Θ_2 = Where the two curves intersect to give the angle beyond which the ship would capsize. Note how the ship capsizes at a slightly smaller angle of heel than if there had been no wind present.

Now consider Figure 25.3. The application of the wind on the side of the ship causes an equal and opposite reaction. This causes an upsetting moment on the ship:

The upsetting moment caused
  by the wind $= P \times \breve{y}$ t m ------------------------------(I)
The righting moment at small
  angles of heel $= W \times GZ$
                                       $= W \times GM_T \times \sin \Theta_1$ ------------------(II)

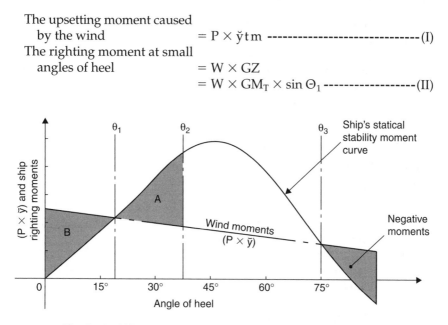

Fig. 25.3   Wind moments and righting moments of a ship.

Equation (I) = Equation (II)

So

$$P \times \check{y} = W \times GM_T \times \sin \Theta_1$$

Hence

$$\sin \Theta_1 = \frac{(P \times \check{y})}{(W \times GM_T)}$$

where Θ_1 = angle of steady heel

Going back, it has been shown that $P \times \check{y} = W \times GZ$

So

$$GZ = \frac{(P \times \check{y})}{W}$$

where GZ = the righting lever

Figure 25.3 shows these moments (I) and (II) superimposed on one view. From Figure 25.3 it can be seen that:

Θ_1 = Angle of steady heel. This occurs with a steady beam wind.
Θ_2 = Angle of lurch, where (area A) = (area B).
        This occurs with a gusting beam wind.
Θ_3 = Angle of heel beyond which the vessel will capsize.

The problems of beam winds are exacerbated on ships with a very large side area, known as 'sail area'. These ships could be:

1. Container vessels, returning with emptied containers stowed five high on the upper deck.
2. VLCCs and ULCCs, due to their very large LBP and depth parameters.
3. Ships sailing in very light ballast condition, with consequent large freeboard values.
4. LNG and LPG ships. In their loaded departure condition, the freeboard divided by the depth moulded is approximately 50%.

Exercise 25

1 (a) With the aid of a statical stability curve, clearly show how wind heeling arm effects on the side of a ship reduce a ship's stability.
   (b) List with explanations, three types of vessels on which side winds could cause these reductions in ship stability.
2 With the aid of a sketch of a statical stability moment curve and a wind moment curve, show the following:
   (a) Angle of steady heel.
   (b) Angle of lurch.
   (c) Angle of heel beyond which the vessel will capsize.

Chapter 26
Icing allowances plus effects on trim and stability

In Arctic ocean conditions, the formation of ice on the upper structures of vessels can cause several problems (see Figures 26.1 and 26.2). Ice build-up can be formed from snowfall, sleet, blizzards, freezing fog and sea-spray in sub-zero temperatures. In the Arctic, the air temperatures can be as low as −40°C in harbour and −30°C at sea.

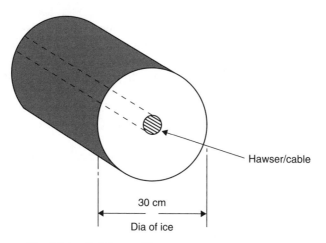

Hawser/cable

30 cm

Dia of ice

Fig. 26.1   Build-up of ice around a hawser/cable.

Icing allowances must be made for:

Rise in G.   Loss of transverse stability.
Increase in weight.   Increased draft due to increased weight.
Loss of freeboard due to increased weight.
Decrease in underkeel clearance.
Contraction of steel due to temperature.

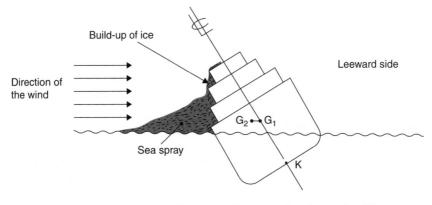

Fig. 26.2   Assymetrical build-up of ice, causing an angle of list.
G_1 moves to G_2.

Increased brittleness in steel structures.
Non-symmetrical formation of ice.
Angle of list.   Angle of loll.
Change of trim.
Impairment of manoeuvrability.
Reduction in forward speed.
Increase in windage area on side of ship.

A 30 cm diameter of ice can form around a hawser or cable. See Figure 26.1. Blocks of ice 100 cm thick have been known to form on the poop deck of a ship in very cold weather zones. Walls of 60 cm of ice forming on the surface of a bridge front have been recorded. In ice-ridden waters the depth of ice may be up to 3 m in depth.

In 1968, three British trawlers rolled over and sank in the North Atlantic. Icing was given as the cause of them capsizing. Only three out of sixty men survived.

Fishing vessels suffer most from the above icing effects. In 1974, the *Gaul* went down in the Barents Sea just off the northern tip of Norway. Thirty-six lives were lost.

These losses indicate not what *could* happen but what *has* happened in icy weather conditions.

Figure 26.3 shows the effect of ice on a typical statical stability curve. The changes that take place are:

Decreased values for the GZ righting levers.
GM_T decreases in value.
GZ_{max} value decreases.
Range of stability decreases.
GZ values and the range of stability will decrease even further if a wind is present on the side of the ship with lesser transverse build-up of ice.

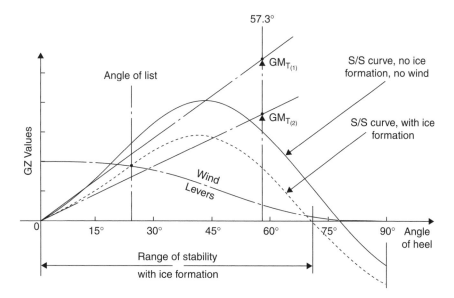

Fig. 26.3   Loss of statical stability due to wind and formation of ice.

IMO suggest the following structures and gears of these vessels must be kept free of ice:

Aerials
Running and navigational lights
Freeing ports and scuppers
Lifesaving craft
Stays, shrouds, masts and rigging
Doors of superstructures and deckhouses
Windlass and hawse holes.

Formation of ice on the upper-works of the vessel must be removed as quickly as possible by:

Cold water pressure.
Hot water and steam.
Break up of ice with ice-crows, axes, ice-picks, metal ice-scrapers, wooden sledgehammers and metal shovels.
Heating of upper structures similar in effect to radiators or central heating arrangements in a house.

In making allowances ships operating in Arctic conditions, Lloyds' Strength Regulations 'request' that transverse framing and beams forward of the collision bulkhead be decreased to about 450 mm maximum, from the more usual 610 mm spacing. In this region, the thickness of the shell plating has to be increased.

After a series of capsizes and vessel losses to fishing boats, the next generation of vessels had fewer structures and gears exposed to the sub-zero temperatures. If it is under cover, it is less likely to freeze.

A final suggestion for these fishing vessels to remember, is not to turn unless sufficient and safe ice allowances have been made.

Exercise 26

1 (a) List the icing allowances a master or mate must make when a vessel is in waters of sub-zero temperatures.
   (b) List four methods for the removal of ice on the upper-works of a vessel operating in sub-zero temperatures.
2 On a sketch of a statical stability curve, show the effects of wind and formation of ice on the following:
   (a) GZ values,
   (b) GM_T value,
   (c) range of stability,
   (d) angle of list.

Chapter 27
Type A, Type B and Type (B-60) vessels plus FL and PL curves (as governed by DfT regulations)

Definitions

Bulkhead deck. This is the uppermost deck to which the transverse watertight bulkheads are carried.

Margin line. This is a line drawn parallel to and 76 mm below the bulkhead deck at side (see Figure 27.1).

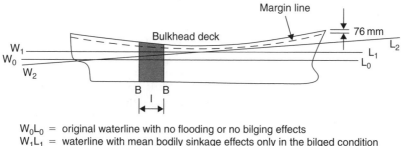

W_0L_0 = original waterline with no flooding or no bilging effects
W_1L_1 = waterline with mean bodily sinkage effects only in the bilged condition
W_2L_2 = waterline with mean bodily sinkage trimming effects in bilged condition, tangential to the margin line
l = length of flooded compartment that has produced the waterline W_2L_2
B = main watertight transverse bulkheads, giving the permissible length of the compartment

Fig. 27.1

Permeability. This is the amount of water that can enter a compartment after the compartment has been bilged. It is denoted as 'μ' and given as a percentage. If the compartment was initially empty, then 'μ' would be 100%.

Floodable length. This is the maximum allowable length of a compartment at any point along the length (with that point as centre), that can be flooded without submerging the margin line. Vessel to be always upright, with no heel.

Floodable length (FL) curve. This is the curve, which at every point in the vessel's length, has an ordinate representing the length of the ship that may be flooded without the margin line being submerged. Vessel to be upright.

Permissible length (PL) curve. This is a lower curve, obtained after the floodable length curve ordinates have been modified for contents within the compartments being considered.

Factor of subdivision (F_S). This is the factor of subdivision. It can range in value from 0.50 to a maximum of 1.00. The 1.00 value signifies that very few passengers are being carried on board. The 0.50 value signifies that a very large number of passengers are being carried on the ship.

By using the following formula, F_S is used to determine the permissible length ordinates.

$$\text{FL ordinates} \times F_S = \text{PL ordinates}$$

Subdivision load line. This is the waterline corresponding to the normal designed waterline. It is drawn parallel to the ship's keel.

Subdivision length (L). This is the length measured between the perpendiculars erected at the ends of the subdivision load line.

Subdivision beam (B). This is the greatest breadth, at or below, the ship's deepest subdivision load line.

Subdivision draft (d). This is the moulded draft to the subdivision load waterline.

Curve of permissible lengths. In any ship, the closer the main transverse bulkheads are, the safer will be the ship. However, too many transverse bulkheads would lead to the vessel being commercially non-viable.

The Regulations Committee suggested that the PL ordinates should be some proportion of the FL ordinates. To achieve this, it was suggested that a factor of subdivision (F_S) be used, where:

$$F_S = A - \frac{\{(A - B)(C_S - 23)\}}{100}$$

$A = 58.2/(L - 60) + 0.18$ mainly for cargo ships
$B = 30.3/(L - 42) + 0.18$ mainly for passenger ships

Cargo–Passenger vessels are vessels that never carry more than 12 passengers.

Passenger–Cargo vessels are vessels that carry more than 12 passengers.

Changing cargo spaces into accommodation spaces will alter the factor of subdivision. It will decrease its value. This will make the permissible lengths smaller and make for a safer ship.

If some compartments are designed to carry cargo on some voyages and passengers on others, then the ship will be assigned more than one subdivisional load line.

Criterion of service numeral (C_s). If the ship's subdivision length is greater than 131 m, then C_S will have as per regulations, a range of values of 23 to 123.

The lower limit of 23 applies for Type 'A' ships (carrying liquid in bulk).
The upper limit of 123 applies for Type 'B' ships.
The regulations state C_S is to be:

$$C_S = 72(M + 2P)/V$$

where:

M = total volume of machinery spaces below the margin line

P = total volume of passenger space and crew space below the margin line. This will obviously take into account the number of passengers and crew onboard ship

V = total volume of ship from keel to the margin line

Worked example 1

Calculate the criterion of service numeral (C_S), when M is 3700 m³, P is 2800 m³ and V is 12 000 m³.

$$C_S = 72(M + 2P)/V = 72(3700 + 5600)/12\,000$$

$$\text{so } C_S = 55.8$$

Worked example 2

Calculate the factor of subdivision, when a ship has a subdivision length (L) of 140 m and a criterion of service numeral of 54.5.

$$A = 58.2/(L - 60) + 0.18 = 58.2/(140 - 60) + 0.18 = 0.908$$
$$B = 30.3/(L - 42) + 0.18 = 30.3/(140 - 42) + 0.18 = 0.489$$

$$F_S = A - \frac{\{(A - B)(C_S - 23)\}}{100}$$

$$F_S = 0.908 - \frac{\{(0.908 - 0.489)(54.5 - 23)\}}{100}$$

$$\text{so } F_S = 0.776$$

Figure 27.2 shows a set of subdivision curves. Observations can be made from these curves to give a greater understanding of why they exist.

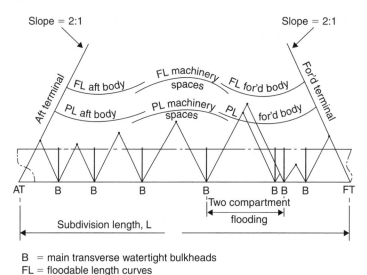

B = main transverse watertight bulkheads
FL = floodable length curves
PL = permissible length curves

Fig. 27.2   Subdivision curves for a passenger liner: diagrammatic sketch.

The triangles all have a height that is equal to the base. Thus, the slope is 2:1. The base in fact is the permissible length of the newly designed compartment.

If need-be, the apex of any of these triangles could go up as far as the PL curve. This would make the compartment have the maximum length within the regulations.

In most cases, the top apex of these shown triangles are not connected to the PL curves. However, in 'two compartment flooding', the regulations do allow the PL curve to be crossed. This is when the adjacent bulkhead sloping line does not extend beyond the FL curve. See illustration of this in Figure 27.2. It is also possible to arrange for 'three compartment flooding'. The resulting smaller length compartments may be used as baggage spaces or storerooms.

Note how the curves of aft body and for'd body do not join those of the machinery space. This is because differences in permeability 'μ' in these localities of the ship. For examples, passenger spaces have a permeability of about 95%, grain spaces have a permeability of 60% to 65% whilst the machinery spaces will have a permeability of 80% to 85%.

Floodable lengths. The basic features that affect the floodable curves for a ship are the block coefficient, sheer ratio, freeboard ratio and permeability.

$$\text{Block coefficient} = \frac{\text{Moulded displacement (excluding bossings)}}{L \times B \times d}$$

$$\text{Sheer ratio} = (\text{Sheer aft or forward})/d$$

$$\text{Freeboard ratio} = \frac{\text{Freeboard to the margin line @ amidships}}{d}$$

$$\text{Permeability} = (\text{Ingress of water}/\text{Volume of compartment}) \times 100/1$$

A DfT 'standard ship' is used as a basis ship. This ship is assigned two permeability values. One is 60% and the other is 100%. Interpolation methods are used to obtain a first estimation of the FL values for the new design being considered. These values are adjusted for sectional area ratios and permeability factors (PF), where:

$$PF = 1.5(100 - \mu)/\mu$$

Summary of procedure steps

1. Determine the subdivision length of the new ship.
2. Calculate the block coefficient, sheer ratios, freeboard ratios and permeability values for the new design.
3. Evaluate the values of 'A' and 'B' coefficients.
4. Determine for the new design, volumes for 'P', 'V' and 'M'.
5. Calculate the criterion of service C_S numeral.
6. Evaluate the factor of subdivision (F_S).
7. Multiply the FL ordinates by F_S to obtain PL ordinates for the aft body, the machinery spaces and the forward body.
8. Plot FL and PL curves, and superimpose the main transverse bulkhead positions together with isosceles triangles having 2:1 slopes as per Figure 27.2.
9. If desired, opt for a 'two compartment flooding' system, as previously described.
10. Adjust and decide upon the *final* positions of main transverse bulkheads for the new design.

Exercise 27

1   Sketch a set of subdivision curves for a passenger ship. Include one example of two compartment flooding. Label the important parts on your diagram.
2   For a passenger ship, the subdivision length is 145 m and the criterion of service numeral is 56.5. Calculate her factor of subdivision (F_S).
3   Define the following floodable and permissible length terms:
    (a) subdivision length (L);
    (b) margin line;
    (c) bulkhead deck;
    (d) factor of subdivision (F_S);
    (e) criterion of service numeral (C_S).
4   Calculate the criterion of service numeral (C_S), when the total volume of machinery spaces below the margin line (M) is 3625 m³, the total volume of passenger space and crew spaces below the margin line (P) is 2735 m³ and the total volume of ship from keel to the margin line (V) is 12 167 m³.

Chapter 28
Load lines and freeboard marks

The link
Freeboard and stability curves are inextricably linked.
With an *increase* in the freeboard:

Righting levers (GZ) are increased.
GM_T increases.
Range of stability increases.
Deck edge immerses later at greater angle of heel.
Dynamical stability increases.
Displacement decreases.
KB decreases.

Overall, both the stability and the safety of the vessel are *improved*.

Historical note
In 1876, Samuel Plimsoll introduced a law into Parliament that meant that ships were assigned certain freeboard markings above which, in particular conditions, they were not allowed to load beyond. Prior to this law a great many ships were lost at sea mainly due to overloading.

In 1930 and in 1966, international conferences modified and expanded these statutory regulations dealing with the safety of ships. These regulations have been further improved over the years by conference meetings every 3 or 4 years up to the present day. One such organisation was the Safety of Life at Sea organisation (SOLAS). In recent years, the IMO has become another important maritime regulatory body.

Definitions
Type 'A' vessel: A ship that is designed to carry only liquid cargoes in bulk, and in which cargo tanks have only small access openings, closed by watertight gasketed covers of steel or equivalent material.

The exposed deck must be one of high integrity.

It must have a high degree of safety against flooding, resulting from the low permeability of loaded cargo spaces and the degree of bulkhead subdivision usually provided.

Type 'B' vessels: All ships that do not fall under the provisions for Type 'A' vessels. For these ships it may be based on:

- The vertical extent of damage is equal to the depth of the ship.
- The penetration of damage is not more than 1/5 of the breadth moulded (B).
- No main transverse bulkhead is damaged.
- Ship's KG is assessed for homogeneous loading of cargo holds, and for 50% of the designed capacity of consumable fluids and stores, etc.

Type (B-60) vessels: The vessel must have an LBP of between 100 and 150 m. It must survive the flooding of any *single* compartment (excluding the machinery space). If greater than 150 m LBP, the machinery space must be considered as a floodable compartment. A typical ship type for a Type (B-60) vessel is a bulk carrier.

Type (B-100) vessels: The vessel must have an LBP of between 100 and 150 m. It must survive the flooding of any two adjacent fore and aft compartments (excluding the machinery space). If greater than 150 m LBP, the machinery space must be considered as a floodable compartment. Such a vessel may be classified as a Type 'A' vessel.

The minimum DfT tabular freeboard values are shown in graphical form on Figure 28.1.

Freeboards of oil tankers and general cargo ships

Oil tankers are permitted to have more Summer freeboard than general cargo ships with a similar LBP. They are considered to be safer ships for the following reasons:

1. They have much *smaller deck openings* in the main deck.
2. They have *greater subdivision,* by the additional longitudinal and transverse bulkheads.
3. Their cargo oil has *greater buoyancy* than grain cargo.
4. They have *more pumps* to quickly control ingress of water after a bilging incident.
5. Cargo oil has a permeability of about 5% whilst grain cargo has a permeability of 60% to 65%. The *lower permeability* will instantly allow less ingress of water following a bilging incident.
6. Oil tankers will have *greater GM values*. This is particularly true for modern double-skin tankers and wide shallow draft tankers.

Tabulated freeboard values

The procedure

When the freeboard for vessel is being assigned, the procedure is to compare a basic Department for Transport (DfT) 'standard ship' with the 'new design' about to enter service. Figure 28.2 shows profiles of these two vessels.

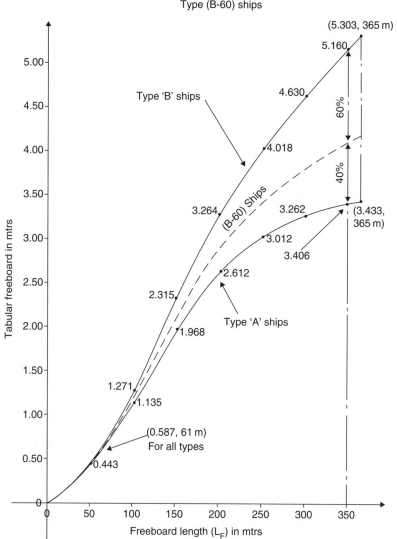

Fig. 28.1   Tabular freeboard for Type 'A', Type 'B' and Type (B-60) Ships.

Both vessels will have the same freeboard length (L_F):

f_1 = the basic or tabular freeboard
f_2 = the final assigned freeboard for the new design

Differences in hull form and structure are considered and compared with these two vessels. If the new design has hull form and structures that would

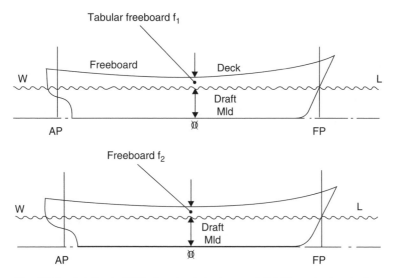

Fig. 28.2   (a) Basic DfT standard design and (b) New design being considered.

increase the danger of operation, then the tabular freeboard is increased by pre-arranged regulations and formulae.

The DfT Tabular freeboard value, based on the freeboard length value, is adjusted or modified for the following six characteristics:

1. Depth D.
2. Block coefficient C_b.
3. Bow height.
4. Length and height of superstructures.
5. Freeboard deck sheer.
6. Structural strength of the new design.

For a standard ship, Depth $= L_F/15$.
For a standard ship, $C_b = 0.680$.

Standard camber is assumed to be parabolic and equal to B.Mld/50.
Freeboard deck sheer is assumed to be parabolic with the sheer forward being twice the deck sheer aft.

$$\text{Standard sheer aft} = (L/3 + 10) \times 25 \text{ mm}$$
$$\text{Standard sheer forward} = (L/3 + 10) \times 50 \text{ mm}$$

Note of caution: for the sheer formulae, L is a ship's LBP in metres (not L_F).

A new vessel can be built to a structural strength of the Lloyds +100A1 standard. If the vessel is indeed built to this classification, then the modification (for strength) to the tabular freeboard is zero.

Whenever the new design has a characteristic that is less safe than the standard DfT standard vessel, the tabular freeboard value will be increased accordingly.

Figure 28.3 shows the bow height measurement to be considered. It is to be measured at the FP, to the uppermost deck exposed to the weather.

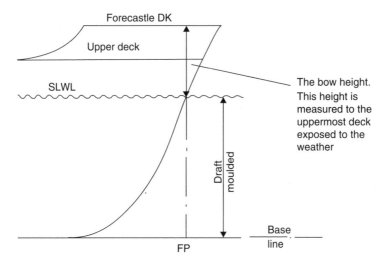

Fig. 28.3   The bow height, as per DfT regulations.

The final assigned statutory freeboard is always measured from the Summer load water line (SLWL) to the *top* of the freeboard deck's stringer plate at amidships. The stringer plate is the outermost line of deck plating. It is the line of deck plating connected to the sheerstrake or gunwhale plate.

The corrections in detail

Depth correction

If Depth D exceeds $L_F/15$, the freeboard is to be increased.

If this is so then the correction $= (D - L_F/15) \times R$
where

$R = L_F/0.48$ if L_F is less than 120 m
$R = 250$ if L_F is 120 m and above

If L_F is 155 m and the Depth D is 11.518 m, then:

$$\text{Depth correction} = (11.518 - 155/15) \times 250 = +296 \text{ mm}$$

If Depth D is less then $L_F/15$, then no reduction is to be made.

C$_b$ correction

If the C$_b$ is greater than the standard 0.680, then the freeboard is to be increased by the following:

$$\text{Correction} = \{(C_b + 0.680)/1.360\} \times \text{Tabular freeboard figure}$$

If the ship's C$_b$ is 0.830 and the Tabular freeboard figure is 2.048 m, then:

C$_b$ correction is $\{(0.830 + 0.680)/1.360\} \times 2.048 = 2.274$ m
Hence addition for actual C$_b$ value = 2.274 − 2.028
$$= +0.226 \text{ m or } +226 \text{ mm}$$

Bow height correction

If the bow height on the actual vessel is *less* than the standard bow height, then the freeboard must be *increased*.

If the bow height on the actual vessel is *greater* than the standard bow height, then there is *no correction* to be made to the freeboard.

The minimum bow height (mBH) for ships is as follows:

If L$_F$ is <250 m, then mBH = 56L$\{1 - L/500\} \times 1.36/(C_b + 0.680)$ mm.
If L$_F$ = 250 m or is >250 m, then mBH = $7000 \times 1.36/(C_b + 0.680)$ mm.

Worked example 1

An oil tanker is 155 m freeboard length with an actual bow height of 6.894 m and a C$_b$ of 0.830. Does the bow height meet with the minimum statutory requirements?

Ship is <250 m so mBH = 56 L$\{1 - L/500\} \times 1.36/(C_b + 0.680)$
mBH = 56 ×155 × $\{1 - 155/500\} \times 1.36/(0.830 + 0.680)$
Hence minimum bow height = 5394 mm or 5.394 m.

The actual bow height is 6.894 m so it is 1.50 m above the minimum statutory limit. No correction to the tabular freeboard is therefore necessary.

Superstructure correction

Where the effective lengths of the superstructure and trunks is **100% × L$_F$,** the freeboard can be reduced by:

350 mm	when L$_F$ is 24 m
860 mm	when L$_F$ is 85 m
1070 mm	when L$_F$ is 122 m and above

These values are shown graphically in Figure 28.4.

However, if less than 100% of the vessel's length is superstructure length, then the following multiple factors should be determined.

Let the actual length of superstructure be denoted as the effective length (E).
Let ratios for E/L$_F$ range from 0 to 1.00.

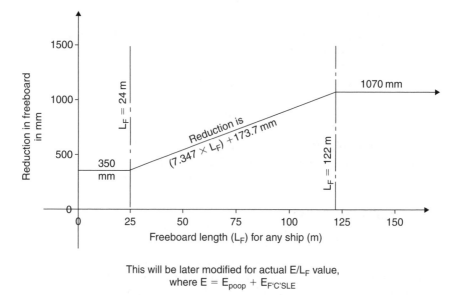

This will be later modified for actual E/L$_F$ value,
where E = E$_{poop}$ + E$_{F'C'SLE}$

Fig. 28.4   Reduction in freeboard for superstructure, when E/L$_F$ is 100%,
for any ship.

First of all, consider *Type 'A' vessels* only (for example, oil tankers).

Table 28.1   Freeboard reduction against effective length/freeboard length
ratios for Type 'A' ships.

E/L$_F$ ratio	0	0.1	0.2	0.3	0.4	0.5	0.6	0.7	0.8	0.9	1.0
Freeboard reduction	0	7%	14%	21%	31%	41%	52%	63%	75.3%	87.7%	100%

Percentages at intermediate ratios of superstructures shall be obtained by linear interpolation.

For E/L$_F$ of 0 to 0.3, the multiple factor is $70 \times$ E/L$_F$ per cent.
For E/L$_F$ of 0.3 to 1.0   factor = $27.78($E/L$_F)^2 + 76.11($E/L$_F) - 3.89$ per cent.

These values are shown graphically in Figure 28.5.
Secondly, consider *Type 'B' vessels* with a forecastle and without a
detached bridge (for example, general cargo ships).

If E/L$_F$ is 0 to 0.3, then the multiple factor = $50($E/L$_F)$ per cent.
If E/L$_F$ is 0.3 to 0.7, then the factor = $175($E/L$_F)^2 - 55($E/L$_F) + 15.75$ per cent.
If E/L$_F$ is 0.7 to 1.0 then the multiple factor = 123.3 E/L$_F - 23.3$ per cent.

These values are shown graphically in Figure 28.6.

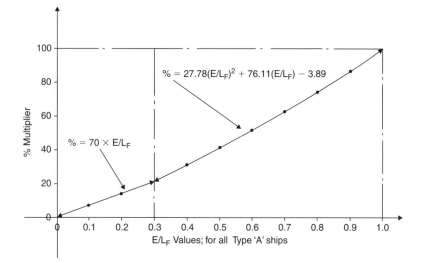

Fig. 28.5   % multipliers for superstructure length 'E' for all Type 'A' ships.

Table 28.2   Freeboard reduction against effective length/freeboard length ratios for Type 'B' vessels.

E/L_F ratio	0	0.1	0.2	0.3	0.4	0.5	0.6	0.7	0.8	0.9	1.0
Freeboard reduction	0	5%	10%	15%	23.5%	32%	46%	63%	75.3%	87.7%	100%

Percentages at intermediate ratios of superstructures shall be obtained by linear interpolation.

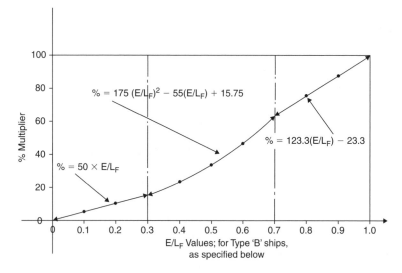

Fig. 28.6   % multipliers for superstructure length 'E' for Type 'B' vessels with forecastle but no bridge.

Worked example 2

An oil tanker is 155 m freeboard length. Superstructure length (E) is 0.51 of the freeboard length (L_F). Estimate the superstructure negative correction to the tabular freeboard.

Oil tanker is greater than 122 m, so the 1st estimate is a reduction of 1070 mm.

Because E/L_F is 0.51 factor $= 27.78(E/L_F)^2 + 76.11(E/L_F) - 3.89$ per cent.
Therefore factor $= 27.78(0.51)^2 + 76.11(0.51) - 3.89$ per cent $= 42.1$ per cent.
So the superstructure correction $= -1070 \times 42.1\% = -450$ mm.

Sheer correction

A vessel with greater than standard sheer will have a reduction in the freeboard.

A vessel with less than standard sheer will have an addition in the freeboard:

$$\text{Sheer correction} = \text{Mean sheer difference} \times (0.75 - S/2) \times L_F$$

The mean sheer difference is the actual mean sheer for the ship relative the mean sheer for the standard ship. For both cases:

$$\text{Mean sheer} = (\text{Aft sheer @ AP} + \text{Forward sheer @ FP})/6 \text{ mm}$$

The denominator of 6 is the sum of Simpson's 1st Rule multipliers (1,4,1) to give a mean sheer value along each vessel's length.

Deck sheer can also be measured at AP, $1/6\,L$, $2/6\,L$, $3/6\,L$, $4/6\,L$, $5/6\,L$ and at FP. L is LBP. These sheers are then put through Simpson's multipliers to obtain an area function.

For both vessels, the mean sheer $= $ area function$/16$

The denominator of 16 is the sum of Simpson's 2nd Rule multipliers (1,3,3,2,3,3,1) to give a mean sheer value along each vessel's length.

S = Total length of enclosed superstructures.

Maximum reduction in freeboard allowed for excess sheer $= 1.25 \times L_F$ mm

Worked example 3

An oil tanker has a freeboard length of 155 m with a total length of enclosed superstructures of 71.278 m. The difference between actual mean sheer and standard ship mean sheer is *minus* 650 mm. Estimate the additional correction to the tabular freeboard figure.

$$\text{The sheer correction} = 650 \times (0.75 - 71.278/2) \times 155$$
$$= +338 \text{ mm}$$

Strength correction

A new vessel can be built to a structural strength of Lloyds +100A1 standard. If the vessel is indeed built to this classification, then the modification (for strength) to the tabular freeboard is zero.

Worked example 4

Consider a tanker of 155 m freeboard length with a Depth Mld of 11.458 m. Calculate the Summer freeboard and the Summer loaded waterline (SLWL) i.e. draft moulded.

From Figure 28.1 at a freeboard length of 155 m, the DfT tabular freeboard from the Type 'A' vessel curve is 2.048 m (recap on previous calculations).

Tabular freeboard	2.048 m
Depth correction	+0.296 m
C_b correction	−0.226 m
Bow height correction – satisfactory	0 m
Superstructure correction	−0.450 m
Freeboard deck sheer correction	+0.338 m
Strength correction – built to Lloyds +100A1 standard	0 m
Moulded Summer freeboard	= 2.458 m
+ Moulded Summer freeboard	= 0.018 m
Final assigned Summer freeboard	= 2.476 m

Summary statement

Tabular freeboard was **2.048 m**.   **Assigned freeboard** was **2.458 m**.

$$\text{Depth Moulded, as given} \quad 11.458 \text{ m} = D$$
$$\text{Summer freeboard (mld)} \quad \underline{2.458 \text{ m}} = f$$
$$\text{Draft Mld or SLWL} \quad \underline{9.000 \text{ m}} = d$$

So Summer load water line = 9.00 m.

Generally, from the stability point of view, the greater the freeboard, the safer the ship will be in day-to-day operations.

The Freeboard Marks

A typical set of freeboard marks is shown in Figure 28.7.

S is the Summer watermark for water of 1.025 t/m^3 density. It is determined by the Department for Transport (DfT) Tabulated Freeboard values, based on the vessel's freeboard length and various corrections. It is placed at the Summer load water line (draft moulded).

T is the Tropical watermark and is 1/48 of the Summer load draft *above* the S mark.

F is the Fresh watermark.

F watermark is W/(4 × TPC$_{SW}$) *or* 1/48 of the Summer load draft *above* the S mark. W and TPC$_{SW}$ are values applicable at the Summer load water line.

TF is the Tropical Fresh watermark and is the (T + F) marks *above* the S mark.

W is the Winter watermark. It is 1/48 of the Summer load draft *below* the S mark.

WNA is the Winter North Atlantic watermark. It is *not* marked on the ship sides for a vessel equal to or more than 100 m freeboard length. If the vessel is less than 100 m floodable length, then the WNA is placed 50 mm *below* the W mark.

The loadlines and freeboard deck line **must** be painted in white or yellow on a dark background, or in black on a light coloured background. The letters on each side of the load Line disc, indicating the assigning authority, should be 115 mm in height and 75 mm in width. See Figure 28.7.

Penalty warning: According to the 1998 load line Regulations, if the appropriate load line on each side of the ship was submerged when it should not have been then:

The owner and master are liable to an additional fine that shall not exceed £1000 *for each complete centimetre* overloaded.

Worked example

(a) Calculate the seasonal allowances and the subsequent drafts for an oil tanker having a SLWL of 9.402 m given that the LBP is 148 m, displacement W is 24 700 t and the TPC$_{SW}$ is 30.2.

(b) Proceed to draw the resulting freeboard marks for this ship. Label your drawing as it would appear on the side of the ship.

T = Tropical fresh water mark = SLWL/48 above SLWL
$$= 9042 \text{ mm}/48 = 188 \text{ mm above SLWL}$$

Hence tropical draft = 9.402 + 0.188 = 9.59 m

F = Fresh water mark = W/(4 × TPC) above SLWL
$$= 24\,700/(4 \times 30.20) = 204 \text{ mm above SLWL}$$

Hence fresh water draft = 9.402 + 0.204 = 9.606 m

TF = Tropical Fresh water mark = (T + F) above SLWL
$$= 0.188 + 0.204 = 0.392 \text{ mm above SLWL}$$

Hence tropical fresh water draft = 9.402 + 0.392 = 9.794 m

W = Winter water mark = SLWL/48 below SLWL
$$= 9042 \text{ mm}/48 = 188 \text{ mm below SLWL}$$

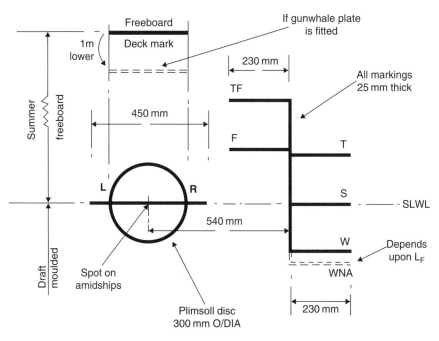

Fig. 28.7   A typical set of freeboard marks.

LR stands for Lloyds Registry based in the UK with offices worldwide.

Alternative classification bodies:

AB = American Bureau of Shipping in USA
NV = Norske Veritas in Norway
BV = Bureau Veritas in France
GL = Germanisches Lloyd of Germany
CA = Common wealth of Australia
NK = Nippon Kaiji Kyokai of Japan

No allowance to be made for the vessel being sagged. Most cargo vessels when fully-loaded will be in a sagging condition.

Hence winter draft = 9.402 − 0.188 = 9.214 m

WNA = Winter North Atlantic water mark. This is *not* marked on the side of this oil tanker because her LBP is greater than 100 m.

Seasonal allowances depend a DfT World zone map (at rear of their freeboard regulations) and on three factors, these three factors being:

1. Time of year.
2. Geographical location of the ship.
3. LBP the ship, relative to a demarcation value of 100 m.

Exercise 28

1 (a) Describe exactly what are Type 'A' vessels, Type 'B' vessels and Type (B-60) vessels.
   (b) Sketch a graph to indicate how the DfT Tabular Freeboard is relative to freeboard length (L_F) for each ship type listed in (a).
2 List the six characteristics that must be considered when modifying the DfT Tabular Freeboard to the ship's final assigned minimum freeboard.
3 Sketch a typical set of freeboard marks for a ship. Show how a ship's seasonal allowances are usually evaluated.
4 Discuss in detail the following corrections for freeboard:
   (a) the freeboard deck sheer;
   (b) the ship's block coefficient;
   (c) the strength of the ship;
   (d) the bow height.
5 An oil tanker is 200 m freeboard length. Superstructure length (E) is 0.60 of the freeboard length (L_F). Estimate the superstructure negative correction to the Tabular freeboard.

Chapter 29
Timber ship freeboard marks

Timber deck cargo is denoted as a cargo of timber carried on an uncovered part of a freeboard or superstructure deck. Timber vessels are allowed less freeboard than other cargo vessels. Certain vessels are assigned timber freeboards but certain additional conditions have to be complied with:

1. The vessel must have a forecastle of at least 0.07 of the vessel's length in extent and of not less than standard height.
2. For a vessel 75 m or less in length the standard height is 1.80 m.
3. For a vessel 125 m in length the standard height is 2.30 m.
4. Intermediate standard heights can be evaluated for the intermediate freeboard lengths ranging from 75 m to 125 m.
5. A poop or raised quarter deck is also required if the vessel's freeboard length is less than 100 m.
6. The double bottom tanks in the midship half-length must have a satisfactory watertight longitudinal subdivision.
7. Timber vessels must have either 1 m high bulwarks with additional strengthening or 1 m high especially strong guardrails.
8. The timber deck cargo shall be compactly stowed, lashed and secured. It shall not interfere in any way with the navigation and necessary work of the ship.

Figure 29.1 shows markings for a vessel operating with two sets of freeboard marks. The markings on the left hand side are the timber markings used when the ship is carrying timber on the upper deck. The right hand side markings are used for conditions of loading without timber stowed on the upper deck.

Additional lumber marks are placed forward of the load line disc and indicated by the 'L' prefix. The decrease in freeboard is allowed because timber ships are considered to provide additional reserve buoyancy. This does not apply for the carriage of wood pulp or other highly absorbent cargoes.

LS is the lumber summer mark. It is determined from the Department for Transport (DfT) tabulated freeboard values, based on the vessel's freeboard length in conjunction with the corrected Type 'B' vessel values.

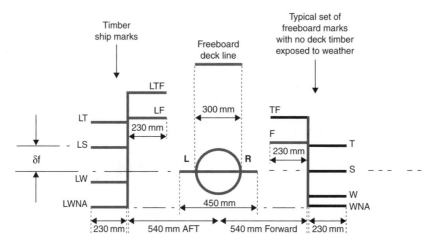

Fig. 29.1   Freeboard marks for a vessel with 2 sets of markings.
Notes. For the distance δf, see graph on Figure 29.2, of 'LS' to 'S' against freeboard length L$_F$. The bow height for the timber ship must be measured *above the 'S' mark* (not the 'LS' mark). The LWNA mark is level with the WNA mark.

LT mark is the lumber tropical mark and is 1/48 of the summer load timber load draft above the LS mark.

LF mark is the lumber fresh watermark. It is calculated in a similar manner to the normal F mark, except that the displacement used in the formula is that of the vessel at her summer timber load draft.

Hence LF mark is $W/(4 \times TPC_{SW})$ *or* 1/48 of the summer load timber load draft above the LS mark.

LTF mark is the lumber tropical fresh watermark and is (LT + LF) marks above the LS mark.

LW mark is the lumber winter watermark and is 1/36 of the summer load timber load draft below the LS mark.

LWNA mark is the lumber winter north atlantic watermark. It is to be horizontally in line with the more usual WNA mark. There is no reduced freeboard allowed on both of these WNA lines.

The load lines and freeboard deck line *must* be painted in white or yellow on a dark background, or in black on a light coloured background. The letters each side of the load line disc indicating the assigning authority should be 115 mm in height and 75 mm in width.

Note that in Figure 29.1 there is a vertical difference between the 'LS' on the left hand side and the 'S' on the right hand side. This is an allowance based on the ratio of length of superstructure used for carrying the timber.

If 100 per cent of the vessel's length is used to carry timber above the freeboard deck, then this reduction in freeboard for carrying timber is as follows:

If a timber ship's freeboard length (L$_F$) is *24 m* or less the reduction is 350 mm. If a timber ship's freeboard length (L$_F$) is *122 m* or more reduction is 1070 mm. For timber ship's freeboard lengths of *24 m* to *122 m*, the reduction in freeboard is:

'LS' mark to 'S' mark is $(7.347 \times L_F) + 173.7$ mm

Figure 29.2 displays this information in graphical form.

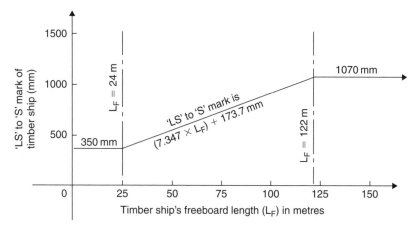

This will be later modified for actual E/L_F value for the timber ship

Fig. 29.2   Distance of 'LS' mark to 'S' mark for a timber ship when 100% of weather deck is used to stow timber.

However, if less than 100 per cent of the vessel's length is used to carry timber above the freeboard deck then Table 29.1 must be consulted. This is denoted as the effective length (E). Consider some ratios for E/L_F.

Table 29.1   Freeboard reduction against effective length/freeboard length ratios for timber ships.

E/L_F ratio	0	0.1	0.2	0.3	0.4	0.5	0.6	0.7	0.8	0.9	1.0
Freeboard reduction	20%	31%	42%	53%	64%	70%	76%	82%	88%	94%	100%

Percentages at intermediate ratios of superstructures shall be obtained by linear interpolation.

If E/L_F of 0 to 0.4   freeboard deduction is $(110 \times E/L_F) + 20$ per cent.
If E/L_F of 0.4 to 1.0   freeboard deduction is $(60 \times E/L_F) + 40$ per cent.

Figure 29.3 portrays this information in graphical form.

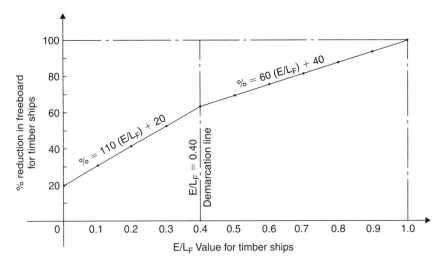

Fig. 29.3   % reduction in freeboard for the superstructures of timber ships.

It is of interest to note that this vertical height 'LS' to 'S' is the approximate mean bodily rise if the timber vessel lost her deck cargo of timber overboard.

Worked example 1

A timber ship has a freeboard length (L_F) of 85 m and has an effective length of superstructure carrying timber of E/L_F of 0.70. Calculate the superstructure allowance, that is, between the 'LS' and the 'S' marks.

$$\text{Reduction} = (\text{allowances for 100\% E}) \times \{(60 \times E/L_F) + 40\}$$
$$= \{(7.347 \times L_F) + 173.7\} \times \{(60 \times 0.70) + 40\}$$
$$= \{(7.347 \times 85) + 173.7\} \times 82\%$$
$$= 654 \text{ mm} \quad \text{say } 0.654 \text{ m.}$$

Carrying timber on the upper deck can create stability problems for those on board ship. They are:

1. The timber may become wet and saturated. This will raise the overall G of the ship thereby possibly decreasing the ship's GM. This leads to a loss in stability. When considering the stability of timber ships the calculations must take into account that because of this saturation:
   (a) Weight of deck timber must be increased by 15 per cent of its dry weight.
   (b) Volume available for reserve buoyancy is only 75 per cent of the total deck timber.
2. The height of the stowed timber can produce a sailing effect, leading to an angle of list situation.
3. Icing effects in very cold weather conditions on the timber will raise the overall G of the ship. Again this will lead to a loss in stability.

4. The IMO suggest that to avoid high strain on the lashing points of the deck timber that the upright GM should not be greater than about 3 per cent of the breadth moulded.

5. As far as statical stability curves are concerned for timber ships, the MCA/UK lumber regulations are as follows:

The upright GM must not be less than 0.05 m (compared to the usual 0.15 m), with the remaining requirements being the same as for any other vessel.

As far as statical stability curves are concerned for timber ships, the IMO lumber regulations are as follows:

The upright GM must not be less than 0.10 m (compared to the usual 0.15 m). Maximum GZ must not be less than 0.25 m (compared to the usual 0.20 m). Area under the statical stability curve, 0° to 40°, must not be less than 0.08 metre radians (compared to the usual 0.09 metre radians).

It also must be remembered that the minimum bow height (as per DfT Regulations) for timber vessels is measured from the ordinary summer load water line. It must NOT be measured from the lumber summer mark.

Exercise 29

1   Certain vessel are assigned timber freeboards with certain conditions. List six of these conditions of assignment.

2   (a) Sketch a typical set of freeboard marks for a timber ship with two sets of markings.
   (b) Explain each of the lumber seasonal allowance markings.

3   A timber ship has a freeboard length (L_F) of 75 m and has an effective length of superstructure carrying timber of E/L_F of 0.60. Calculate the superstructure allowance, that is, between the 'LS' and the 'S' marks.

4   (a) List four problems that could occur when carrying timber on a deck that is exposed to the weather.
   (b) As per DfT requirements, for timber ships, from where is the bow height measured?
   (c) List three IMO stability requirements for timber ships.

Chapter 30
IMO Grain Rules for the safe carriage of grain in bulk

The intact stability characteristics of any ship carrying bulk grain must be shown to meet, throughout the voyage, the following criteria relating to the moments due to grain shift:

1. The angle of heel due to the shift of grain shall not be greater than 12° or in the case of ships constructed on or after 1 January 1994 the angle at which the deck edge is immersed, whichever is the lesser. For most grain carrying ships, the deck edge will not be immersed at or below 12°, so this 12° is normally the lesser of the two angles.
2. In the statical stability diagram, the net or residual area between the heeling arm curve and the righting arm curve up to the angle of heel of maximum difference between the ordinates of the two curves, or 40° or the angle of flooding (θ_1), whichever is the least, shall in all conditions of loading be not less than 0.075 metre-radians.
3. The initial metacentric height, after correction for free surface effects of liquids in tanks, shall not be less than 0.30 m.

Before loading bulk grain the Master shall, if so required by the contracting government of the country of the port of loading, demonstrate the ability of the ship at all stages of any voyage to comply with the stability criteria required by this section.

After loading, the Master shall ensure that the ship is upright before proceeding to sea.

Consider Figure 30.1. The IMO Grain Code stipulates that λ_0 and λ_{40} values to graphically determine the angle of list in the event of a shift of grain where

$$\lambda_0 = \frac{\text{Assumed volumetric heeling moment due to transverse shift}}{\text{Stowage factor} \times \text{Displacement}}$$

$$\lambda_{40} = 0.8 \times \lambda_0$$

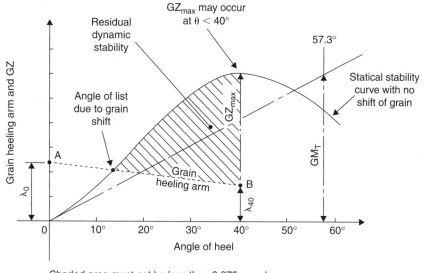

Shaded area must not be less than 0.075 m rad
Angle of list shall not be greater than 12°
$\lambda_{40} = \lambda_0 \times 0.80$ A–B is a straight line
GM_T (fluid) is not to be less than 0.30 m

Fig. 30.1   IMO grain heeling stability requirements.

This gives a grain heeling line.

Stowage Factor = volume per unit weight of grain cargo. It can be represented by three values, 1.25, 1.50 and 1.75 m³/tonne.

Displacement = weight of ship, fuel fresh water, stores, cargo, etc.

The righting arm shall be derived from cross-curves which are sufficient in number to accurately define the curve for the purpose of these requirements and shall include cross-curves at 12° and 40°.

The angle of list occurs at the intersection of the GZ curve and the grain heeling line. The angle of list shown in Figure 30.1 may also be approximated in the following manner:

The approximate angle of heel = ahm/(mpm × SF) × 12°
ahm = actual transverse volumetric heeling moments in m⁴
SF = appropriate stowage factor in m³/tonne, say 1.50 to 1.70

mpm = maximum permissible grain heeling moments in tm. These are supplied to the ship by the Naval Architect in the form of a table of Fluid KG against displacement W. Refer to the Datasheet in the worked examples section.

Adjustments:

If the holds are full with grain then the multiple shall be 1.00 × each ahm.

A multiple of 1.06 is applied to each ahm in holds and tween decks filled to the hatchway.

A multiple of 1.12 is applied to each ahm in holds and tween decks partially filled full width grain compartments.

A study of the worked examples will give an understanding of exactly how these adjustments are made.

Studies of grain cargoes after shifting has taken place have lead to authorities to make the following stipulations with regard to the transverse slope 'α' of grain across the ship.

$\alpha = 25°$ to the horizontal assumed for full width partially filled compartments. This is at the ends of the compartment situated forward and aft of the hatchway. α also equals 25° to the horizontal for hatchway trunks (see Figure 30.2).

$\alpha = 15°$ to the horizontal assumed for full compartments where the grain stow is situated abreast of the hatchway (see Figure 30.2).

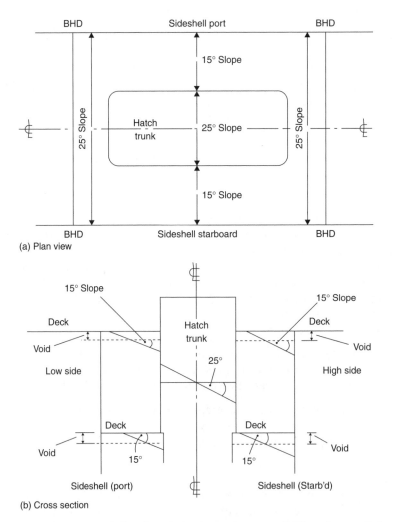

(a) Plan view

(b) Cross section

Fig. 30.2　Assumed shifts of grain in degrees, as per IMO grain regulations for hatch trunks.

Worked example 1

A vessel has loaded grain, stowage factor $1.65\,\mathrm{m}^3/\mathrm{tonne}$ to a displacement of $13\,000$ tonne. In the loaded condition the effective KG is $7.18\,\mathrm{m}$.

All grain spaces are full, except No. 2 tween deck, which is partially full. The tabulated transverse volumetric heeling moments are as follows:

No. 1 hold	$1008\,\mathrm{m}^4$
No. 2 hold	$1211\,\mathrm{m}^4$
No. 3 hold	$1298\,\mathrm{m}^4$
No. 4 hold	$1332\,\mathrm{m}^4$
No. 1 TD	$794\,\mathrm{m}^4$
No. 2 TD	$784\,\mathrm{m}^4$
No. 3 TD	$532\,\mathrm{m}^4$

The value of the Kg used in the calculation of the vessel's effective KG were as follows:

for lower holds, the centroid of the space;

for tween decks, the actual Kg of the cargo.

(a) Using Datasheet Q.1 *Maximum Permissible Grain Heeling Moments Table*, determine the vessel's ability to comply with the statutory grain regulations.
(20) marks
(b) Calculate the vessel's approximate angle of heel in the event of a shift of grain assumed in the grain regulations. (5) marks
(c) State the stability criteria in the current International Grain Code. (10) marks

Using the *Maximum Permissible Grain Heeling Moments Table*, for a KG of $7.18\,\mathrm{m}$ and a $13\,000\,\mathrm{t}$ displacement, by direct interpolation procedure it can be determined that:

The maximum permissible grain heeling moment = $3979\,\mathrm{tm}$, say mpm
The actual volumetric heeling moments (ahm) for this vessel are:

No. 1 hold	$1008\,\mathrm{m}^4 \times 1.00 = 1008$
No. 2 hold	$1211\,\mathrm{m}^4 \times 1.00 = 1211$
No. 3 hold	$1298\,\mathrm{m}^4 \times 1.00 = 1298$
No. 4 hold	$1332\,\mathrm{m}^4 \times 1.00 = 1332$
No. 1 TD	$794\,\mathrm{m}^4 \times 1.06 =\ \ 842$
No. 2 TD	$784\,\mathrm{m}^4 \times 1.12 =\ \ 878$
No. 3 TD	$532\,\mathrm{m}^4\ \times 1.06 =\ \ \underline{564}$
	Total ahm = $\overline{7133\,\mathrm{m}^4}$

The factor 1.00 for the four holds is because, stated in the question, in the full holds the centroid was at the centre of the space.

The factor 1.06 is because, in the question, for the two full tween deck cargo spaces the Kg was at the actual Kg.

The factor of 1.12 is because, in the question, No. 2 tween deck is given as being partially full.

TS002-72   STABILITY

DATASHEET Q.1

N.B. This Datasheet must be returned with your examination answer book

TABLE OF MAXIMUM PERMISSIBLE GRAIN HEELING MOMENTS (tm)

FLUID KG (metres)

Displacement tonne	6.50	6.60	6.70	6.80	6.90	7.00	7.10	7.20	7.30	7.40
14 500	6141	5820	5499	5179	4858	4537	4217	3896	3575	3255
14 000	5957	5647	5338	5028	4719	4409	4099	3790	3480	3171
13 500	5924	5625	5327	5028	4730	4431	4132	3834	3535	3237
13 000	5934	5647	5359	5072	4784	4497	4209	3922	3634	3347
12 500	5891	5614	5338	5062	4785	4509	4232	3956	3679	3403
12 000	5857	5591	5326	5061	4795	4530	4265	3999	3734	3468
11 500	5893	5639	5385	5130	4876	4622	4368	4113	3859	3605
11 000	5944	5701	5457	5214	4971	4728	4484	4241	3998	3755
10 500	5948	5716	5484	5251	5019	4787	4555	4323	4090	3858
10 000	5940	5719	5498	5276	5055	4834	4613	4392	4171	3950
9500	5961	5751	5541	5331	5121	4911	4701	4491	4281	4071
9000	6027	5828	5629	5430	5231	5032	4833	4634	4435	4236
8500	6127	5939	5751	5563	5375	5187	4999	4811	4623	4435
8000	6210	6033	5856	5679	5502	5325	5148	4971	4795	4618
7500	6252	6087	5921	5755	5589	5423	5257	5091	4926	4760
7000	6343	6189	6034	5879	5724	5569	5415	5260	5105	4950
6500	6550	6406	6262	6118	5975	5831	5687	5543	5400	5256
6000	6832	6699	6566	6434	6301	6168	6035	5903	5770	5637
5500	7120	6998	6877	6755	6633	6512	6390	6268	6147	6025
5000	7320	7209	7099	6988	6877	6767	6656	6546	6435	6325

Candidate's Name ....

Examination Centre ....

Let stowage factor be denoted as SF.

$$\text{The approximate angle of heel} = \text{ahm}/(\text{mpm} \times \text{SF}) \times 12°$$
$$\text{Approximate angle of heel} = 7133/(1.65 \times 3979) \times 12°$$
$$\text{Hence approximate angle of heel} = \underline{13.04°}$$

This angle of heel is *not* acceptable because it does not comply with the IMO regulations because it is greater than 12°.

Worked example 2

A vessel has loaded grain, stowage factor $1.65\,\text{m}^3/\text{tonne}$ to a displacement of 13 000 tonne. In the loaded condition the effective KG is 7.18 m.

All grain spaces are full, except No. 2 tween deck, which is partially full.
The tabulated transverse volumetric heeling moments are as follows:

No. 1 hold $851\,\text{m}^4$
No. 2 hold $1022\,\text{m}^4$
No. 3 hold $1095\,\text{m}^4$
No. 4 hold $1124\,\text{m}^4$
No. 1 TD $669\,\text{m}^4$
No. 2 TD $661\,\text{m}^4$
No. 3 TD $448\,\text{m}^4$

The value of the Kg used in the calculation of the vessel's effective KG were as follows:
for lower holds, the centroid of the space;
for tween decks, the actual Kg of the cargo.

(a) Using Datasheet Q.1 *Maximum Permissible Grain Heeling Moments Table*, determine the vessel's ability to comply with the statutory grain regulations.
(20) marks
(b) Calculate the vessel's approximate angle of heel in the event of a shift of grain assumed in the grain regulations. (5) marks
(c) State the stability criteria in the current International Grain Code.
(10) marks

Using the *Maximum Permissible Grain Heeling Moments Table*, for a KG of 7.18 m and a 13 000 tonnes displacement, by direct interpolation procedure it can be determined that:
The maximum permissible grain heeling moment = 3979 tm, say mpm
The actual volumetric heeling moments (ahm) for this vessel are:

No. 1 hold $851\,\text{m}^4 \times 1.00 = 851$
No. 2 hold $1022\,\text{m}^4 \times 1.00 = 1022$
No. 3 hold $1095\,\text{m}^4 \times 1.00 = 1095$
No. 4 hold $1124\,\text{m}^4 \times 1.00 = 1124$
No. 1 TD $669\,\text{m}^4 \times 1.06 = 709$
No. 2 TD $661\,\text{m}^4 \times 1.12 = 740$
No. 3 TD $448\,\text{m}^4 \times 1.06 = \underline{475}$
Total ahm = $\underline{6016\,\text{m}^4}$

The factor 1.00 for the four holds is because, stated in the question, in the full holds the centroid was at the centre of the space.

The factor 1.06 is because, in the question, for the two full tween deck cargo spaces the Kg was at the actual Kg.

The factor of 1.12 is because, in the question, No. 2 tween deck is given as being partially full.

Let stowage factor be denoted as SF.

$$\text{The approximate angle of heel} = \text{ahm}/(\text{mpm} \times \text{SF}) \times 12°$$
$$\text{Approximate angle of heel} = 6016/(1.65 \times 3979) \times 12°$$
$$\text{Hence approximate angle of heel} = \underline{11°}$$

This angle of heel is quite acceptable because it complies with the IMO regulations in that it is less than 12°.

Worked example 3

A vessel has loaded grain, stowage factor $1.60\,\text{m}^3/\text{tonne}$ to a displacement of 13 674 tonne. In this loaded condition, the fluid GM_T is 0.90 m.

All grain spaces are full, except No. 2 tween deck, which is partially full.

The tabulated transverse volumetric heeling moments are as follows:

No. 1 hold	$774\,\text{m}^4$
No. 2 hold	$929\,\text{m}^4$
No. 3 hold	$995\,\text{m}^4$
No. 4 hold	$1022\,\text{m}^4$
No. 1 TD	$608\,\text{m}^4$
No. 2 TD	$601\,\text{m}^4$
No. 3 TD	$407\,\text{m}^4$

The value of the Kg used in the calculation of the vessel's effective KG were as follows:

for lower holds, the centroid of the space;
for tween decks, the actual Kg of the cargo.

The righting levers for GZ in metres at angles of heel in degrees are as shown in the table:

Angle of heel	0	5	10	15	20	25	30	35	40	45	50
GZ ordinate	0	0.09	0.22	0.35	0.45	0.51	0.55	0.58	0.59	0.58	0.55

(a) Use λ_0 and λ_{40} and the tabulated GZ values to graphically determine the angle of list in the event of a shift of grain.
(b) Calculate the enclosed area between the GZ curve and the grain heeling arm line.

The actual volumetric heeling moments (ahm) for this vessel are:

No. 1 hold	$774\,m^4 \times 1.00 =$	774	
No. 2 hold	$929\,m^4 \times 1.00 =$	929	
No. 3 hold	$995\,m^4 \times 1.00 =$	995	
No. 4 hold	$1022\,m^4 \times 1.00 =$	1022	
No. 1 TD	$608\,m^4 \times 1.06 =$	644	
No. 2 TD	$601\,m^4 \times 1.12 =$	673	
No. 3 TD	$407\,m^4 \times 1.06 =$	431	

$$\text{Total ahm} = 5468\,m^4$$

The factor 1.00 for the four holds is because, in the question, in the full holds the centroid was at the centre of the space.

The factor 1.06 is because, in the question, for the two full tween deck cargo spaces the Kg was at the actual Kg.

The factor of 1.12 is because, in the question, No. 2 tween deck is given as being partially full.

Let stowage factor be denoted as SF.

Let W = vessel's displacement in tonnes.

$$\lambda_0 = \text{ahm}/(W \times SF)$$
$$\lambda_0 = 5468/(13\,674 \times 1.60) = 0.25 \text{ metres}$$

This value of 0.25 m is plotted on Figure 30.3 at angle of heel of 0°.

Now according to IMO Grain Regulations:

$$\lambda_{40} = \lambda_0 \times 0.80$$

So

$$\lambda_{40} = 0.25 \times 0.80 = 0.20 \text{ m.}$$

This value of 0.20 m is then plotted on Figure 30.3 at angle of heel of 40°.

The next step is to plot a curve of GZ against angle of heel (to give part of a statical stability curve) and then to superimpose a line connecting the λ_0 and λ_{40} values. This sloping line is a line of grain heeling arm values. Do not forget also to plot the GM_T value at 57.3°. This assists in establishing the guidance triangle that runs into the 0-0 axis.

At the intersection of these two lines the value for the angle of list may be lifted off. Figure 30.3 shows this angle of list to be *11*°.

Area enclosed between 11° and 40° angles of heel:

Ordinate	Simpson's multiplier	Area function
0	1	0
0.31	4	1.24
0.39	1	0.39
		$1.63 = \Sigma_1$

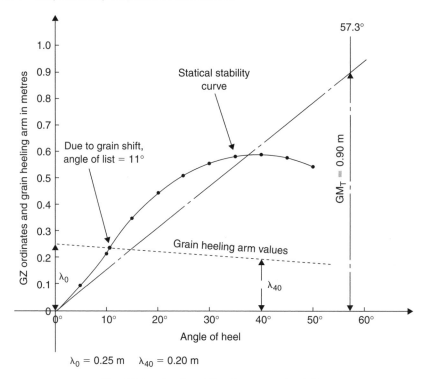

$\lambda_0 = 0.25\,\text{m}$ $\lambda_{40} = 0.20\,\text{m}$

Fig. 30.3   Angle of list due to grain shift.

Area enclosed $= 1/3 \times \text{CI} \times \Sigma_1$
$= 1/3 \times 14.5°/57.3° \times 1.63 = 0.137\,\text{m radians}$

where CI = common interval of $(40° - 11°)/2$
This well above the IMO stipulated minimum value of 0.075 m radians.

Exercise 30

1   A vessel has loaded grain, stowage factor $1.55\,\text{m}^3/\text{tonne}$ to a displacement of 13 500 tonne. In the loaded condition the effective KG is 7.12 m.
All grain spaces are full, except No. 3 tween deck, which is partially full.
The tabulated transverse volumetric heeling moments are as follows:

No. 1 hold	$810\,\text{m}^4$
No. 2 hold	$1042\,\text{m}^4$
No. 3 hold	$1075\,\text{m}^4$
No. 4 hold	$1185\,\text{m}^4$
No. 1 TD	$723\,\text{m}^4$
No. 2 TD	$675\,\text{m}^4$
No. 3 TD	$403\,\text{m}^4$

The value of the Kg used in the calculation of the vessel's effective KG were as follows:

for lower holds, the centroid of the space;

for tween decks, the actual Kg of the cargo.

(a) Using Datasheet Q.1, determine the vessel's ability to comply with the statutory grain regulations.

(b) Calculate the vessel's approximate angle of heel in the event of a shift of grain assumed in the grain regulations.

2 State the stability criteria in the current IMO International Grain Code.

3 A vessel has loaded grain, stowage factor $1.69\,m^3$/tonne to a displacement of 13 540 tonne.

All grain spaces are full, except No. 2 tween deck, which is partially full.

The tabulated transverse volumetric heeling moments are as follows:

No. 1 hold	$762\,m^4$
No. 2 hold	$941\,m^4$
No. 3 hold	$965\,m^4$
No. 4 hold	$1041\,m^4$
No. 1 TD	$618\,m^4$
No. 2 TD	$615\,m^4$
No. 3 TD	$414\,m^4$

The value of the Kg used in the calculation of the vessel's effective KG were as follows:

for lower holds, the centroid of the space;

for tween decks, the actual Kg of the cargo.

Calculate the grain heeling arms λ_0, λ_{40} and λ_{20} in metres.

4 (a) In grain carrying ships, what is an angle of repose?

(b) For a grain carrying ship, discuss the assumed angles of repose suggested by IMO, in way of a hatch trunk. Show these assumed angles of repose in an elevation view and in a plan view.

Chapter 31
Angle of loll

When a ship with negative initial metacentric height is inclined to a small angle, the righting lever is negative, resulting in a capsizing moment. This effect is shown in Figure 31.1(a) and it can be seen that the ship will tend to heel still further.

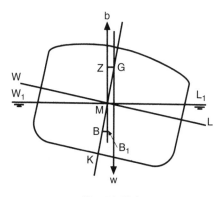

Fig. 31.1(a)

At a large angle of heel the centre of buoyancy will have moved further out the low side and the force of buoyancy can no longer be considered to act vertically upwards though M, the initial metacentre. If, by heeling still further, the centre of buoyancy can move out far enough to lie vertically under G the centre of gravity, as in Figure 31.1(b), the righting lever and thus the righting moment, will be zero.

The angle of heel at which this occurs is referred to as the *angle of loll* and may be defined as the angle to which a ship with negative initial metacentric height will lie at rest in still water.

If the ship should now be inclined to an angle greater than the angle of loll, as shown in Figure 31.1(c), the righting lever will be positive, giving a moment to return the ship to the angle of loll.

From this it can be seen that the ship will oscillate about the angle of loll instead of the upright.

The curve of statical stability for a ship in this condition of loading is illustrated in Figure 31.2. Note from the figure that the GZ at the angle of

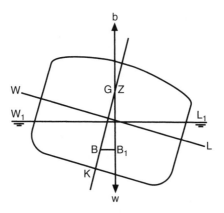

Fig. 31.1(b)

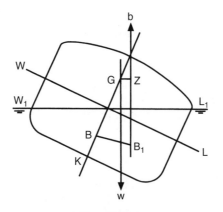

Fig. 31.1(c)

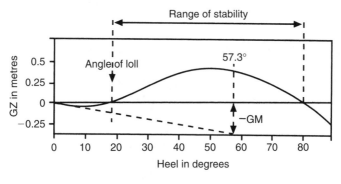

Fig. 31.2

loll is zero. At angles of heel less than the angle of loll the righting levers are negative, whilst beyond the angle of loll the righting levers are positive up to the angle of vanishing stability.

Note how the range of stability in this case is measured from the angle of loll and not from the 'o–o' axis.

To calculate the angle of loll

When the vessel is 'wall-sided' between the upright and inclined water-lines, the GZ may be found using the formula:

$$GZ = \sin\theta(GM + \tfrac{1}{2}BM\tan^2\theta)$$

At the angle of loll:

$$GZ = 0$$
$$\therefore \text{ either } \sin\theta = 0$$

or

$$(GM + \tfrac{1}{2}BM\tan^2\theta) = 0$$

If

$$\sin\theta = 0$$

then

$$\theta = 0$$

But then angle of loll cannot be zero, therefore:

$$(GM + \tfrac{1}{2}BM\tan^2\theta) = 0$$
$$\tfrac{1}{2}BM\tan^2\theta = -GM$$
$$BM\tan^2\theta = -2GM$$
$$\tan^2\theta = \frac{-2GM}{BM}$$
$$\tan\theta = \sqrt{\frac{-2GM}{BM}}$$

The angle of loll is caused by a negative GM, therefore:

$$\tan\theta = \sqrt{\frac{-2(-GM)}{BM}}$$

or

$$\tan\theta = \sqrt{\frac{2GM}{BM}}$$

where

$$\theta = \text{the angle of loll}$$
$$GM = \text{a negative initial metacentric height}$$
$$BM = \text{the BM when upright}$$

Example

Will a homogeneous log 6 m × 3 m × 3 m and relative density 0.4 float in fresh water with a side perpendicular to the waterline? If not, what will be the angle of loll?

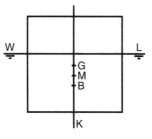

Fig. 31.3

Since the log is homogeneous the centre of gravity must be at half-depth, i.e. KG = 1.5 m.

$$\frac{\text{Draft of log}}{\text{Depth of log}} = \frac{\text{SG of log}}{\text{SG of water}}$$

$$\text{Draft of log} = \frac{3 \times 0.4}{1}$$

$$d = 1.2 \,\text{m}$$

$$\text{KB} = \frac{1}{2} \,\text{draft}$$

$$\text{KB} = 0.6 \,\text{m}$$

$$\text{BM} = \text{B}^2/12\text{d}$$

$$= \frac{3 \times 3}{12 \times 1.2}$$

$$\text{BM} = 0.625 \,\text{m}$$

$$+$$

$$\underline{\text{KB} = 0.600 \,\text{m}}$$

$$\text{KM} = 1.225 \,\text{m}$$

$$\underline{\text{KG} = 1.500 \,\text{m}}$$

$$\text{GM} = -0.275 \,\text{m}$$

Therefore the log is unstable and will take up an angle of loll.

$$\tan\theta = \sqrt{\frac{2\text{GM}}{\text{BM}}}$$

$$= \sqrt{\frac{0.55}{0.625}} = 0.9381$$

$$\theta = 43° \, 10'$$

Ans. The angle of loll = 43° 10'

Question: What exactly is angle of list and angle of loll? List the differences/characteristics.

Angle of list

'G', the centroid of the loaded weight, has *moved off the centre line* due to a shift of cargo or bilging effects, say to the port side.

GM is positive, i.e. 'G' is below 'M'. In fact GM will *increase* at the angle of list compared to GM when the ship is upright. The ship is in *stable equilibrium*.

In still water conditions the ship will remain at this *fixed* angle of heel. She will list to one side only, i.e. the same side as movement of weight.

In heavy weather conditions the ship will roll about this angle of list, say 3° P, but will not stop at 3° S. See comment below.

To bring the ship back to upright, load weight on the other side of the ship, for example if she lists 3° P add weight onto starboard side of ship.

Angle of loll

KG = KM so *GM is zero*. 'G' remains *on the centre line* of the ship.

The ship is in *neutral equilibrium*. She is in a *more dangerous situation* than a ship with an angle of list, because once 'G' goes above 'M' she will capsize.

Angle of loll may be *3° P or 3° S* depending upon external forces such as wind and waves acting on her structure. She may suddenly flop over from 3° P to 3° S and then back again to 3° P.

To improve this condition 'G' must be brought below 'M'. This can be done by moving weight downwards towards the keel, adding water ballast in double-bottom tanks or removing weight above the ship's 'G'. Beware of free surface effects when moving, loading and discharging liquids.

With an angle of list or an angle of loll the calculations must be carefully made *prior* to any changes in loading being made.

Exercise 31

1   Will a homogeneous log of square cross-section and relative density 0.7 be stable when floating in fresh water with two opposite sides parallel to the waterline? If not, what will be the angle of loll?

2   A box-shaped vessel 30 m × 6 m × 4 m floats in salt water on an even keel at 2 m, draft F and A. KG = 3 m. Calculate the angle of loll.

3   A ship is upright and is loaded with a full cargo of timber with timber on deck. During the voyage the ship develops a list, even though stores, fresh water and bunkers have been consumed evenly from each side of the centre line. Discuss the probable cause of the list and the method which should be used to bring the ship to the upright.

4   A ship loaded with a full cargo of timber and timber on deck is alongside a quay and has taken up an angle of loll away from the quay. Describe the correct method of discharging the deck cargo and the precautions which must be taken during this process.

Chapter 32
True mean draft

In previous chapters it has been shown that a ship trims about the centre of flotation. It will now be shown that, for this reason, a ship's true mean draft is measured at the centre of flotation and may not be equal to the average of the drafts forward and aft. It only does when LCF is at average ⋓.

Consider the ship shown in Figure 32.1(a) which is floating on an even keel and whose centre of flotation is FY aft of amidships. The true mean draft is KY, which is also equal to ZF, the draft at the centre of flotation.

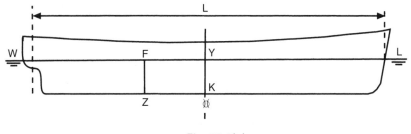

Fig. 32.1(a)

Now let a weight be shifted aft within the ship so that she trims about F as shown in Figure 32.1(b).

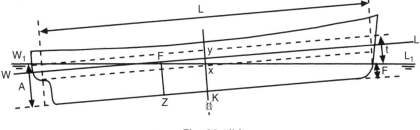

Fig. 32.1(b)

The draft at the centre of flotation (ZF) remains unchanged.

Let the new draft forward be F and the new draft aft be A, so that the trim (A − F) is equal to 't'.

Since no weights have been loaded or discharged, the ship's displacement will not have changed and the true mean draft must still be equal to

KY. It can be seen from Figure 32.1(b) that the average of the drafts forward and aft is equal to KX, the draft amidships. Also,

$$ZF = KY = KX + XY$$

or

$$\text{True mean draft} = \text{Draft amidships} + \text{Correction}$$

Referring to Figure 32.1(b) and using the property of similar triangles:

$$\frac{XY}{FY} = \frac{t}{L} \qquad XY = \frac{t \times FY}{L}$$

or

$$Correction\ FY = \frac{Trim \times FY}{Length}$$

where FY is the distance of the centre of flotation from amidships.

It can also be seen from the figure that, when a ship is trimmed by the stern and the centre of flotation is aft of amidships, the correction is to be added to the mean of the drafts forward and aft. Also by substituting forward for aft and aft for forward in Figure 32.1(b), it can be seen that the correction is again to be added to the mean of the drafts forward and aft when the ship is trimmed by the head and the centre of flotation is forward of amidships.

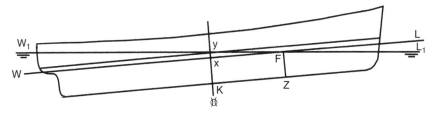

Fig. 32.1(c)

Now consider the ship shown in Figure 32.1(c), which is trimmed by the stern and has the centre of flotation forward of amidships. In this case:

$$ZF = KY = KX - XY$$

or

$$\text{True mean draft} = \text{Draft amidships} - \text{Correction}$$

The actual correction itself can again be found by using the above formula, but in this case the correction is to be subtracted from the mean of the drafts forward and aft. Similarly, by substituting forward for aft and aft for forward in this figure, it can be seen that the correction is again to be subtracted from the average of the drafts forward and aft when the ship is trimmed by the head and the centre of flotation is aft of amidships.

A general rule may now be derived for the application of the correction to the draft amidships in order to find the true mean draft.

Rule

When the centre of flotation is in the *same direction* from amidships as the maximum draft, the correction is to be *added* to the mean of the drafts. When the centre of flotation is in the *opposite direction* from amidships to the maximum draft, the correction is to be *subtracted.*

Example

A ship's minimum permissible freeboard is at a true mean draft of 8.5 m. The ship's length is 120 m, centre of flotation being 3 m aft of amidships. TPC = 50 tonnes. The present drafts are 7.36 m F and 9.00 m A. Find how much more cargo can be loaded.

$$\text{Draft forward} = 7.36 \text{ m}$$
$$\text{Draft aft} = \underline{9.00 \text{ m}}$$
$$\text{Trim} = 1.64 \text{ m by the stern}$$
$$\text{Correction} = \frac{t \times FY}{L} = \frac{1.64 \times 3}{120}$$
$$\text{Correction} = 0.04 \text{ m}$$
$$\text{Draft forward} = 7.36 \text{ m}$$
$$\text{Draft aft} = 9.00 \text{ m}$$
$$\text{Sum} = 16.36 \text{ m}$$
$$\text{Average} = \text{Draft amidships} = 8.18 \text{ m}$$
$$\text{Correction} = +0.04 \text{ m}$$
$$\text{True mean draft} = 8.22 \text{ m}$$
$$\text{Load mean draft} = \underline{8.50 \text{ m}}$$
$$\text{Increase in draft} = 0.28 \text{ m or 28 cm}$$
$$\text{Cargo to load} = \text{Increase in draft required} + \text{TPC}$$
$$= 28 \times 50$$

Ans. Cargo to load = 1400 tonnes

Effect of hog and sag on draft amidships

When a ship is neither hogged nor sagged the draft amidships is equal to the mean of the drafts forward and aft. In Figure 32.1(d) the vessel is shown in hard outline floating without being hogged or sagged. The draft forward is F, the draft aft is A, and the draft amidships (KX) is equal to the average of the drafts forward and aft.

Now let the vessel be sagged as shown in Figure 32.1(d) by the broken outline. The draft amidships is now K_1X, which is equal to the mean of the drafts forward and aft (KX), plus the sag (KK_1). The amount of hog or sag must therefore be taken into account in calculations involving the draft amidships. The depth of the vessel amidships from the keel to the deck line (KY or K_1Y_1) is constant being equal to the draft amidships plus the freeboard.

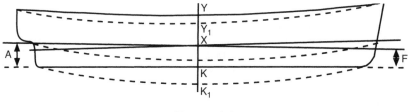

Fig. 32.1(d)

Example

A ship is floating in water of relative density 1.015. The present displacement is 12 000 tonnes, KG 7.7 m, KM 8.6 m. The present drafts are F 8.25 m, A 8.65 m, and the present freeboard amidships is 1.06 m. The Summer draft is 8.53 m and the Summer freeboard is 1.02 m FWA 160 mm TPC 20. Assuming that the KM is constant, find the amount of cargo (KG 10.0 m) which can be loaded for the ship to proceed to sea at the loaded Summer draft. Also find the amount of the hog or sag and the initial GM on departure.

Summer freeboard	1.02 m	Present mean freeboard	1.06 m
Summer draft	+ 8.53 m	Depth Mld	9.55 m
Depth Mld	9.55 m	Present draft amidships	8.49 m
		Average of drafts F and A	8.45 m
		Ship is sagged by	0.04 m

$$\text{Dock water allowance (DWA)} = \frac{(1025 - \rho_{DW})}{25} \times \text{FWA} = \frac{10}{25} \times 160 = 64\,\text{mm}$$
$$= 0.064\,\text{m}$$

$$\text{TPC in dock water} = \frac{RD_{DW}}{RD_{SW}} \times TPC_{SW} = \frac{1.015}{1.025} \times 20$$
$$= 19.8\,\text{tonnes}$$

$$\begin{aligned}
\text{Summer freeboard} &= 1.020\,\text{m} \\
\text{DWA} &= 0.064\,\text{m} \\
\text{Min. permissible freeboard} &= 0.956\,\text{m} \\
\text{Present freeboard} &= 1.060\,\text{m} \\
\text{Mean sinkage} &= 0.104\,\text{m or } 10.4\,\text{cm} \\
\text{Cargo to load} &= \text{Sinkage} \times TPC_{DW} = 10.4 \times 19.8 \\
\text{Cargo to load} &= 205.92\,\text{tonnes}
\end{aligned}$$

$$GG_1 = \frac{w \times d}{W + w} = \frac{205.92 \times (10 - 7.7)}{12\,000 + 205.92} = \frac{473.62}{12\,205.92}$$

$$\therefore \ \text{Rise of G} = 0.039\,\text{m}$$
$$\text{Present GM } (8.6 - 7.7) = 0.900\,\text{m}$$
$$\underline{\text{GM on departure} = 0.861\,\text{m}}$$

and ship has a sag of 0.04 m.

Exercise 32

1 The minimum permissible freeboard for a ship is at a true mean draft of 7.3 m. The present draft is 6.2 m F and 8.2 m A. TPC = 10. The centre of flotation is 3 m aft of amidships. Length of the ship 90 m. Find how much more cargo may be loaded.

2 A ship has a load salt water displacement of 12 000 tonnes, load draft in salt water 8.5 m, length 120 m, TPC 15 tonnes, and centre of flotation 2 m aft of amidships. The ship is at present floating in dock water of density 1015 kg per cu. m at drafts of 7.2 m F and 9.2 m A. Find the cargo which must yet be loaded to bring the ship to the maximum permissible draft.

3 Find the weight of the cargo the ship in Question 2 could have loaded had the centre of flotation been 3 m forward of amidships instead of 2 m aft.

4 A ship is floating in dock water of relative density 1.020. The present displacement is 10 000 tonnes, KG 6.02 m, KM 6.92 m. Present drafts are F 12.65 m and A 13.25 m. Present freeboard 1.05 m. Summer draft 13.10 m and Summer freeboard is 1.01 m. FWA 150 mm. TPC 21. Assuming that the KM is constant find the amount of cargo (KG 10.0 m) which can be loaded for the ship to sail at the load Summer draft. Find also the amount of the hog or sag and the initial metacentric height on departure.

Chapter 33
The inclining experiment plus fluctuations in a ship's lightweight

It has been shown in previous chapters that, before the stability of a ship in any particular condition of loading can be determined, the initial conditions must be known. This means knowing the ship's lightweight, the VCG or KG at this lightweight, plus the LCG for this lightweight measured from amidships. For example, when dealing with the height of the centre of gravity above the keel, the initial position of the centre of gravity must be known before the final KG can be found. It is in order to find the KG for the light condition that the Inclining Experiment is performed.

The experiment is carried out by the builders when the ship is as near to completion as possible; that is, as near to the light condition as possible. The ship is forcibly inclined by shifting weights a fixed distance across the deck. The weights used are usually concrete blocks, and the inclination is measured by the movement of plumb lines across specially constructed battens which lie perfectly horizontal when the ship is upright. Usually two or three plumb lines are used and each is attached at the centreline of the ship at a height of about 10 m above the batten. If two lines are used then one is placed forward and the other aft. If a third line is used it is usually placed amidships. For simplicity, in the following explanation only one weight and one plumb line is considered.

The following conditions are necessary to ensure that the KG obtained is as accurate as possible:

1. There should be little or no wind, as this may influence the inclination of the ship. If there is any wind the ship should be head on or stern on to it.
2. The ship should be floating freely. This means that nothing outside the ship should prevent her from listing freely. There should be no barges or lighters alongside; mooring ropes should be slacked right down, and

there should be plenty of water under the ship to ensure that at no time during the experiment will she touch the bottom.

3. Any loose weights within the ship should be removed or secured in place.
4. There must be no free surfaces within the ship. Bilges should be dry. Boilers and tanks should be completely full or empty.
5. Any persons not directly concerned with the experiment should be sent ashore.
6. The ship must be upright at the commencement of the experiment.
7. A note of 'weights on' and 'weights off' to complete the ship each with a VCG and LCG₵.

When all is ready and the ship is upright, a weight is shifted across the deck transversely, causing the ship to list. A little time is allowed for the ship to settle and then the deflection of the plumb line along the batten is noted. If the weight is now returned to its original position the ship will return to the upright. She may now be listed in the opposite direction. From the deflections the GM is obtained as follows:

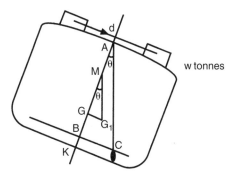

Fig. 33.1

In Figure 33.1 let a mass of 'w' tonnes be shifted across the deck through a distance 'd' metres. This will cause the centre of gravity of the ship to move from G to G_1 parallel to the shift of the centre of gravity of the weight. The ship will then list to bring G_1 vertically under M, i.e. to θ degrees list. The plumb line will thus be deflected along the batten from B to C. Since AC is the new vertical, angle BAC must also be θ degrees.

In triangle ABC,

$$\cot \theta = \frac{AB}{BC}$$

In triangle GG_1M,

$$\tan \theta = \frac{GG_1}{GM}$$

$$\therefore \frac{GM}{GG_1} = \frac{AB}{BC}$$

Or

$$GM = GG_1 \times \frac{AB}{BC}$$

But

$$GG_1 = \frac{w \times d}{W}$$

$$\therefore GM = \frac{w \times d}{W} \times \frac{AB}{BC}$$

Hence

$$GM = \frac{w \times d}{W \tan \theta}$$

In this formula AB, the length of the plumb line and BC, the deflection along the batten can be measured. 'w' the mass shifted, 'd' the distance through which it was shifted, and 'W' the ship's displacement, will all be known. The GM can therefore be calculated using the formula.

The naval architects will already have calculated the KM for this draft and hence the present KG is found. By taking moments about the keel, allowance can now be made for weights which must be loaded or discharged to bring the ship to the light condition. In this way the light KG is found.

Example 1

When a mass of 25 tonnes is shifted 15 m transversely across the deck of a ship of 8000 tonnes displacement, it causes a deflection of 20 cm in a plumb line 4 m long. If the KM = 7 m, calculate the KG.

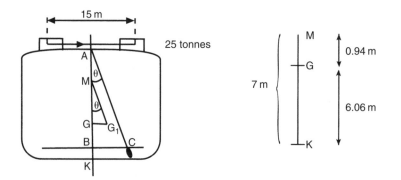

Fig. 33.2

$$\frac{GM}{GG_1} = \frac{AB}{BC} = \frac{1}{\tan \theta}$$

$$\therefore \tan \theta \ GM = GG_1$$

$$GM = \frac{w \times d}{W} \times \frac{1}{\tan \theta}$$

$$= \frac{25 \times 15}{8000} \times \frac{4}{0.2}$$

$$GM = 0.94 \text{ m}$$

$$KM = 7.00 \text{ m}$$

Ans. KG = 6.06 m as shown in the sketch.

Example 2

When a mass of 10 tonnes is shifted 12 m, transversely across the deck of a ship with a GM of 0.6 m it causes 0.25 m deflection in a 10 m plumb line. Calculate the ship's displacement.

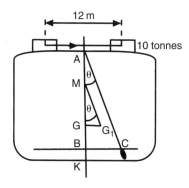

Fig. 33.3

$$GM = \frac{w \times d}{W} \times \frac{1}{\tan \theta}$$

$$W = \frac{w \times d}{GM} \times \frac{1}{\tan \theta}$$

$$= \frac{10 \times 12 \times 10}{0.6 \times 0.25}$$

$$W = 8000.$$

Ans. Displacement = 8000 tonnes

Summary

Every **new** ship should have an Inclining Experiment. However, some ship-owners do not request one if their ship is a sister-ship to one or more in the company's fleet.

Fluctuations in a ship's lightweight over a period of time

Lightweight is made up of the steel weight plus the wood and outfit weight plus the machinery weight. The lightweight of a ship is the weight of the ship when completely *empty*. There will be no deadweight items on board.

Over the years in service, there will be increases in the lightweight due to:

Accretion of paintwork.	Formation of oxidation or rust.
Build up of cargo residue.	Sediment in bottom of oil tanks.
Mud in bottom of ballast tanks.	Dunnage.
Gradual accumulation up of rubbish.	Lashing material.

Retrofits on accommodation fittings and in navigational aids.
Barnacle attachment or animal growth on the shell plating.
Vegetable growth on shell plating.
Additional engineers' spares and electricians' spares.
Changing a bulbous bow to a soft-nosed bow.
Major ship surgery such as an addition to ship's section at amidships.

Each item in the above list will change the weight of an empty ship. It can also be accumulative. One example of increase in lightweight over a period of years is the 'Herald of Free Enterprise,' which capsized in 1987. At the time of capsize it was shown that the lightweight had increased by 270 t, compared to when newly built.

Regular dry-docking of the ship will decrease the animal and vegetable growth on the shell plating. It has been known to form as much as 5 cm of extra weight around the hull.

Regular tank-cleaning programmes will decrease the amount of oil sediment and mud in the bottom of tanks. Regular routine inspections should also decrease the accumulation of rubbish.

Over years in service, there will also be decreases in the lightweight due to:

Oxidation or corrosion of the steel shell plating, and steel decks exposed to the sea and to the weather.
Wear and tear on moving parts.
Galvanic corrosion at localities having dissimilar metals joined together.

Corrosion and loss of weight is prevalent and vulnerable in the boot-topping area of the side-shell of a vessel, especially in way of the machinery spaces. Feedback has shown that the side-shell thickness can decrease over the years from being 18 mm thickness to being only 10 mm in thickness. This would result in an appreciable loss of weight.

Wear and tear occurs on structures such as masts and derricks, windlass, winches, hawse pipes and capstans.

These additions and reductions will all have their own individual centres of gravity and moments of weight. The result will be an overall change in the lightweight itself, plus a new value for the KG corresponding to this new lightweight.

One point to also consider is the weight that crew and passengers bring onto ships. It is usually denoted as 'effects'. Strictly speaking, these 'effects' are part of the deadweight. However, some people bring more onto the ship than average. Some leave the ship without all their belongings. It is difficult as time goes by to keep track of all these small additions. Conjecture suggests that ships with large numbers of persons will experience greater additions. Examples are passenger liners, RO-RO ships and Royal Naval vessels.

It has been documented that the lightweight of a vessel can amount to an average addition of 0.5% of the lightweight for each year of the ship's life. As a ship gets older it becomes heavier.

Major ship surgery can involve the insertion of a length of structure at amidships. However, it can involve raising a deck to give greater tween deck height. There is also on record a ship that was cut right down through the centreline (stern to bow), pulled apart and structure was added in to increase the breadth moulded. Each installation of course changed each ship's lightweight.

Returning to the line on bows, there was one notable example where the ship-owner requested the removal of a bulbous bow. It was replaced with a soft-nosed bow. The modified ship actually gave an increase in service speed for similar input of engine power. This increase in ship performance was mainly due to the fact that this vessel had a low service speed (about 12 knots) and a small summer deadweight. After this retrofit, obviously there would be a change in lightweight.

These notes indicate that sometimes the lightweight will increase, for example due to plate renewal or animal and vegetable growth. Other times it will decrease, for example due to wear and tear or build up of corrosion. There will be fluctuations. It would seem judicial to plan for an inclining experiment perhaps every 5 years. This will re-establish for the age of the ship exactly the current lightweight. Passenger liners are required to do just this.

Exercise 33

1 A ship of 8000 tonnes displacement has KM = 7.3 m and KG = 6.1 m. A mass of 25 tonnes is moved transversely across the deck through a distance of 15 m. Fined the deflection of a plumb line which is 4 m long.

2 As a result of performing the inclining experiment it was found that a ship had an initial metacentric height of 1 m. A mass of 10 tonnes, when shifted 12 m transversely, had listed the ship $3\frac{1}{2}$ degrees and produced a deflection of 0.25 m in the plumb line. Find the ship's displacement and the length of the plumb line.

3 A ship has KM = 6.1 m and displacement of 3150 tonnes. When a mass of 15 tonnes, already on board, is moved horizontally across the deck through a distance of 10 m it causes 0.25 m deflection in an 8 m long plumb line. Calculate the ship's KG.

4   A ship has an initial GM = 0.5 m. When a mass of 25 tonnes is shifted transversely a distance of 10 m across the deck, it causes a deflection of 0.4 m in a 4 m plumb line. Find the ship's displacement.

5   A ship of 2304 tonnes displacement has an initial metacentric height of 1.2 m. Find the deflection in a plumb line which is suspended from a point 7.2 m above a batten when a mass of 15 tonnes, already on board, is shifted 10 m transversely across the deck.

6   During the course of an inclining experiment in a ship of 4000 tonnes displacement, it was found that, when a mass of 12 tonnes was moved transversely across the deck, it caused a deflection of 75 mm in a plumb line which was suspended from a point 7.5 m above the batten. KM = 10.2 m. KG = 7 m. Find the distance through which the mass was moved.

7   A box-shaped vessel 60 m × 10 m × 3 m is floating upright in fresh water on an even keel at 2 m draft. When a mass of 15 tonnes is moved 6 m transversely across the deck a 6 m plumb line is deflected 20 cm. Find the ship's KG.

8   The transverse section of a barge is in the form of a triangle, apex downwards. The ship's length is 65 m, breadth at the waterline 8 m, and the vessel is floating upright in salt water on an even keel at 4 m draft. When a mass of 13 tonnes is shifted 6 m transversely it causes 20 cm deflection in a 3 m plumb line. Find the vessel's KG.

9   A ship of 8000 tonnes displacement is inclined by moving 4 tonnes transversely through a distance of 19 m. The average deflections of two pendulums, each 6 m long, was 12 cm 'Weights on' to complete this ship were 75 t centred at KG of 7.65 m 'Weights off' amounted to 25 t centred at KG of 8.16 m.
   (a) Calculate the GM and angle of heel relating to this information, for the ship as inclined.
   (b) From Hydrostatic Curves for this ship as inclined, the KM was 9 m. Calculate the ship's final Lightweight and VCG at this weight.

Chapter 34
The calibration book plus soundings and ullages

The purpose of a calibration book is to give volumes, displacements and centre of gravity at pre-selected tank levels. The Naval Architect calculates the contents of a tank at, say, 0.01 m to 0.20 m intervals of height within the tank. For an example of 0.10 m sounding height intervals (see Table 34.1). This information is supplied by the shipbuilder to each ship, for use by masters and mates.

A Sounding is the vertical distance between the base of the tank and the surface of the liquid. A sounding pipe is a plastic pipe of about 37 mm diameter, down which a steel sounding tape is lowered (see Figures 34.1 and 34.2 and Table 34.1).

A Ullage is the vertical distance between the surface of the liquid and the top of the ullage plug or top of the sounding pipe (see Figure 34.1 and Tables 34.2 and 34.3).

For tanks containing fresh water, distilled water, water ballast, etc. it is advisable to take soundings. These tanks can be filled up to 100 per cent capacity.

For tanks containing cargo oil, oil fuel, diesel oil, lubricating oil, etc. it is advisable to take ullage readings. Oil tanks are usually filled to a maximum of 98 per cent full capacity. This is to make allowance for expansion due to heat.

Methods for reading soundings and ullages

A reading may be taken by using:

(i) Steel measuring tape with a weight attached to its end (see Figure 34.2).
(ii) Calibrated glass tube (see Figure 34.3).
(iii) Whessoe gastight tank gauge (see Figure 34.4).
(iv) Saab Radar tank gauge (see Figure 34.5).

Consider an oil fuel tank (see Table 34.1). The master or mate can take a sounding in the tank on a particular day. If it happens to be 0.60 m then the contents of the tank are 3.30 cu. metres.

Table 34.1   Sample of a calibrated oil fuel tank.

Sounding in metres	Volume in cu. metres
0.10	0.60
0.20	0.90
0.30	1.40
0.40	1.90
0.50	2.60
0.60	3.30
0.70	4.10
0.80	5.00
0.90	6.10
1.00	7.20

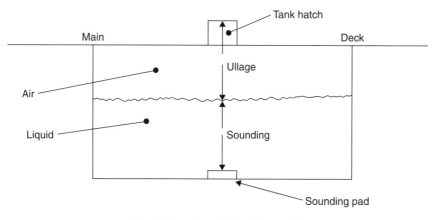

Fig. 34.1   A partially filled tank.

(i) The steel measuring tape is marked with white chalk and lowered into the tank of liquid until the user hears a dull sound as the end weight makes contact with the sounding pad.

 The scale on the tape is then checked to the level where the chalk has been washed off and where the chalk has remained in dry condition. At this interface, the vertical measurement lifted off. This is the sounding depth for the tank (see Figure 34.2 and Table 34.1).

(ii) A calibrated tube can be fitted on the outside of a small tank. This may be a fresh water tank, distilled water tank, lubricating oil tank or heavy oil tank. The observer can quickly read the height on the calibrated glass tube, thus obtaining the ullage or sounding in the tank (see Figure 34.3 and Table 34.3).

(iii) When carrying poisons, corrosives, highly inflammable products and cargoes that react with air, means of taking ullages and samples without

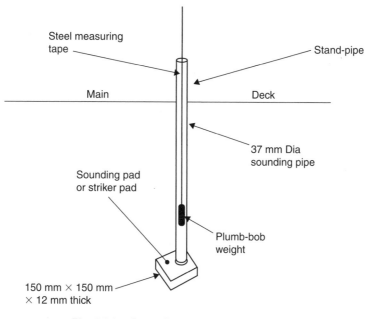

Fig. 34.2   Soundings, using a steel tape.

Table 34.2   Consider a cargo oil tank. Sample of a calibrated cargo oil tank. Volumes and weights.

Ullage in metres	*Volume in m³*	*Oil @ 0.860 m³ per tonne*	*Oil @ 0.880 m³ per tonne*	*Oil @ 0.900 m³ per tonne*
24.30	4011.73	3450	3530	3611
24.32	4012.72	3451	3531	3611
24.34	4013.61	3452	3532	3612
24.36	4014.37	3452	3532	3613
24.38	4015.01	3453	3533	3614
24.40	4015.55	3453	3534	3614
24.42	4015.97	3454	3534	3614
24.44	4016.28	3454	3534	3615
24.46	4016.47	3454	3534	3615
24.48	4016.55	3454	3535	3615
24.50	4016.58	3454	3535	3615

Weights have been rounded off to the nearest tonne.

releasing any vapour must be provided. Such a system is the 'Whessoe' gastight tank gauge (see Figure 34.4).

Remote reading temperature sensors and high level alarms are also commonly fitted on modern ships. When these alarms are set off the pumps stop automatically.

Table 34.3 Consider a fresh water tank. Calibrated FW tank. Weights, vertical and longitudinal moments.

Sight glass tube Ullage reading (m)	Weight of fresh Water in tonnes	Vertical moment Weight $\times$ kg (tm)	Longitudinal moment Weight $\times$ lcg (tm)
0.20	6.90	23.10	68.90
0.40	5.70	18.50	57.00
0.60	4.60	14.30	45.30
0.80	3.50	10.50	33.80
1.00	2.40	6.90	22.80
1.20	1.50	4.00	13.20
1.40	0.80	2.00	6.90
1.60	0.30	0.80	2.70
1.80	0.10	0.10	0.50

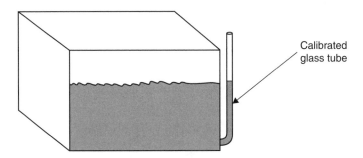

Calibrated glass tube

Fig. 34.3   Calibration, using a glass sounding tube.

The distance of the float from the tank top is read on the 'Whessoe' ullage meter. This is moved by the measuring tape passing over a sprocket in the gauge head, which in turn moves the counter drum whose rotation is transmitted to the ullage meter. The ullage for the tank is then read off the screen, as shown in Figure 34.4.

(iv) The Saab radar cased cargo tank and ballast monitoring system (see Figure 34.5). The tank gauge unit measures the distance to the surface of the liquid in the tank using a continuous radar signal. This signal can have a measurement range of 0 to 60 m. The reflected signal is received and processed via tank gauge electronics. When converted, the calculated values can then be sent to a communications centre; for example to officers working on the bridge.

This modern system has several very useful advantages. They are as follows:

• From the bridge or other central control, it is easy to check and recheck the contents in any tank containing liquid.

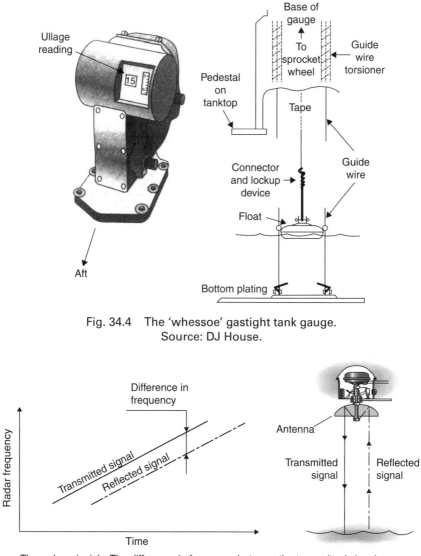

Fig. 34.4   The 'whessoe' gastight tank gauge.
Source: DJ House.

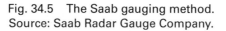

The radar principle. The difference in frequency between the transmitted signal
and the reflected signal is directly proportional to the ullage.

Fig. 34.5   The Saab gauging method.
Source: Saab Radar Gauge Company.

- Continuous minute by minute readings of liquid level in tank.
- Radar waves are extremely robust to any condition in the tank.
- Radar waves are generally not affected by the atmosphere above the product in the tank.
- The only part located inside the tank is the antenna without any moving parts.

- High accuracy and high reliability.
- The electronics can be serviced and replaced during closed tank conditions.

The Bosun on board the ship used to check all tanks at least once a day. This was to ensure that water had not entered the ship via the hull plating and that liquid had not leaked into an adjacent tank. With modern calibration systems continuous assessment is now possible.

Cofferdams must also be checked to ensure no seepage has occurred. A cofferdam is a drainage tank placed between compartments carrying dissimilar liquids; for example, it may be a water ballast tank and an oil fuel tank. Cofferdams should not be used to carry any liquids so they do not usually have filling pipes fitted.

Adjustments for angle of heel and trim

If the sounding pipe or ullage plug is not transversely in line with the tank's centreline then an adjustment to the ullage reading must be made.

1. If vessel has a list to *Port* and the ullage plug is a horizontal distance 'L' to the port side of the tank's vcg, then the *positive* correction to the ullage reading is:

$$+L \times \tan \theta \text{ metres} \qquad \text{where } \theta = \text{Angle of list.}$$

2. If vessel has a list to *Port* and the ullage plug is a horizontal distance 'L' to the Starboard side of the tank's vcg, then the *negative* correction to the ullage reading is:

$$-L \times \tan \theta \text{ metres} \qquad \text{where } \theta = \text{Angle of list.}$$

3. If vessel has a list to *Starboard* and the ullage plug is a horizontal distance 'L' to the port side of the tank's vcg, then the *negative* correction to the ullage reading is:

$$-L \times \tan \theta \text{ metres} \qquad \text{where } \theta = \text{Angle of list.}$$

4. If vessel has a list to *Starboard* and the ullage plug is a horizontal distance 'L' to the Starboard side of the tank's vcg, then the *positive* correction to the ullage reading is:

$$+L \times \tan \theta \text{ metres} \qquad \text{where } \theta = \text{Angle of list.}$$

Worked example 1

A ship has an ullage plug that is 1.14 m to the port side of a tank's vcg. Estimate the adjustment to the ullage readings in cms, when this ship has angles of heel to port up to 3° at intervals of 0.5°.

Angle of heel $\theta(°)$	0	0.5	1.0	1.5	2.0	2.5	3.0
$+114 \times \tan \theta$ (cm)	0	+1	+2	+3	+4	+5	+6

The table for Worked Example 1 shows the transverse adjustments are 0 to 6 cm. Consequently, they do need to be considered by the master and mate on-board ship.

If the sounding pipe or ullage plug is not longitudinally in line with the tank's lcg then an adjustment to the ullage reading must be made.

1. If vessel is trimming by the *Stern* and the ullage plug is a horizontal distance 'L' *Forward* of the tank's lcg, then the *negative* correction to the ullage reading is:

$$-L \times \tan \theta \text{ metres} \qquad \text{where } \tan \theta = \text{trim}/\text{LBP}$$

2. If vessel is trimming by the *Stern* and the ullage plug is a horizontal distance 'L' *Aft* of the tank's lcg, then the *positive* correction to the ullage reading is:

$$+L \times \tan \theta \text{ metres} \qquad \text{where } \tan \theta = \text{trim}/\text{LBP}$$

3. If vessel is trimming by the *Bow* and the ullage plug is a horizontal distance 'L' *Forward* of the tank's lcg, then the *positive* correction to the ullage reading is:

$$+L \times \tan \theta \text{ metres} \qquad \text{where } \tan \theta = \text{trim}/\text{LBP}$$

4. If vessel is trimming by the *Bow* and the ullage plug is a horizontal distance 'L' *Aft* of the tank's lcg, then the *negative* correction to the ullage reading is:

$$-L \times \tan \theta \text{ metres} \qquad \text{where } \tan \theta = \text{trim}/\text{LBP}$$

Worked example 2

A vessel is 120 m LBP and has an ullage pipe 2.07 m aft of a tank's lcg. Calculate the adjustments to the ullages readings when the ship is trimming zero metres, 0.3 m, 0.6 m, 0.9 m and 1.2 m by the stern.

Trim in metres	0	0.3	0.6	0.9	1.2
Trim angle (°)	0	0.0025	0.0050	0.0075	0.0100
$+207 \times \tan \theta$ (cm)	0	+0.009	+0.018	+0.027	+0.036

The table for Worked Example 2 indicates the adjustments for trim are very small and can in practice, be ignored by the master and mate on board the ship.

Adjustments for temperature

Oil expands when heated and consequently, its density *decreases* with a *rise* in temperature. This means that the density in t/m^3 must be adjusted to give a better reading. The change of relative density due to a change of one degree in temperature is known as the relative density coefficient. For most oils, this lies between 0.00025 and 0.00050 per one degree Centigrade. The following worked example shows how this is done.

Worked example

A sample of oil has a density of 0.8727 at 16° Centigrade. Its expansion coefficient is 0.00027 per degree C. Proceed to calculate its density at 26°C.

$$\text{Difference in temperatures} = 26° - 16° = 10°\text{C}.$$
$$\text{Change in density} = 10 \times 0.00027 = 0.0027$$
$$\text{Density at } 26° = 0.8727 - 0.0027 = 0.870 \, \text{t/m}^3.$$

Worked example

A reworking of previous example with degrees F.

A sample oil has a density of 0.8727 at 60.8° Fahrenheit. Its expansion coefficient is 0.00015 per degree F. Proceed to calculate its density at 78.8°F.

$$\text{Difference in temperatures} = 78.8° - 60.8° = 18°\text{C}.$$
$$\text{Change in density} = 18 \times 0.00015 = 0.0027$$
$$\text{Density at } 78.8° = 0.8727 - 0.0027 = 0.870 \, \text{t/m}^3 \text{ (as before)}$$

A tank with a zero sounding will not always have a zero volume reading. This is because at the base of a tank there could be a 12 mm thick sounding pad to conserve the bottom of the ship's tank.

Exercise 34

1 A ship 120 m long is trimmed 1.5 m by the stern. A double bottom tank is 15 m long × 20 m × 1 m and has the sounding pipe situated at the after end of the tank. Find the sounding that will indicate that the tank is full.

2 A ship 120 m long is trimmed 2 m by the stern. A double bottom tank 36 m × 15 m × 1 m is being filed with oil fuel of relative density 0.96. The sounding pipe is at the after end of the tank and the present sounding is 1.2 m. Find how many tonnes of oil are yet required to fill this tank and also find the sounding when the tank is full.

Chapter 35
Drydocking and grounding

When a ship enters a drydock she must have a positive initial GM, be upright, and trimmed slightly, usually by the stern. On entering the drydock the ship is lined up with her centreline vertically over the centreline of the keel blocks and the shores are placed loosely in position. The dock gates are then closed and pumping out commences. The rate of pumping is reduced as the ship's stern post nears the blocks. When the stern lands on the blocks the shores are hardened up commencing from aft and gradually working forward so that all of the shores will be hardened up in position by the time the ship takes the blocks overall. The rate of pumping is then increased to quickly empty the dock.

As the water level falls in the drydock there is no effect on the ship's stability so long as the ship is completely waterborne, but after the stern lands on the blocks the draft aft will decrease and the trim will change by the head. This will continue until the ship takes the blocks overall throughout her length, when the draft will then decrease uniformly forward and aft.

The interval of time between the stern post landing on the blocks and the ship taking the blocks overall is referred to as the *critical period*. During this period part of the weight of the ship is being borne by the blocks, and this creates an upthrust at the stern which increases as the water level falls in the drydock. The upthrust causes a virtual loss in metacentric height and it is essential that positive effective metacentric height be maintained throughout the critical period, or the ship will heel over and perhaps slip off the blocks with disastrous results.

The purpose of this chapter is to show the methods by which the effective metacentric height may be calculated for any instant during the drydocking process.

Figure 35.1 shows the longitudinal section of a ship during the critical period. 'P' is the upthrust at the stern and 'l' is the distance of the centre of flotation from aft. The trimming moment is given by $P \times l$. But the trimming moment is also equal to MCTC $\times$ change of trim.

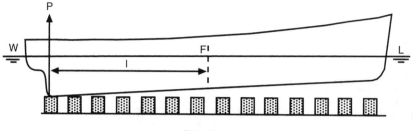

Fig. 35.1

Therefore

$$P \times 1 = MCTC \times t$$

or

$$P = \frac{MCTC \times t}{1}$$

where
P = the upthrust at the stern in tonnes
t = the change of trim since entering the drydock in centimetres
1 = the distance of the centre of flotation from aft in metres

Now consider Figure 35.2 which shows a transverse section of the ship during the critical period after she has been inclined to a small angle (θ degrees) by a force external to the ship. For the sake of clarity the angle of heel has been magnified. The weight of the ship (W) acts downwards through the centre of gravity (G). The force P acts upwards through the keel (K) and is equal to the weight being borne by the blocks. For equilibrium the force of buoyancy must now be (W − P) and will act upwards through the initial metacentre (M).

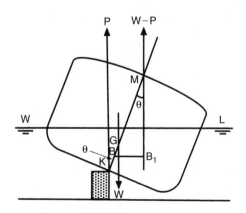

Fig. 35.2

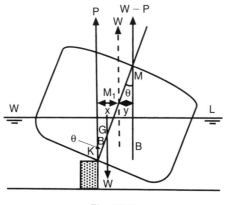

Fig. 35.3

There are, thus, three parallel forces to consider when calculating the effect of the force P on the ship's stability. Two of these forces may be replaced by their resultant (see Figure 1.5) in order to find the effective metacentric height and the moment of statical stability.

Method (a)

In Figure 35.3 consider the two parallel forces P and $(W - P)$. Their resultant W will act upwards through M_1 such that:

$$(W - P) \times y = P \times X$$

or

$$(W - P) \times MM_1 \times \sin \theta = P \times KM_1 \times \sin \theta$$
$$(W - P) \times MM_1 = P \times KM_1$$
$$W \times MM_1 - P \times MM_1 = P \times KM_1$$
$$W \times MM_1 = P \times KM_1 + P \times MM_1$$
$$= P (KM_1 + MM_1)$$
$$= P \times KM$$
$$MM_1 = \frac{P \times KM}{W}$$

There are now two forces to consider: W acting upwards through M_1 and W acting downwards through G. These produce a righting moment of $W \times GM_1 \times \sin \theta$.

Note also that the original metacentric height was GM but has now been reduced to GM_1. Therefore MM_1 is the virtual loss of metacentric height due to drydocking.

Or

$$\text{Virtual loss of GM (MM}_1) = \frac{P \times KM}{W}$$

Method (b)

Now consider the two parallel forces W and P in Figure 35.4. Their result-ant (W − P) acts downwards through G_1 such that:

$$W \times y = P \times X$$

or

$$W \times GG_1 \times \sin\theta = P \times KG_1 \times \sin\theta$$
$$W \times GG_1 = P \times KG_1$$
$$= P(KG + GG_1)$$
$$= P \times KG + P \times GG_1$$
$$W \times GG_1 - P \times GG_1 = P \times KG$$
$$GG_1(W - P) = P \times KG$$
$$GG_1 = \frac{P \times KG}{W - P}$$

There are now two forces to consider: (W − P) acting upwards through M and (W − P) acting downwards through G_1. These produce a righting moment of (W − P) × G_1M × sin θ.

The original metacentric height was GM but has now been reduced to G_1M. Therefore GG_1 is the virtual loss of metacentric height due to drydocking.

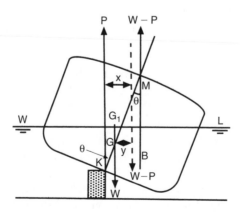

Fig. 35.4

Or

$$\text{Virtual loss of GM (GG}_1\text{)} = \frac{P \times KG}{W - P}$$

Example 1

A ship of 6000 tonnes displacement enters a drydock trimmed 0.3 m by the stern. KM = 7.5 m, KG = 6 m and MCTC = 90 tonnes m. The centre of flotation is 45 m from aft. Find the effective metacentric height at the critical instant before the ship takes the blocks overall.

Note. Assume that the trim at the critical instant is zero.

$$P = \frac{MCTC \times t}{1}$$

$$= \frac{90 \times 30}{45}$$

$$P = 60 \text{ tonnes}$$

Method (a)

$$\text{Virtual loss of GM (MM}_1\text{)} = \frac{P \times KM}{W}$$

$$= \frac{60 \times 7.5}{6000}$$

$$= 0.075 \text{ m}$$

$$\text{Original GM} = 7.5 - 6.0 = 1.500 \text{ m}$$

Ans. New GM = 1.425 m

Method (b)

$$\text{Virtual loss of GM} = \frac{P \times KG}{W - P}$$

$$GG_1 = \frac{60 \times 6}{5940}$$

$$= 0.061 \text{ m}$$

$$\text{Original GM} = 1.500 \text{ m}$$

Ans. New GM = 1.439 m

From these results it would appear that there are two possible answers to the same problem, but this is not the case. The ship's ability to return to the upright is indicated by the righting moment and not by the effective metacentric height alone.

To illustrate this point, calculate the righting moments given by each method when the ship is heeled to a small angle ($\theta°$)

Method (a)

$$\text{Righting moment} = W \times GM_1 \times \sin\theta$$
$$= 6000 \times 1.425 \times \sin\theta$$
$$= (8550 \times \sin\theta) \text{ tonnes m}$$

Method (b)

$$\text{Righting moment} = (W - P) \times G_1M \times \sin\theta$$
$$= 5940 \times 1.439 \times \sin\theta$$
$$= (8549 \times \sin\theta) \text{ tonnes metres}$$

Thus each of the two methods used gives a correct indication of the ship's stability during the critical period.

Example 2

A ship of 3000 tonnes displacement is 100 m long, has KM = 6 m, KG = 5.5 m. The centre of flotation is 2 m aft of amidships and MCTC = 40 tonnes m. Find the maximum trim for the ship to enter a drydock if the metacentric height at the critical instant before the ship takes the blocks forward and aft is to be not less than 0.3 m.

$$KM = 6.0\,m$$
$$KG = 5.5\,m$$
$$\overline{\text{Original GM} = 0.5\,m}$$
$$\text{Virtual GM} = 0.3\,m$$
$$\overline{\text{Virtual loss} = 0.2\,m}$$

Method (a)

$$\text{Virtual loss of GM } (MM_1) = \frac{P \times KM}{W}$$

or

$$P = \frac{\text{Virtual loss} \times W}{KW}$$
$$= \frac{0.2 \times 3000}{6}$$
$$\overline{\text{Maximum P} = 100 \text{ tonnes}}$$

But

$$P = \frac{MCTC \times t}{l}$$

or

$$Maximum \ t = \frac{P \times l}{MCTC}$$

$$= \frac{100 \times 48}{40}$$

Ans. Maximum trim = 120 cm by the stern

Method (b)

$$Virtual \ loss \ of \ GM \ (GG_1) = \frac{P \times KG}{W - P}$$

$$0.2 = \frac{P \times 5.5}{3000 - P}$$

$$600 - 0.2P = 5.5P$$

$$5.7P = 600$$

$$Maximum \ P = \frac{600}{5.7} = 105.26 \ tonnes$$

But

$$P = \frac{MCTC \times t}{l}$$

or

$$Maximum \ t = \frac{P \times l}{MCTC}$$

$$= \frac{105.26 \times 48}{40}$$

Ans. Maximum trim = 126.3 cm by the stern

There are, therefore, two possible answers to this question, depending on the method of solution used. The reason for this is that although the effective metacentric height at the critical instant in each case will be the same, the righting moments at equal angles of heel will not be the same.

Example 3

A ship of 5000 tonnes displacement enters a drydock trimmed 0.45 m by the stern. KM = 7.5 m, KG = 6.0 m and MCTC = 120 tonnes m. The centre of flotation is 60 m from aft. Find the effective metacentric height at the critical instant before the ship takes the blocks overall, assuming that the transverse metacentre rises 0.075 m.

$$P = \frac{MCTC \times t}{1}$$

$$= \frac{120 \times 45}{60}$$

$$P = 90 \text{ tonnes}$$

Method (a)

$$\text{Virtual loss } (MM_1) = \frac{P \times KM}{W}$$

$$= \frac{90 \times 7.575}{5000}$$

$$= 0.136 \text{ m}$$

$$\text{Original KM} = 7.500 \text{ m}$$

$$\text{Rise of M} = \underline{0.075 \text{ m}}$$

$$\text{New KM} = 7.575 \text{ m}$$

$$\text{KG} = \underline{6.000 \text{ m}}$$

$$\text{GM} = 1.575 \text{ m}$$

$$\text{Virtual loss } (MM_1) = \underline{0.136 \text{ m}}$$

Ans. New GM = 1.439 m

Method (b)

$$\text{Virtual loss } (GG_1) = \frac{P \times KG}{W - P}$$

$$= \frac{90 \times 6.0}{4910}$$

$$= 0.110 \text{ m}$$

$$\text{Old KG} = 6.000 \text{ m}$$

$$\text{Virtual loss } (GG_1) = \underline{0.110 \text{ m}}$$

$$\text{New KG} = 6.110 \text{ m}$$

$$\text{New KM} = 7.575 \text{ m}$$

Ans. New GM = 1.465 m

The virtual loss of GM after taking the blocks overall

When a ship takes the blocks overall, the water level will then fall uniformly about the ship, and for each centimetre fallen by the water level

P will be increased by a number of tonnes equal to the TPC. Also, the force P at any time during the operation will be equal to the difference between the weight of the ship and the weight of water she is displacing at that time.

Example

A ship of 5000 tonnes displacement enters a drydock on an even keel. $KM = 6\,m$, $KG = 5.5\,m$ and $TPC = 50$ tonnes. Find the virtual loss of metacentric height after the ship has taken the blocks and the water has fallen another 0.24 m.

$$P = TPC \times \text{Reduction in draft in cm}$$
$$= 50 \times 24$$
$$P = 1200 \text{ tonnes}$$

Method (a)

$$\text{Virtual loss (MM}_1) = \frac{P \times KM}{W}$$
$$= \frac{1200 \times 6}{5000}$$

Ans. Virtual loss = 1.44 m

Method (b)

$$\text{Virtual loss (GG}_1) = \frac{P \times KG}{W - P}$$
$$= \frac{1200 \times 5.5}{3800}$$

Ans. Virtual loss = 1.74 m

Note to students

In the DfT examinations, when sufficient information is given in a question, either method of solution may be used. It has been shown in this chapter that both are equally correct. In some questions, however, there is no choice, as the information given is sufficient for only one of the methods to be used. It is therefore advisable for students to learn both of the methods.

Example

A ship of 8000 tonnes displacement takes the ground on a sand bank on a falling tide at an even keel draft of 5.2 metres, KG 4.0 metres. The predicted depth of water over the sand bank at the following low water is 3.2 metres. Calculate the GM at this time assuming that the KM will then be 5.0 metres and that the mean TPC is 15 tonnes.

$$P = TPC \times \text{Fall in water level (cm)} = 15 \times (520 - 320)$$
$$= 15 \times 200$$
$$P = 3000 \text{ tonnes}$$

Method (a)

$$\text{Virtual loss of GM (MM}_1) = \frac{P \times KM}{W}$$

$$= \frac{3000 \times 5}{8000}$$

$$= 1.88 \text{ m}$$

$$\text{Actual KM} = 5.00 \text{ m}$$
$$\text{Virtual KM} = 3.12 \text{ m}$$
$$KG = 4.00 \text{ m}$$

Ans. New GM = −0.88 m

Method (b)

$$\text{Virtual loss of GM (GG}_1) = \frac{P \times KG}{W - P}$$

$$= \frac{3000 \times 4}{5000}$$

$$= 2.40 \text{ m}$$

$$KG = 4.00 \text{ m}$$
$$\text{Virtual KG} = 6.40 \text{ m}$$
$$KM = 5.00 \text{ m}$$

Ans. New GM = −1.40 m

Note that in example, this vessel has developed a negative GM. Consequently she is *unstable*. She would capsize if transverse external forces such as wind or waves were to remove her from zero angle of heel. Suggest a change of loading to reduce KG and make GM a positive value greater than DfT minimum of 0.15 m.

Exercise 35

1 A ship being drydocked has a displacement of 1500 tonnes. TPC = 5 tonnes, KM = 3.5 m, GM = 0.5 m, and has taken the blocks fore and aft at 3 m draft. Find the GM when the water level has fallen another 0.6 m.

2 A ship of 4200 tonnes displacement has GM 0.75 m and present drafts 2.7 m F and 3.7 m A. She is to enter a drydock. MCTC = 120 tonnes m. The after

keel block is 60 m aft of the centre of flotation. At 3.2 m mean draft KM = 8 m. Find the GM on taking the blocks forward and aft.

3 A box-shaped vessel 150 m long, 10 m beam and 5 m deep, has a mean draft in salt water of 3 m and is trimmed 1 m by the stern, KG = 3.5 m. State whether it is safe to drydock this vessel in this condition or not, and give reasons for your answer.

4 A ship of 6000 tonnes displacement is 120 m long and is trimmed 1 m by the stern. KG = 5.3 m, GM = 0.7 m and MCTC = 90 tonnes m. Is it safe to drydock the ship in this condition? (Assume that the centre of flotation is amidships.)

5 A ship of 4000 tonnes displacement, 126 m long, has KM = 6.7 m, KG = 6.1 m. The centre of flotation is 3 m aft of amidships, MCTC = 120 tonnes m. Find the maximum trim at which the ship may enter a drydock if the minimum GM at the critical instant is to be 0.3 m.

Chapter 36
Liquid pressure and thrust plus centres of pressure

Pressure in liquids

When a fluid is in equilibrium the stress across any surface in it is normal to the surface and the pressure intensity at any point of a fluid at rest is the same in all directions. The pressure intensity in a homogeneous liquid at rest under gravity increases uniformly with depth; i.e.

$$P = w \times g \times D$$

where

P = pressure intensity
w = density of the liquid
g = acceleration due to gravity

and

D = depth below the surface

Total thrust and resultant thrust

If the thrust on each element of area of a surface immersed in a fluid is found, the scalar sum of all such thrusts is called the 'total thrust' whilst their vector sum is called the 'resultant thrust'. When the surface is plane then the total thrust is equal to the resultant thrust.

To find the resultant thrust

In Figure 36.1, G represents the centroid of an area which is immersed, though not necessarily vertical, in a liquid, and Z represents the depth of the centroid of the area below the surface. Let w be the mass density of the liquid. If an element (dA) of the area, whose centroid is Z_1 below the surface, is considered, then:

$$\text{Thrust on the element dA} = \text{Pressure intensity} \times \text{Area}$$
$$= w \times g \times Z_1 \times dA$$

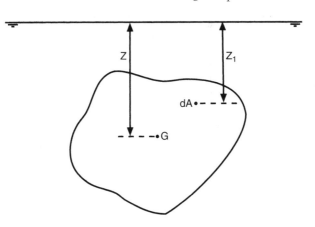

Fig. 36.1

$$\text{Resultant thrust on the whole area are} = \int w \times g \times Z_1 \, dA$$
$$= wg \times \int Z_1 \, dA$$

but

$$\int Z_1 \, dA = ZA$$

and

$$w \times g \int Z_1 \, dA = wgZA$$

∴ Resultant thrust = Density × g × Depth of centroid × Area

It should be noted that this formula gives only the magnitude of the resultant thrust. It does not indicate the point at which the resultant thrust may be considered to act.

The centre of pressure

The centre of pressure is the point at which the resultant thrust on an immersed surface may be considered to act. Its position may be found as follows:

(i) *For a rectangular lamina immersed with one side in the surface of the liquid.* In Figure 36.2, each particle of the strip GH is approximately at the same depth and therefore, the pressure intensity is nearly the same on each particle. Hence, the resultant thrust on each strip will act at its mid-point. The resultant of the thrusts on all of the strips in the lamina will act at a point on EF.

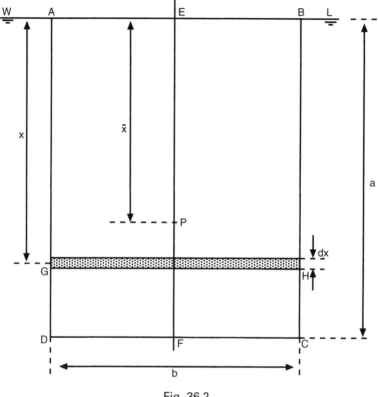

Fig. 36.2

Let w = Mass density of the liquid

The area of the elementary strip = b dx

The depth of the centroid of the strip below the surface is x if the plane of the rectangle is vertical, or x sin θ if the plane is inclined at an angle θ to the horizontal.

The thrust on the strip = $wgbx \sin \theta \, dx$

$$\text{The resultant thrust on the lamina} = \int_{O}^{a} wgbx \sin \theta \, dx$$

$$= \frac{wgba^2}{2} \sin \theta$$

$$= a \times b \times \tfrac{1}{2} \times a \times \sin \theta \times w \times g$$

$$= \text{Area} \times \text{Depth of centroid} \times g \times \text{Density}$$

$$\begin{array}{l}\text{The moment of the thrust} \\ \text{on the strip about AB}\end{array} = \text{wgbx}^2 \sin\theta\,\text{dx}$$

$$\begin{array}{l}\text{The moment of the total} \\ \text{thrust about AB}\end{array} = \int_O^a \text{wgbx}^2 \sin\theta$$

$$= \text{wgb}\frac{a^2}{3}\sin\theta$$

Let $\bar{x}$ be the distance of the centre of pressure (P) from AB, then:

$$\bar{x} \times \text{Total thrust} = \text{Total moment about AB}$$

$$\bar{x} \times \text{wgb}\frac{a^2}{2}\sin\theta = \text{wgb}\frac{a^3}{3}\sin\theta$$

or

$$\bar{x} = \frac{2}{3}a \text{ (unless } \sin\theta = 0)$$

(ii)   For any plane area immersed in a liquid.

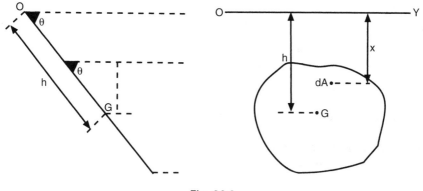

Fig. 36.3

In Figure 36.3, let OY be the line in which the plane cuts the surface of the liquid. Let the plane be inclined at an angle θ to the horizontal.

Let

h = the x co-ordinate of the centroid (G)
w = the mass density of the liquid

$$\text{Depth of the element dA} = x\sin\theta$$
$$\text{Thrust on dA} = \text{wgx}\sin\theta\,\text{dA}$$
$$\text{Moment of thrust about OY} = \text{wgx}^2\sin\theta\,\text{dA}$$
$$\text{Moment of total thrust about OY} = \int \text{wgx}^2\sin\theta\,\text{dA}$$
$$= \text{wg}\sin\theta\int x^2\,\text{dA}$$

$$\text{Total thrust on A} = \text{wgA} \times \text{Depth of centroid}$$
$$= \text{wgAh} \sin \theta$$
$$\text{Moment of total thrust about OY} = \text{wgAh} \sin \theta \cdot \bar{x}$$
$$\therefore \text{wgAh} \sin \theta \cdot \bar{x} = \text{wg} \sin \theta \int x^2 \, dA$$

or

$$\bar{x} = \frac{\int x^2 \, dA}{hA} \text{(unless } \sin \theta = 0)$$

Let I_{OY} be the second moment of the area about OY, then:

$$I_{OY} = \int x^2 \, dA$$

and

$$\bar{x} = \frac{I_{OY}}{hA}$$

or

$$\bar{x} = \frac{\text{Second moment of area about the waterline}}{\text{First moment of area about the waterline}}$$

Centres of pressure by Simpson's Rules

Using horizontal ordinates
Referring to Figure 36.4:

$$\text{Thrust on the element} = \text{wgxy} \, dx$$
$$\text{Moment of the thrust about OY} = \text{wgx}^2 \text{y} \, dx$$
$$\text{Moment of total thrust about OY} = \int \text{wgx}^2 \text{y} \, dx$$
$$= \text{wg} \int x^2 \text{y} \, dx$$
$$\text{Total thrust} = \text{wgA} \times \text{Depth of centroid}$$
$$= \text{wg} \int y \, dx \frac{\int xy \, dx}{\int y \, dx}$$
$$= \text{wg} \int xy \, dx$$
$$\text{Moment of total thrust about OY} = \text{Total thrust} \times \bar{x}$$

where

$\bar{x} = $ Depth of centre of pressure below the surface

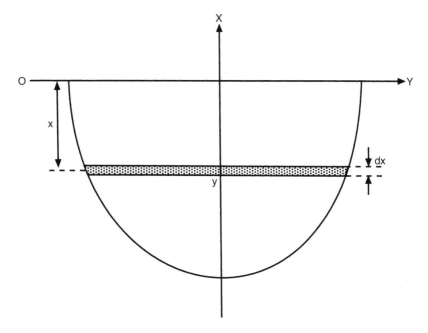

Fig. 36.4

$\therefore$ Moment of total thrust about OY $= wg\int xy\,dx \times \bar{x}$

or

$$wg\int xy\,dx \times \bar{x} = wg\int x^2y\,dx$$

and

$$\bar{x} = \frac{\int x^2y\,dx}{\int xy\,dx}$$

The value of the expression $\int x^2y\,dx$ can be found by Simpson's Rules using values of the product x^2y as ordinates, and the value of the expression $\int xy\,dx$ can be found in a similar manner using values of the product xy as ordinates.

Example 1

A lower hold bulkhead is 12 metres deep. The transverse widths of the bulkhead, commencing at the upper edge and spaced at 3-m intervals, are as follows:

15.4, 15.4, 15.4, 15.3 and 15 m, respectively

Find the depth of the centre of pressure below the waterplane when the hold is flooded to a depth of 2 metres above the top of the bulkhead.

Ord.	SM	Area func.	Lever	Moment func.	Lever	Inertia func.
15.4	1	15.4	0	0	0	0
15.4	4	61.6	1	61.6	1	61.6
15.4	2	30.8	2	61.6	2	123.2
15.3	4	61.2	3	183.6	3	550.8
15.0	1	15.0	4	60.0	4	240.0
		$184.0 = \Sigma_1$		$366.8 = \Sigma_2$		$975.6 = \Sigma_3$

$$\text{Area} = \frac{1}{3} \times h \times \Sigma_1 = \frac{3}{3} \times 184.0$$
$$= 184 \text{ sq. m}$$

Referring to Figure 36.5:

$$CG = \frac{\Sigma_2}{\Sigma_1} \times h$$
$$= \frac{366.8}{184} \times 3$$
$$= 5.98 \text{ m}$$
$$+CD = 2.00 \text{ m}$$
$$\bar{z} = 7.98 \text{ m}$$
$$I_{OZ} = \frac{1}{3} \times h^3 \times \Sigma_3 = \frac{1}{3} \times 3^3 \times 975.6$$
$$= 8780 \text{ m}^4$$
$$I_{CG} = I_{OZ} - A(CG)^2, \text{ i.e. parallel axis theorem}$$
$$I_{WL} = I_{CG} + A\bar{z}^2$$
$$= I_{OZ} - A(CG^2 - \bar{z}^2)$$
$$I_{WL} = 8780 - 184(5.98^2 - 7.98^2)$$
$$= 13\,928 \text{ m}^4$$
$$\bar{y} = \frac{I_{WL}}{A\bar{z}}$$
$$= \frac{13\,928}{184 \times 7.98}$$
$$\bar{y} = 9.5 \text{ m}$$

Ans. The centre of pressure is 9.5 m below the waterline.

Using vertical ordinates
Referring to Figure 36.6:

$$\text{Thrust on the element} = wg\frac{y}{2} y \, dx$$
$$= \frac{wgy^2}{2} dx$$

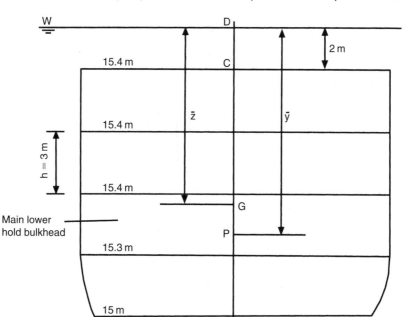

Fig. 36.5

$$\text{Moment of the thrust about OX} = \frac{wgy^2}{2}\,dx \times \frac{2}{3}y$$

$$= \frac{wgy^3}{3}\,dx$$

$$\text{Moment of total thrust about OX} = \frac{w}{3} \times g \times \int y^3\,dx$$

$$\text{Total thrust} = wgA \times \text{Depth at centre of gravity}$$

$$= wg\!\int y\,dx \,\frac{\frac{1}{2}\int y^2\,dx}{\int y\,dx}$$

$$= \frac{w}{2}g\!\int y^2\,dx$$

Let $\bar{y}$ be the depth of the centre of pressure below the surface, then:

$$\text{Moment of total thrust about OX} = \text{Total thrust} \times \bar{y}$$

$$\frac{wg}{3}\int y^3\,dx = \frac{wg}{2}\int y^2\,dx \times \bar{y}$$

or

$$\bar{y} = \frac{\frac{1}{3}\int y^3\,dx}{\frac{1}{2}\int y^2\,dx}$$

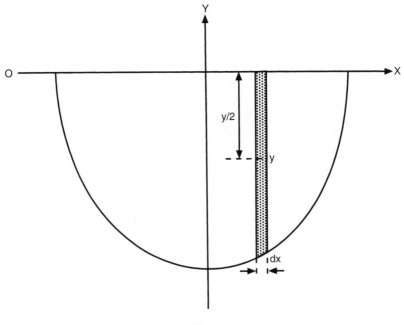

Fig. 36.6

The values of the two integrals can again be found using Simpson's Rules.

Example 2

The breadth of the upper edge of a deep tank bulkhead is 12 m. The vertical heights of the bulkhead at equidistant intervals across it are 0, 3, 5, 6, 5, 3 and 0 m, respectively. Find the depth of the centre of pressure below the waterline when the tank is filled to a head of 2 m above the top of the tank.

$$\text{Area} = \tfrac{1}{3} \times CI \times \Sigma_1$$
$$\text{Area} = \tfrac{1}{3} \times 2 \times 68$$
$$= 45\tfrac{1}{3}\,\text{sq. m}$$

Ord.	SM	Area func.	Ord.	Moment func.	Ord.	Inertia func.
0	1	0	0	0	0	0
3	4	12	3	36	3	108
5	2	10	5	50	5	250
6	4	24	6	144	6	864
5	2	10	5	50	5	250
3	4	12	3	36	3	108
0	1	0	0	0	0	0
		$68 = \Sigma_1$		$316 = \Sigma_2$		$1580 = \Sigma_3$

Referring to Figure 36.7:

$$CG = \frac{\Sigma_2}{\Sigma_1} \times \frac{1}{2}$$

$$= \frac{316}{68} \times \frac{1}{2}$$

$$= 2.324 \text{ m}$$

$$CD = \underline{2.000 \text{ m}}$$

$$DG = \underline{4.324 \text{ m}} = \bar{z}$$

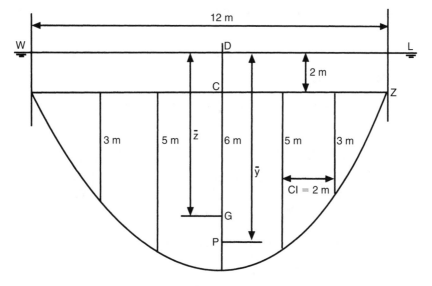

Fig. 36.7

$$I_{OZ} = \tfrac{1}{9} \times CI \times \Sigma_3$$

$$= \tfrac{1}{9} \times 2 \times 1580 = 351 \text{ m}^4$$

$$I_{CG} = I_{OZ} - A(CG)^2$$

$$I_{WL} = I_{CG} + A\bar{z}^2$$

$$= I_{OZ} - A(CG)^2 + A\bar{z}^2$$

$$= I_{OZ} - A(CG^2 - \bar{z}^2)$$

$$= 351 - 45.33(2.324^2 - 4.324^2)$$

$$I_{WL} = 953.75 \text{ m}^4$$

$$\bar{y} = \frac{I_{WL}}{A\bar{z}}$$

$$= \frac{953.75}{45.33 \times 4.324}$$

$$\bar{y} = 4.87 \text{ m}$$

Ans. The centre of pressure is 4.87 m below the waterline.

Summary

When using Simpson's Rules to estimate the area of a bulkhead under liquid pressure together with the VCG and centre of pressure the procedure should be as follows:

1. Make a sketch from the given information.
2. Make a table and insert the relevant ordinates and multipliers.
3. Calculate the area of bulkhead's plating.
4. Estimate the ship's VCG below the stipulated datum level.
5. Using the parallel axis theorem, calculate the requested centre of pressure.
6. Remember: sketch, table, calculation.

Exercise 36

1 A fore-peak tank bulkhead is 7.8 m deep. The widths at equidistant intervals from its upper edge to the bottom are as follows:

16, 16.6, 17, 17.3, 16.3, 15.3 and 12 m, respectively

Find the load on the bulkhead and the depth of the centre of pressure below the top of the bulkhead when the fore peak is filled with salt water to a head of 1.3 m above the crown of the tank.

2 A deep tank transverse bulkhead is 30 m deep. Its width at equidistant intervals from the top to the bottom is:

20, 20.3, 20.5, 20.7, 18, 14 and 6 m, respectively

Find the depth of the centre of pressure below the top of the bulkhead when the tank is filled to a head of 4 m above the top of the tank.

3 The transverse end bulkhead of a deep tank is 18 m wide at its upper edge. The vertical depths of the bulkhead at equidistant intervals across it are as follows:

0, 3.3, 5, 6, 5, 3.3 and 0 m, respectively

Find the depth of the centre of pressure below the top of the bulkhead when the tank is filled with salt water to a head of 2 m above the top of the bulkhead. Find also the load on the bulkhead.

4 A fore-peak bulkhead is 18 m wide at its upper edge. Its vertical depth at the centreline is 3.8 m. The vertical depths on each side of the centreline at 3 m intervals are 3.5, 2.5 and 0.2 m, respectively. Calculate the load on the bulkhead and the depth of the centre of pressure below the top of the bulkhead when the fore-peak tank is filled with salt water to a head of 4.5 m above the top of the bulkhead.

5 The vertical ordinates across the end of a deep tank transverse bulkhead measured downwards from the top at equidistant intervals, are:

$$4, 6, 8, 9.5, 8, 6 \text{ and } 4\,\text{m, respectively}$$

Find the distance of the centre of pressure below the top of the bulkhead when the tank is filled with salt water.

6 A square plate of side 'a' is vertical and is immersed in water with an edge of its length in the free surface. Prove that the distance between the centres of pressure of the two triangles into which the plate is divided by a diagonal, is $\frac{a\sqrt{13}}{8}$.

Chapter 37
Ship squat in open water and in confined channels

What exactly is ship squat?

When a ship proceeds through water, she pushes water ahead of her. In order not to have a 'hole' in the water, this volume of water must return down the sides and under the bottom of the ship. The streamlines of return flow are speeded up under the ship. This causes a drop in pressure, resulting in the ship dropping vertically in the water.

As well as dropping vertically, the ship generally trims forward or aft. The overall decrease in the static underkeel clearance, forward or aft, is called ship squat. It is *not* the difference between the draughts when stationary and the draughts when the ship is moving ahead.

If the ship moves forward at too great a speed when she is in shallow water, say where this static even-keel underkeel clearance is 1.0 to 1.5 m, then grounding due to excessive squat could occur at the bow or at the stern.

For full-form ships such as supertankers or OBO vessels, grounding will occur generally at the *bow*. For fine-form vessels such as passenger liners or container ships the grounding will generally occur at the stern. This is assuming that they are on even keel when stationary. It must be *generally*, because in the last two decades, several ship types have tended to be shorter in LBP and wider in breadth moulded. This has led to reported groundings due to ship squat at the bilge strakes at or near to amidships when slight rolling motions have been present.

Why has ship squat become so important in the last 40 years? Ship squat has always existed on smaller and slower vessels when underway. These squats have only been a matter of centimetres and thus have been inconsequential.

However, from the mid-1960s and into the new millennium, ship size has steadily grown until we have supertankers of the order of 350 000 tonnes dwt and above. These supertankers have almost outgrown the ports they visit, resulting in small static even-keel underkeel clearances of 1.0 to 1.5 m.

Alongside this development in ship size has been an increase in service speed on several ships, for example container ships, where speeds have gradually increased from 16 knots up to about 25 knots.

Ship design has seen tremendous changes in the 1980s, 1990s and 2000s. In oil tanker design we have the *Jahre Viking* with a dwt of 564 739 tonnes and an LBP of 440 m. This is equivalent to the length of five football pitches. In 2002, the biggest container ship to date, the 'Hong Kong Express' came into service with a dwt of 82 800 tonnes, a service speed of 25.3 kts, an LBP of 304 m, Br. Mld of 42.8 m and a draft moulded of 13 m.

As the static underkeel clearances have decreased and as the service speeds have increased, ship squats have gradually increased. They can now be of the order of 1.50 to 1.75 m, which are of course by no means inconsequential.

Recent ship groundings

To help emphasise the dangers of excessive squat only one has to remember the grounding of these *nine* vessels in recent years.

Herald of Free Enterprise	Ro-Ro vessel at Zeebrugge	06/03/1987
QE11	Passenger liner at Massachusetts	07/08/1992
Sea Empress	Supertanker at Milford Haven	15/02/1996
Napoleon Bonaparte	Passenger liner at Marseille	05/02/1999
Don Raul	37 000 t bulk carrier at Pulluche, Chile	31/03/2001
Capella Voyager	Oil tanker at Marsden Point, N. Zealand	14/04/2003
Tasman Spirit	87 500 t oil tanker at Karachi Harbour	27/07/2003
Thor Guardian	Timber ship at River Plate	24/06/2004
Iran Noor	Oil tanker at Ningbo Port, China	05/07/2004

I currently, have a database of 44 vessels that have recently gone aground attributable to excessive ship squat and other reasons. There may well be others not officially reported.

In the United Kingdom, over the last 20 years the D.Tp. have shown their concern by issuing four 'M' notices concerning the problems of ship squat and accompanying problems in shallow water. These alert all mariners to the associated dangers.

Signs that a ship has entered shallow water conditions can be one or more of the following:

1. Wave making increases, especially at the forward end of the ship.
2. Ship becomes more sluggish to manoeuvre. A pilot's quote, 'almost like being in porridge'.

3. Draught indicators on the bridge or echo-sounders will indicate changes in the end draughts.
4. Propeller rpm indicator will show a decrease. If the ship is in 'open water' conditions, i.e. without breadth restrictions, this decrease may be up to 15 per cent of the service rpm in deep water. If the ship is in a confined channel, this decrease in rpm can be up to 20 per cent of the service rpm.
5. There will be a drop in speed. If the ship is in open water conditions this decrease may be up to 35 per cent. If the ship is in a confined channel such as a river or a canal then this decrease can be up to 75 per cent.
6. The ship may start to vibrate suddenly. This is because of the entrained water effects causing the natural hull frequency to become resonant with another frequency associated with the vessel.
7. Any rolling, pitching and heaving motions will all be reduced as the ship moves from deep water to shallow water conditions. This is because of the cushioning effects produced by the narrow layer of water under the bottom shell of the vessel.
8. The appearance of mud could suddenly show in the water around the ship's hull say in the event of passing over a raised shelf or a submerged wreck.
9. Turning circle diameter (TCD) increases. TCD in shallow water could increase 100 per cent.
10. Stopping distances and stopping times increase, compared to when a vessel is in deep waters.
11. Effectiveness of the rudder decreases.

What are the factors governing ship squat?

The main factor is ship speed V_k. Squat varies approximately with the speed squared. In other words, we can take as an example that if we halve the speed we quarter the squat. In this context, speed V_k is the ship's speed relative to the water; in other words, effect of current/tide speed with or against the ship must be taken into account.

Another important factor is the block coefficient C_b. Squat varies directly with C_b. Oil tankers will therefore have comparatively more squat than passenger liners.

The blockage factor 'S' is another factor to consider. This is the immersed cross-section of the ship's midship section divided by the cross-section of water within the canal or river see Figure 37.1. If the ship is in open water the width of influence of water can be calculated. This ranges from about 8.25b for supertankers, to about 9.50b for general cargo ships, to about 11.75 ship-breadths for container ships.

The presence of another ship in a narrow river will also affect squat, so much so, that squats can *double in value* as they pass/cross the other vessel.

Formulae have been developed that will be satisfactory for estimating maximum ship squats for vessels operating in confined channels and in open water conditions. These formulae are the results of analysing about 600 results some measured on ships and some on ship models. Some of the empirical formulae developed are as follows:

Let

$$b = \text{breadth of ship}$$

$$B = \text{breadth of river or canal}$$

$$H = \text{depth of water}$$

$$T = \text{ship's even-keel static draft}$$

$$C_b = \text{block co-efficient}$$

$$V_k = \text{ship speed relative to the water or current}$$

$$\text{CSA} = \text{cross-sectional area (see Figure 37.1)}$$

Let

$$S = \text{blockage factor} = \text{CSA of ship/CSA of river or canal}$$

If ship is in open water conditions, then the formula for B becomes

$$B = \{7.7 + 20(1 - C_b)^2\} \cdot b, \text{ known as the 'width of influence'}$$

$$\text{Blockage factor} = S = \frac{b \times T}{B \times H}$$

$$\text{Maximum squat} = \delta_{max}$$

$$= \frac{C_b \times S^{0.81} \times V_k^{2.08}}{20} \text{ metres, for } open\ water$$
$$\text{and } confined\ channels$$

Two short-cut formulae relative to the previous equation are:

$$\text{Maximum squat} = \frac{C_b \times V_k^2}{100} \text{ metres for } open\ water \text{ conditions } only,$$

with H/T of 1.1 to 1.4.

$$\text{Maximum squat} = \frac{C_b \times V_k^2}{50} \text{ metres for } confined\ channels,$$

where S = 0.100 to 0.265.

A worked example showing how to predict maximum squat and how to determine the remaining underkeel clearance is shown at the end of this chapter. It shows the use of the more detailed formula and then compares the answer with the short-cut method.

These formulae have produced several graphs of maximum squat against ship's speed V_k. One example of this is in Figure 37.2, for a 250 000 t. dwt supertanker. Another example is in Figure 37.3, for a container vessel having shallow water speeds up to 18 knots.

Figure 37.4 shows the maximum squats for merchant ships having C_b values from 0.500 up to 0.900, in open water and in confined channels. Three items of information are thus needed to use this diagram. First, an idea of the ship's C_b value, secondly the speed V_k and thirdly to decide if the ship is in open water or in confined river/canal conditions. A quick graphical prediction of the maximum squat can then be made.

In conclusion, it can be stated that if we can predict the maximum ship squat for a given situation then the following advantages can be gained:

1. The ship operator will know which speed to reduce to in order to ensure the safety of his/her vessel. This could save the cost of a very large repair bill. It has been reported in the technical press that the repair bill for the *QEII* was *$13 million*, plus an estimation for lost passenger booking of *$50 million*!!

   In Lloyd's lists, the repair bill for the *Sea Empress* had been estimated to be in the region of *$28 million*. In May 1997, the repairs to the *Sea Empress* were completed at Harland & Wolff Ltd of Belfast, for a reported cost of *£20 million*. Rate of exchange in May 1997 was the order of £1 = $1.55. She was then renamed the *Sea Spirit*.

2. The ship officers could load the ship up an extra few centimetres (except of course where load-line limits would be exceeded). If a 100 000 tonne dwt tanker is loaded by an extra 30 cm or an SD14 general cargo ship is loaded by an extra 20 cm, the effect is an extra 3 per cent onto their dwt. This gives these ships extra earning capacity.

3. If the ship grounds due to excessive squatting in shallow water, then apart from the large repair bill, there is the time the ship is 'out of service'. Being 'out of service' is indeed very costly because loss of earnings can be greater than *£100 000* per *day*.

4. When a vessel goes aground there is always a possibility of leakage of oil resulting in compensation claims for oil pollution and fees for cleanup operations following the incident. These costs eventually may have to be paid for by the shipowner.

These last four paragraphs illustrate very clearly that not knowing about ship squat can prove to be very costly indeed. Remember, in a marine court hearing, ignorance is not acceptable as a legitimate excuse!!

Summarising, it can be stated that because maximum ship squat can now be predicted, it has removed the 'grey area' surrounding the phenomenon. In the past ship pilots have used 'trial and error', 'rule of thumb' and years of experience to bring their vessels safely in and out of port.

Empirical formulae quoted in this study, modified and refined by the author over a period of 33 years' research on the topic, give *firm guidelines*. By maintaining the ship's trading availability a shipowner's profit margins

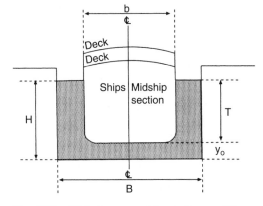

Fig. 37.1   Ship in a canal in static condition.

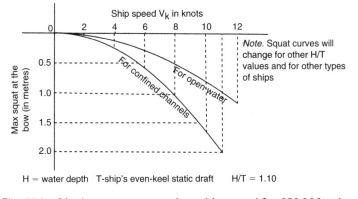

H = water depth   T-ship's even-keel static draft   H/T = 1.10

Fig. 37.2   Maximum squats against ship speed for 250 000 t. dwt
supertanker.

are not decreased. More important still, this research can help prevent loss of life as occurred with the *Herald of Free Enterprise* grounding.

It should be remembered that the quickest method for reducing the danger of grounding due to ship squat is to *reduce the ship's speed*. 'Prevention is better than cure' and *much cheaper*.

$$A_S = \text{cross-section of ship at amidships} = b \times T$$
$$A_C = \text{cross-section of canal} = B \times H$$
$$\text{Blockage factor} = S = \frac{A_S}{A_C} = \frac{b \times T}{B \times H}$$
$$y_o = \text{Static underkeel clearance}$$

$\dfrac{H}{T}$ range is 1.10 to 1.40.

Blockage factor range is 0.100 to 0.265.

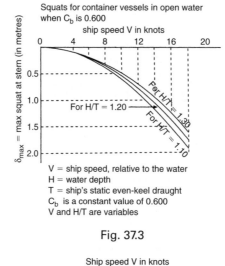

V = ship speed, relative to the water
H = water depth
T = ship's static even-keel draught
C_b is a constant value of 0.600
V and H/T are variables

Fig. 37.3

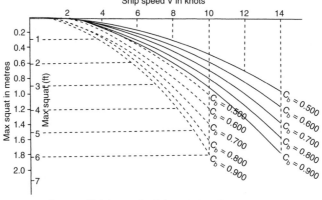

--- Denotes ship is in a confined channel where S = 0.100 to 0.265
— Denotes ship is in open water, where H/T = 1.10 to 1.40

Ship type	Typical C_b, fully loaded	Ship type	Typical C_b, fully loaded
ULCC	0.850	General cargo	0.700
Supertanker	0.825	Passenger liner	0.575 to 0.625
Oil tanker	0.800	Container ship	0.575
Bulk carrier	0.775 to 0.825	Coastal tug	0.500

Fig. 37.4   Maximum ship squats in confined channels and in open water conditions.

Width of influence $= F_B = \dfrac{\text{Equivalent 'B'}}{b}$ in open water.

V_k = speed of ship relative to the water, in knots.

'B' $= 7.7 + 20(1 - C_b)^2$ ship breadths

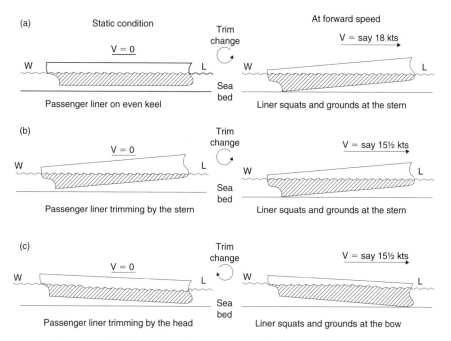

Fig. 37.5   The link between static trim and location of maximum squat.

Ship squat for ships with static trim

Each ship to this point has been assumed to be on *even keel* when static. For a given forward speed the maximum ship squat has been predicted. Based on the C_b, the ship will have this maximum squat at the bow, at the stern or right away along the length of the ship. See Figure 37.5(a).

However, some ships will have *trim by the bow* or *trim by the stern* when they are stationary. This static trim will decide where the maximum squat will be located when the ship is underway.

Tests on ship-models and from full size squat measurements have shown that:

1. If a ship has static trim by the stern when static, then when underway she will have a maximum squat (as previously estimated) at the stern. The ship will have dynamic trim in the same direction as the static trim. In other words, when underway she will have increased trim and could possibly go aground at the stern. See Figure 37.5(b).

    This is due to streamlines under the vessel at the stern moving faster than under the vessel at the bow. Cross-sectional area is less at the stern than under the bow and this causes a greater suction at the stern. Vessel trims by the stern. In hydraulics, this is known as the Venturi effect.

2. If a ship has static trim by the bow when static, then when underway she will have a maximum squat (as previously estimated) at the bow. The ship will have dynamic trim in the same direction as the static trim. See Figure 37.5(c). In other words, when underway she will have increased trim and

could possibly go aground at the bow. The *Herald of Free Enterprise* grounding is a prime example of this trimming by the bow when at forward speed.

Note of caution. Some masters on oil tankers trim their vessels by the stern before going into shallow waters. They believe that full-form vessels trim by the bow when underway. In doing so they believe that the ship will level out at even keel when at forward speed. This does not happen!! Maximum squat at the bow occurs when the tankers are on even keel when static, or when they have trim by the bow when static.

Worked example – ship squat for a supertanker

Question:

A supertanker operating in open water conditions is proceeding at a speed of 11 knots. Her $C_b = 0.830$, static even-keel draft $= 13.5\,m$ with a static under-keel clearance of $2.5\,m$. Her breadth moulded is $55\,m$ with LBP of $320\,m$.

Calculate the maximum squat for this vessel at the given speed via *two* methods, and her remaining ukc (underkeel clearance) at V_k of 11 kts.

Answer:

$$\text{Width of influence} = \{7.7 + 20(1 - C_b)^2\} \times b = \text{'B'}$$
$$\therefore \ \text{'B'} = \{7.7 + 20(1 - 0.830)^2\} \times 55$$

$$\therefore \ \underline{\text{'B'} = 455\,m}$$

i.e. artificial boundaries in open water or wide rivers

$$\text{Blockage factor, S} = \frac{b \times T}{\text{'B'} \times H} = \frac{55 \times 13.5}{455 \times (13.5 + 2.5)} = \underline{0.102}$$
$$\text{(water depth)}$$

Method 1

$$\text{Maximum squat} = \frac{C_b \times S^{0.81} \times V_k^{2.08}}{20} = \delta_{max}$$

$$\therefore \ \delta_{max} = 0.830 \times 0.102^{0.81} \times 11^{2.08} \times \frac{1}{20}$$

$$\therefore \ \underline{\delta_{max} = 0.96\,m}$$

at the bow, because $C_b > 0.700$.

Method 2
Simplified approx. formula

$$\delta_{max} = \frac{C_b \times V_k^2}{100}\ \text{metres}$$

$$\therefore \ \delta_{max} = \frac{0.830 \times 11^2}{100} = 1.00\,m$$

i.e. slightly above previous answer, so overpredicting on the *safe side*.

$$\text{Average } \delta_{max} = \frac{0.96 + 100}{2} = 0.98 \text{ m} \quad y_o = 2.50 \text{ m}$$

Hence, remaining underkeel clearance at bow $= y_o - \delta_{max} = y_2$.

$$\therefore \; y_2 = 2.500 - 0.98 = \underline{1.52 \text{ m}} \; @ \; V_k \text{ of 11 kts.}$$

Ship squat in wide and narrow rivers

As previously noted, when a ship is in shallow waters and at forward speed there is a danger she will go aground due to ship squat. The danger is greater when the ship is in a river or canal than when in open water. This is due to interaction effects with the adjacent banks and the sides of the moving vessel. The narrower the river, the greater will be the ship squats.

Because of erosion of riverbanks and interaction with moored ships in a river, the forward speeds of ships are a lot less than at sea or in open water conditions. Some port authorities request maximum speeds of only 4 knots in rivers. Others may allow slightly higher transit speeds, commensurate with ship size.

Another restriction often imposed on incoming ships is a minimum static ukc. Some port authorities require at least 10% of the ship's static mean draft. Others demand at least 1.0 m while others still can request a static ukc of at least 1.25 m.

In the interests of safety, I favour the minimum requirement to be irrespective of a ship's static draft and to be in units of metres. After all, 10% of a ballast draft is less than 10% of a draft in a fully loaded condition.

Note of caution. Whilst Figure 37.4 shows C_b values for fully loaded conditions, it must be realised that ships do go aground at drafts less than their draft moulded.

To predict maximum ship squat in river conditions, I have produced a diagram that involves 'K' coefficients. See Figure 37.6. For this study, the value of 'K' will range only from 1.0 to 2.0. If 'K' < 1 on the diagram, then use 'K' = 1. If 'K' is >2 on the diagram, then use 'K' = 2.

The parameters associated with this diagram are H/T and B/b, where:

H = depth of water in a rectangular cross-sectional shaped river (m)
T = ship's static even-keel draft (m)
B = breadth of water in a rectangular cross-sectional shaped river (m)
b = breadth moulded of ship in transit (m)

Assume first of all that each ship when stationary is on *even keel*. This appears to be a sensible pre-requisite prior to entering shallow waters.

In this final section I have concentrated on shallow waters where the static H/T ranges from 1.10 to 1.30. It is in this range of 10% to 30% ukc that there is greater chance of a ship going aground. At greater than 30% ukc, likelihood of touchdown is greatly decreased.

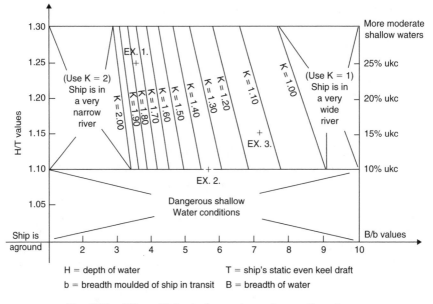

Fig. 37.6   'K' coefficients for rectangular section rivers.

Procedure for using Figure 37.6

The following procedure should be used:

(a) Determine and record the values of H, T, B and b.
(b) Calculate the values H/T and B/b.
(c) Enter Figure 37.6, and plot the intersection point of H/T with B/b. See the results of the final three worked examples.
(d) Using this intersection point and the 'K' contour lines, determine the value of 'K' appropriate to the river condition under consideration.

Having now obtained a value for 'K', the next step is to link it with two more variables. They are the block coefficient C_b and the ship speed V is measured in knots. C_b will depend on the ship being considered and her condition of loading. As shown in Chapter 9, C_b values will decrease with a decrease in draft.

V is the ship speed relative to the water. The tidal speed and the direction of current must always be taken into account on the bridge by the master, mate or ship pilot.

Calculation of maximum squat

To obtain the maximum squat for these ships in a river, the squat predicted for open water conditions must be multiplied by the previously determined 'K' value.

Hence

Maximum squat in a river or a canal = 'K' $\times$ ($C_b \times V^2/100$) metres

Worked example 1

A container ship has a C_b of 0.575 and is proceeding upriver at a speed (V) of 6 kts. This river of rectangular cross-section has an H/T of 1.25 and a B/b of 3.55. When static, the ship was on even keel. Calculate the maximum squat at this speed and where it will occur.

Using Figure 37.6, at the point of intersection of 1.25 and 3.55, the 'K' value lifted off is 1.752.

So

$$\text{Maximum squat} = \text{'K'} \times (C_b \times V^2/100) \text{ metres}$$
$$= 1.752 \times (0.575 \times 6^2/100) = 0.36 \text{ m.}$$

This 0.36 m will be located at the stern, because the $C_b < 0.700$ and the ship when static was on even keel.

Worked example 2

A general cargo ship has a C_b of 0.700 and is proceeding upriver at a speed (V) of 5 kts. This river of rectangular cross-section has an H/T of 1.10 and a B/b of 5.60. When static, the ship was on even keel. Calculate the maximum squat at this speed and where it will occur.

Using Figure 37.6, at the point of intersection of 1.10 and 5.60, the 'K' value lifted off is 1.374.

So

$$\text{Maximum squat} = \text{'K'} \times (C_b \times V^2/100) \text{ metres}$$
$$= 1.374 \times (0.700 \times 5^2/100) = 0.24 \text{ m}$$

This 0.24 m will be located from the stern to amidships to the bow, because the C_b is 0.700 and the ship when static was on even keel.

Worked example 3

A supertanker has a C_b of 0.825 and is proceeding upriver at a speed (V) of 7 kts. This river of rectangular cross-section has an H/T of 1.15 and a B/b of 7.25. When static, the ship had trim by the stern. Calculate the maximum squat at this speed and where it will occur.

Using Figure 37.6, at the point of intersection of 1.15 and 7.25, the 'K' value lifted off is 1.120.

So

$$\text{Maximum squat} = \text{'K'} \times (C_b \times V^2/100) \text{ metres}$$
$$= 1.120 \times (0.825 \times 7^2/100) = 0.45 \text{ m}$$

This 0.45 m will be located at the stern, because when static, this ship had trim by the stern.

Conclusions

By using this procedure, it becomes possible to quickly and accurately predict the maximum ship squat and its location. This procedure caters for all types of merchant ships operating in confined channels such as a river or in a canal. It will be suitable for typical speeds of transits of merchant ships operating in rivers of any size.

The golden rule again is, if there is danger of going aground, then simply reduce the ship speed.

Exercise 37

1 A container ship is operating in open water river conditions at a forward speed of 12.07 kts. Her C_b is 0.572 with a static even-keel draft of 13 m. Breadth moulded is 32.25 m. If the depth of water is 14.5 m; calculate the following:
   (a) Width of influence for this wide river.
   (b) Blockage factor.
   (c) Maximum squat, stating with reasoning where it occurs.
   (d) Dynamical underkeel clearance corresponding to this maximum squat.

2 A vessel has the following particulars:
   Br. Mld is 50 m. Depth of water in river is 15.50 m. C_b is 0.817. Width of water in river is 350 m. Static even-keel draft is 13.75 m.
   (a) Prove this ship is operating in a 'confined channel' situation.
   (b) Draw a graph of maximum squat against ship speed for a range of speeds up to 10 kts.
   (c) The pilot decides that the dynamical underkeel clearance is not to be less than 1.00 m.
   Determine graphically and mathematically the maximum speed of transit that this pilot must have in order to adhere to this pre-requisite.

3 A 75 000 tonne dwt oil tanker has the following particulars:
   Br.Mld is 37.25 m. Static even-keel draft is 13.5 m. C_b is 0.800. Depth of water is 14.85 m. Width of river is 186 m. Calculate the forward speed at which, due to ship squat, this vessel would just go aground at the bow.

4 A passenger liner has a C_b of 0.599. She is proceeding upriver at a speed (V) of 7.46 kts. This river of rectangular cross-section has an H/T of 1.25 and a B/b of 4.36. When static, ship was on even keel. Calculate her maximum squat at this speed and state whereabout it will occur. Use Figure 37.6.

Chapter 38
Interaction, including two case studies

What exactly is interaction?

Interaction occurs when a ship comes too close to another ship or too close to, say, a river or canal bank. As ships have increased in size (especially in breadth moulded), interaction has become very important to consider. In February 1998, the Marine Safety Agency (MSA) issued a Marine Guidance note 'Dangers of Interaction', alerting Owners, Masters, Pilots and Tug-Masters on this topic.

Interaction can result in one or more of the following characteristics:

1. If two ships are on a passing or overtaking situation in a river the squats of both vessels could be doubled when their amidships are directly in line.
2. When they are directly in line each ship will develop an angle of heel and the smaller ship will be drawn bodily towards the larger vessel.
3. Both ships could lose steerage efficiency and alter course without change in rudder helm.
4. The smaller ship may suddenly veer off course and head into the adjacent riverbank.
5. The smaller ship could veer into the side of the larger ship or worse still be drawn across the bows of the larger vessel, bowled over and capsized.

In other words there is:

(a) a ship to ground interaction,
(b) a ship to ship interaction,
(c) a ship to shore interaction.

What causes these effects of interaction? The answer lies in the pressure bulbs that exist around the hull form of a moving ship model or a moving ship (see Figure 38.1). As soon as a vessel moves from rest, hydrodynamics produce the shown positive and negative pressure bulbs. For ships with greater parallel body such as tankers these negative bulbs will be comparatively longer in length. When a ship is stationary in water of zero current speed these bulbs disappear.

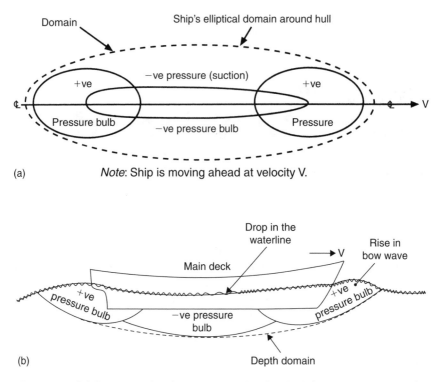

Fig. 38.1   (a) Pressure distribution around ship's hull (not drawn to scale).
(b) Pressure bulbs around a ship's profile when at forward speed.

Note the elliptical domain that encloses the vessel and these pressure bulbs. This domain is very important. When the domain of one vessel interfaces with the domain of another vessel then interaction effects will occur. Effects of interaction are increased when ships are operating in shallow waters.

Ship to ground (squat) interaction

In a report on measured ship squats in the St Lawrence seaway, A.D. Watt stated: 'meeting and passing in a channel also has an effect on squat. It was found that when two ships were moving at the low speed of 5 knots that squat increased up to double the normal value. At higher speeds the squat when passing was in the region of one and a half times the normal value.' Unfortunately, no data relating to ship types, gaps between ships, blockage factors, etc. accompanied this statement.

Thus, at speeds of the order of 5 knots the squat increase is +100 per cent whilst at higher speeds, say 10 knots, this increase is +50 per cent. Figure 38.2 illustrates this passing manoeuvre. Figure 38.3 interprets the percentages given in the previous paragraph.

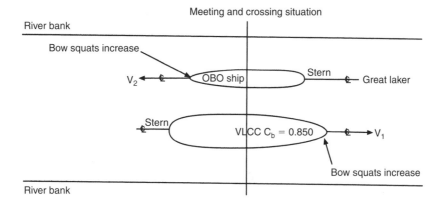

Fig. 38.2 Amidships (⊗) of VLCC directly in line with amidships of OBO ship in St Lawrence seaway.

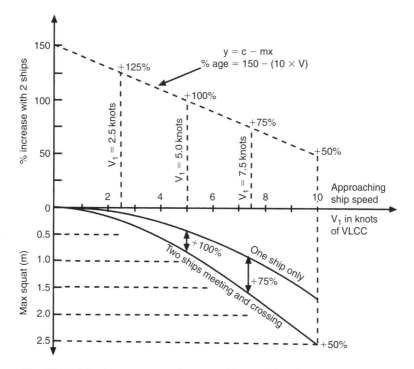

Fig. 38.3 Maximum squats for one ship, and for the same ship with another ship present.

How may these squat increases be explained? It has been shown in the chapter on Ship Squat that its value depends on the ratio of the ship's cross-section to the cross-section of the river. This is the blockage factor 'S'. The presence of a second ship meeting and crossing will of course increase the blockage factor. Consequently the squat on each ship will increase.

Maximum squat is calculated by using the equation:

$$\delta_{max} = \frac{C_b \times S^{0.81} \times V_k^{2.08}}{20} \text{ metres}$$

Consider the following example.

Example 1

A supertanker has a breadth of 50 m with a static even-keel draft of 12.75 m. She is proceeding along a river of 250 m and 16 m depth rectangular cross-section. If her speed is 5 kts and her C_b is 0.825, calculate her maximum squat when she is on the centreline of this river.

$$S = \frac{b \times T}{B \times H} = \frac{50 \times 12.75}{250 \times 16} = 0.159$$

$$\delta_{max} = \frac{0.825 \times 0.159^{0.81} \times 5^{2.08}}{20} = 0.26 \text{ m}$$

Example 2

Assume now that this supertanker meets an oncoming container ship also travelling at 5 kts (see Figure 38.4). If this container ship has a breadth of 32 m a C_b of 0.580, and a static even-keel draft of 11.58 m, calculate the maximum squats of both vessels when they are transversely in line as shown.

$$S = \frac{(b_1 \times T_1) + (b_2 \times T_2)}{B \times H}$$

$$S = \frac{(50 \times 12.75) + (32 \times 11.58)}{250 \times 16} = 0.252$$

Supertanker:

$$\delta_{max} = \frac{0.825 \times 0.252^{0.81} \times 5^{2.08}}{20}$$

$$= 0.38 \text{ m at the bow}$$

Container ship:

$$\delta_{max} = \frac{0.580 \times 0.252^{0.81} \times 5^{2.08}}{20}$$

$$= 0.27 \text{ m at the stern}$$

The maximum squat of 0.38 m for the supertanker will be at the bow because her C_b is greater than 0.700. Maximum squat for the container ship will be at the stern, because her C_b is less than 0.700. As shown this will be 0.27 m.

If this container ship had travelled alone on the centreline of the river then her maximum squat at the stern would have only been 0.12 m. Thus the presence of the other vessel has more than doubled her squat.

Clearly, these results show that the presence of a second ship does increase ship squat. Passing a moored vessel would also make blockage effect and

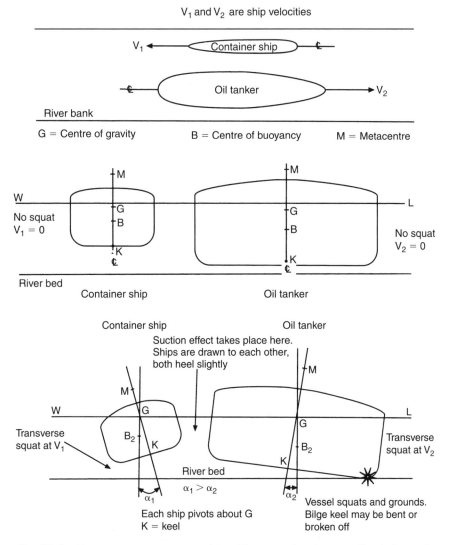

Fig. 38.4 Transverse squat caused by ships crossing in a confined channel.

squat greater. These values are not qualitative but only illustrative of this phenomenon of interaction in a ship to ground (squat) situation. Nevertheless, they are supportive of A.D. Watt's statement.

Ship to ship interaction

Consider Figure 38.5 where a tug is overtaking a large ship in a narrow river. Three cases have been considered:

Case 1. The tug has just come up to aft port quarter of the ship. The domains have become in contact. Interaction occurs. The positive bulb of the ship

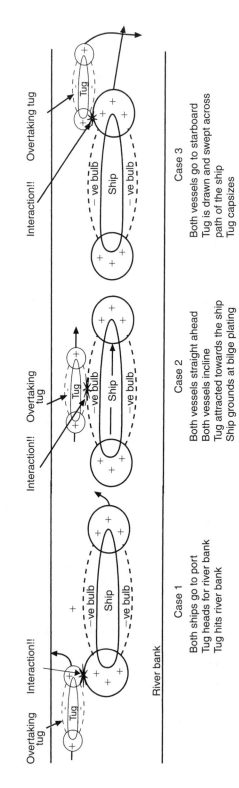

Fig. 38.5 Ship to ship interaction in a narrow river during an overtaking manoeuvre.

reacts with the positive bulb of the tug. Both vessels veer to port side. Rate of turn is greater on the tug. There is a possibility of the tug veering off into the adjacent riverbank as shown in Figure 38.5.

Case 2. The tug is in danger of being drawn bodily towards the ship because the negative pressure (suction) bulbs have interfaced. The bigger the differences between the two deadweights of these ships the greater will be this transverse attraction. Each ship develops an angle of heel as shown. There is a danger of the ship losing a bilge keel or indeed fracture of the bilge strakes occurring. This is 'transverse squat', the loss of underkeel clearance at forward speed. Figure 38.4 shows this happening with the tanker and the container ship.

Case 3. The tug is positioned at the ship's forward port quarter. The domains have become in contact via the positive pressure bulbs (see Figure 38.5). Both vessels veer to the starboard side. Rate of turn is greater on the tug. There is great danger of the tug being drawn across the path of the ship's heading and bowled over. This has actually occurred with resulting loss of life.

Note how in these three cases that it is the smaller vessel, be it a tug, a pleasure craft or a local ferry involved, that ends up being the casualty!!

Figures 38.6 and 38.7 give further examples of ship to ship interaction effects in a river.

Methods for reducing the effects of interaction in Cases 1 to 5

Reduce speed of both ships and then if safe increase speeds after the meeting crossing manoeuvre time slot has passed. Resist the temptation to go for the order 'increase revs'. This is because the forces involved with interaction vary as the *speed squared*. However, too much a reduction in speed produces a loss of steerage because rudder effectiveness is decreased. This is even more so in shallow waters, where the propeller rpm decrease for similar input of deep water power. Care and vigilance are required.

Keep the distance between the vessels as large as practicable bearing in mind the remaining gaps between each ship side and nearby riverbank.

Keep the vessels from entering another ship's domain, for example crossing in wider parts of the river.

Cross in deeper parts of the river rather than in shallow waters, bearing in mind those increases in squat.

Make use of rudder helm. In Case 1, starboard rudder helm could be requested to counteract loss of steerage. In Case 3, port rudder helm would counteract loss of steerage.

Ship to shore interaction

Figures 38.8 and 38.9 show the ship to shore interaction effects. Figure 38.8 shows the forward positive pressure bulb being used as a pivot to bring a ship alongside a riverbank.

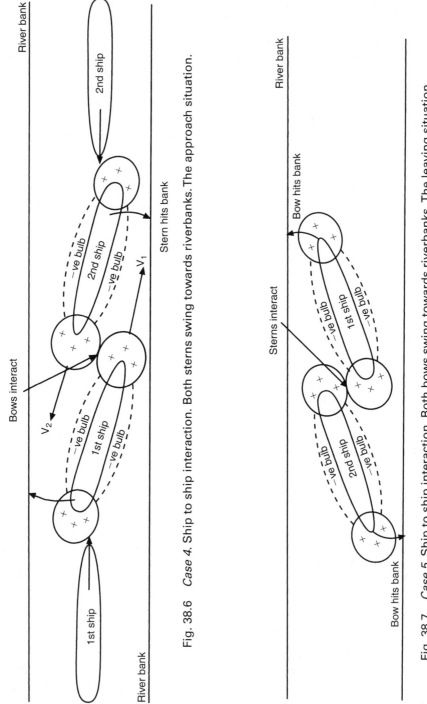

Fig. 38.6   *Case 4.* Ship to ship interaction. Both sterns swing towards riverbanks. The approach situation.

Fig. 38.7   *Case 5.* Ship to ship interaction. Both bows swing towards riverbanks. The leaving situation.

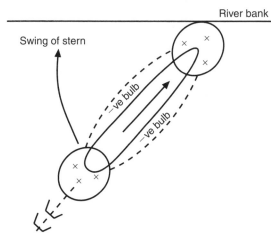

Fig. 38.8   Ship to bank interaction. Ship approaches slowly and pivots on forward positive pressure bulb.

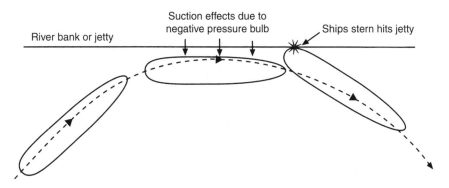

Fig. 38.9   Ship to bank interaction. Ship comes in at too fast a speed. Interaction causes stern to swing towards riverbank and then hits it.

Figure 38.9 shows how the positive and negative pressure bulbs have caused the ship to come alongside and then to veer away from the jetty. Interaction could in this case cause the stern to swing and collide with the wall of this jetty.

Summary

An understanding of the phenomenon of interaction can avert a possible marine accident. Generally a reduction in speed is the best preventive procedure. This could prevent on incident leading to loss of sea worthiness, loss of income for the shipowner, cost of repairs, compensation claims and maybe loss of life.

A collision due to interaction? An example of directional stability

CASE STUDY 1

On 23 October 1970, the *Pacific Glory* (a 77 648 tonne dwt oil tanker) collided with the *Allegro* (a 95 455 tonne dwt oil tanker). The collision took place just south of the Isle of Wight.

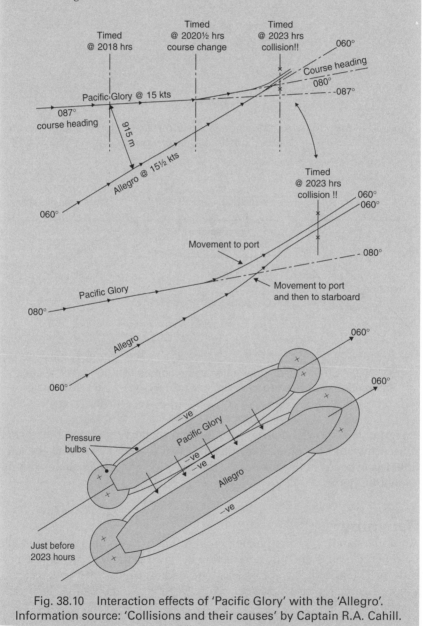

Fig. 38.10   Interaction effects of 'Pacific Glory' with the 'Allegro'.
Information source: 'Collisions and their causes' by Captain R.A. Cahill.

Both vessels were travelling from west to east up the English Channel. It was approximately 2000 hrs as the vessels approached each other. It was a clear night with good visibility of about 10 nautical miles. Both oil tankers had full loads of cargo oil.

The *Pacific Glory* on transit from Nigeria to Rotterdam, had been to Brixham to pick up a North Sea pilot. This vessel had a service speed of 15 kts and was fitted with diesel machinery. It was on a course of 087° that changed at 2020½ hrs to a heading of 080°.

The *Allegro* had been on route from Libya to the Fawley Oil Terminal at Southampton. It had a service speed of 15.5 kts, was fitted with steam turbine machinery and was proceeding on a heading of 060°. The *Allegro* was larger, longer and faster than the *Pacific Glory*.

At 2018 hrs, the vessels were 915 m apart. Figure 38.10 shows the converging courses of these two ships prior to their collision at 2023 hrs. The top diagram shows the movements in the 5 minutes before the collision. The middle diagram provides details 1 to 2 minutes before the collision. The lower diagram shows the *Pacific Glory* being drawn towards the larger *Allegro*.

Just moments before the collision, the vessels must have been very nearly on parallel headings, each one close to 060°.

The three diagrams in Figure 38.10 clearly show belated attempts to steer clear of one another. Interaction between the parallel bodies of the vessels caused them to attract and then to collide.

CASE STUDY 2

On 11 May 1972 at 0500 hrs, the *Royston Grange* (a 10 262 tonne dwt reefer) collided with the *Tien Chee* (a 19 700 tonne dwt Liberian oil tanker). The collision took place in the Indio Channel of Montevideo in the River Plate.

The *Royston Grange* was outward bound, on route to Montevideo and then on to London. It had a static mean draft of 7 m and a trim by the stern of 0.30 m. On board were grains, butter and refrigerated meats (see Figure 38.11(a)).

The *Tien Chee* was inward bound, on route to Buenos Aires and had a mean draft of 9 m with a trim of 0.30 m by the stern. It was almost fully loaded with cargo oil and forward speed just prior to the collision was 12 kts.

Both vessels were travelling at forward speed and each had their pilot aboard. Both vessels were also on an approaching manoeuvre, passing port to port, and were in shallow waters. They were both in the southern part of the channel and it was at low tide.

Just after their bows crossed, the *Royston Grange* suddenly veered to port and crashed into the amidships port side of the *Tien Chee*. The photograph, showing extensive bow damage to the reefer (see Figure 38.11(b)) suggests that the vessels must have been travelling towards each other at a fast speed.

Analysis and investigation into the collision has drawn the following conclusions:

1. The *Royston Grange* hit the *Tien Chee* at an angle of 40° (see Figure 38.12). The first point of contact was at No. 7 cargo wing tank port. Damage then occurred at Nos. 8, 9 and 10 wing tanks of the *Tien Chee*. As shown in the

Fig. 38.11(a)

Royston Grange

Damaged
bow

Fig. 38.11(b)

photograph substantial structural damage to the bow structure of the *Royston Grange* occurred and 800 tonnes of oil were spilled.

2. The pilot on the bridge of the *Tien Chee* made a claim to the Master on 11/5/72, that there would be an extra 0.60 m of water than was shown for low water. He convinced the ship's Master that this would be sufficient to proceed upriver without the need to wait and the Master was persuaded.

3. The *Tien Chee* had been delayed and was 3 hrs late in arriving at the river. The Master was in a hurry to take the ship upriver because of this delay, resulting in human error of judgement.

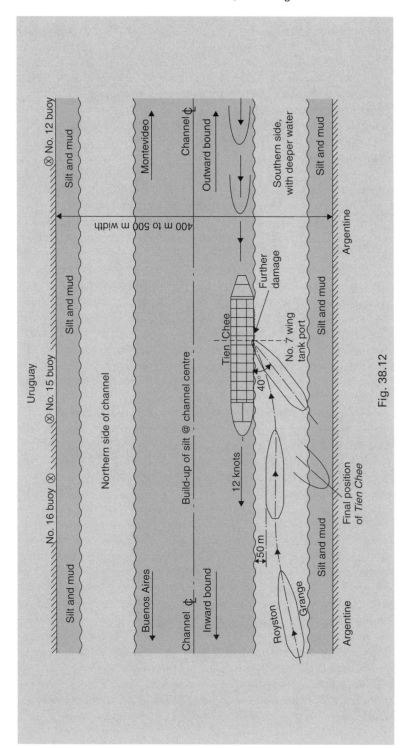

Fig. 38.12

4. The approach speed of the *Tien Chee* was too fast. This can be considered another human error of judgement.
5. The Tien Chee was ploughing through silt and mud on the riverbed.
6. At 12 kts speed the *Tien Chee* had squatted about 0.90 m at the stern and would have increased the trim by the stern. The total trim would then be about 0.66 m, thereby drawing the stern further into the mud.
7. The bow of the *Royston Grange* was drawn into the port side of the *Tien Chee* (see Figure 38.12).
8. Just before the collision both vessels were drawn as close as 50 m on a parallel course. Much too close!!
9. The negative pressure bulb of the *Tien Chee's* side-shell was much larger than the negative pressure bulb of the *Royston Grange's* side-shell.
10. If the forward speed (not given) of the *Royston Grange* had been 6 to 10 kts (average 8 kts) then the closing speed would have been a very dangerous speed of 20 kts (see Figure 38.12).
11. A northerly wind present at the time of the collision would have decreased the water depth. This would have increased squat and interaction.
12. The approach manoeuvre took place in the southern part of this channel in a confined channel condition (see Figure 38.13).
13. The build up of silt, due to poor local dredging arrangements, meant there was an appreciable build up of mud and silt on the riverbed. The bottom saucers of silt had been built up by the transit of many vessels going up and downriver. This silt and mud would increase the blockage factor and would increase squat and interaction effects.
14. When the vessel collided it was at low tide and was in a narrow part of the river. In these conditions, rudder helm is less effective.
15. The deadweight of the *Tien Chee* was 19 700 t. The deadweight of the *Royston Grange* was only 10 262 t. This gives a ratio of almost 2:1. The greater the difference of deadweight between ships in this situation, the greater will be the effects of interaction.
16. The mean drafts of the *Tien Chee* and the *Royston Grange* were in the ratio of about 1.25:1.00. Interaction increases with greater ratios of drafts, so this would definitely contribute to the collision.
17. The maintenance (dredging arrangements) by the Argentine and Uruguay port authorities left a lot to be desired. To say the least, it allegedly was very suspect (see Figure 38.13).
18. The *Tien Chee* was loaded so much so that there was evidence of transitting through mud when at 12 kts forward speed. The ship would have experienced the greater squat effects with its H/T of 1.10 (see Figure 38.13).
19. In this river the greatest water depth is not on the centreline of the channel but rather in the southern part of the channel (see Figure 38.13).
20. The *Royston Grange* was turned to port because the large negative pressure bulb of the *Tien Chee* was moving upriver whilst the smaller negative pressure bulb was moving downriver and at the same time towards it. This set

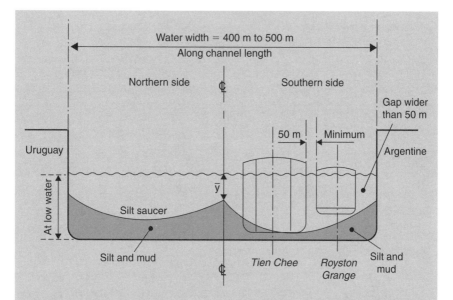

Observations

1. Note how $\bar{y}$ at channel ₵ is *not* the deepest water depth.
2. Water depth in Southern side is deeper than in Northern side.
3. When on parallel courses, these vessels came to being only 50 m apart.
4. Silt 'saucers' were scoured out by many vessels in transit along this channel.
5. *Tien chee* was travelling through silt and mud.
6. *Royston Grange* was very much in a confined channel situation.

Fig. 38.13   Cross-section through the Indio Channel of Montevideo.

up an anticlockwise couple that moved the *Royston Grange* from being parallel, to being at 40° angle to the *Tien Chee* (see Figure 38.14).

21. The parallel body of the *Tien Chee* would have been 50% to 60% of its LBP. This represents a long wall of moving water going upriver at 12 kts, almost like a moving riverbank. The greater the percentage, the greater the effect of interaction.

Due to the butter cargo the *Royston Grange* was carrying, the oil vapours from the *Tien Chee* and the sparking from the collision contact, the reefer was incinerated. Both vessels were later scrapped.

The total complement (74) of people aboard the *Royston Grange* lost their lives. Eight people aboard the *Tien Chee* also died.

Summary and conclusions

These case studies on interaction effects has clearly shown the perils of allowing the domain of one ship to enter the domain of another ship.

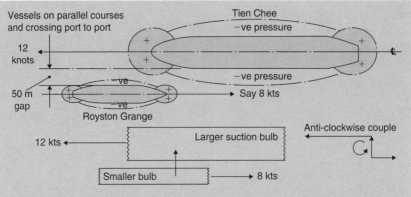

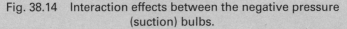

Observations

1. Interaction effects would exist between the port negative bulbs of the vessels.
2. The Royston Grange would be drawn bodily towards the Tien Chee, because she was much smaller than the Tien Chee
3. The ACW couple would also cause the Royston Grange to veer ACW and collide with the Tien Chee at the reported angle of 40°.

Fig. 38.14   Interaction effects between the negative pressure (suction) bulbs.

If one ship is *overtaking* another ship in a narrow river and the amidships are coming in line then it is possible that the smaller vessel will be drawn bodily towards the other vessel.

However, if the vessels are on an *approaching* manoeuvre and transversely close together, then as well as this suction effect the smaller ship may suddenly veer, turn and then run into the side of the other vessel.

Exercise 38

1   A river is 150 m wide and has a 12 m depth of water. A passenger liner having a breadth of 30 m, a static even-keel draft of 10 m and a C_b of 0.625 is proceeding along this river at 8 kts. It meets an approaching general cargo vessel having a breadth of 20 m, a static even-keel draft of 8 m and a C_b of 0.700, moving at 7 kts.

   Estimate the maximum squats for each vessel when their amidships are transversely in line.

2   With the aid of sketches, define interaction and describe how its effects may be reduced. Show clearly how interaction and transverse squat are inter-related.

3   With the aid of sketches, show the domains of a moving ship in plan view and in profile view.

Chapter 39
Heel due to turning

When a body moves in a circular path there is an acceleration towards the centre equal to v^2/r where v represents the velocity of the body and r represents the radius of the circular path. Theforce required to produce this acceleration, called a 'Centripetal' force, is equal to $\dfrac{Mv^2}{r}$, where M is the mass of the body.

In the case of a ship turning in a circle, the centripetal force is produced by the water acting on the side of the ship away from the centre of the turn. The force is considered to act at the centre of lateral resistance which, in this case, is the centroid of the underwater area of the ship's side away from the centre of the turn. The centroid of this area is considered to be at the level of the centre of buoyancy. For equilibrium there must be an equal and opposite force, called the 'Centrifugal' force, and this force is considered to act at the centre of mass (G).

When a ship's rudder is put over to port, the forces on the rudder itself will cause the ship to develop a small angle of heel initially to port, say α_1°.

However, the underwater form of the ship and centrifugal force on it cause the ship to heel to starboard, say α_2°.

In this situation α_2° is always greater than α_1°. Consequently for port rudder helm, the final angle of heel due to turning will be to starboard and vice versa.

It can be seen from Figure 39.1 that these two forces produce a couple which tends to heel the ship away from the centre of the turn, i.e.

$$\text{Heeling couple} = \frac{Mv^2}{r} \times B_1Z$$

Equilibrium is produced by a righting couple equal to $W \times GZ$, where W is equal to the weight of the ship, the weight being a unit of force, i.e. $W = Mg$.

$$\therefore \quad MgGZ = \frac{Mv^2}{r} \times B_1Z$$

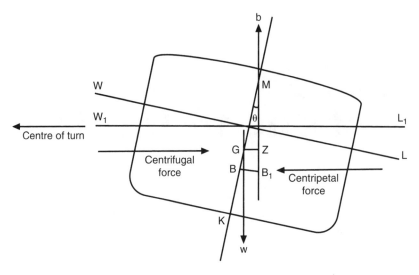

Fig. 39.1

or

$$GZ = \frac{v^2}{gr} \times B_1Z$$

but at a small angle

$$GZ = GM \sin \theta$$

and

$$B_1Z = BG \cos \theta$$
$$\therefore\ GM \sin \theta = \frac{v^2}{gr} BG \cos \theta$$

and

$$\tan \theta = \frac{v^2 \times BG}{grGM}$$

Example
A ship turns to port in a circle of radius 100 m at a speed of 15 knots. The GM is 0.67 m and BG is 1 m. If g = 9.81 m/sec^2 and 1 knot is equal to 1852 km/hour, find the heel due to turning.

$$\text{Ship speed in m/sec} = 15 \times \frac{1852}{3600}$$

$$v = 7.72 \text{ m/sec}$$

$$\tan \theta = \frac{v^2 \times BG}{grGM}$$

$$= \frac{7.72^2 \times 1.0}{9.81 \times 100 \times 0.67}$$

$$\tan \theta = 0.0907$$

Ans. Heel = 5° 11′ to starboard, due to Contrifugal forces only.

In practice, this angle of heel will be slightly smaller. Forces on the rudder will have produced an angle of heel, say 1° 17′ to port. Consequently the overall angle of heel due to turning will be:

$$\text{Heel} = 5° 11′ - 1° 17′ = 3° 54′ \text{ or } 3.9° \text{ to starboard}$$

Exercise 39

1 A ship's speed is 12 knots. The helm is put hard over and the ship turns in a circle of radius 488 m. GM = 0.3 m and BG = 3 m. Assuming that 1 knot is equal to 1852 km/hour, find heel due to turning.

2 A ship steaming at 10 knots turns in a circle of radius 366 m. GM = 0.24 m. BM = 3.7 m. Calculate the heel produced.

3 A ship turns in a circle of radius 100 m at a speed of 15 knots. BG = 1 m. Find the heel if the GM = 0.6 m.

4 A ship with a transverse metacentric height of 0.40 m has a speed of 21 kts. The centre of gravity is 6.2 m above keel whilst the centre of lateral resistance is 4.0 m above keel. The rudder is put hard over to port and the vessel turns in a circle of 550 m radius.

Considering only the centrifugal forces involved, calculate the angle of heel as this ship turns at the given speed.

Chapter 40
Rolling, pitching and heaving motions

A ship will not normally roll in still water but if a study be made of such rolling some important conclusions may be reached. For this study it is assumed that the amplitude of the roll is small and that the ship has positive initial metacentric height. Under the conditions rolling is considered to be simple harmonic motion so it will be necessary to consider briefly the principle of such motion.

Let XOY in Figure 40.1 be a diameter of the circle whose radius is 'r' and let OA be a radius vector which rotates about O from position OY at a constant angular velocity of 'w' radians per second. Let P be the projection of the point A on to the diameter XOY. Then, as the radius vector rotates, the point P will oscillate backwards and forwards between Y and X. The motion of the point P is called 'simple harmonic'.

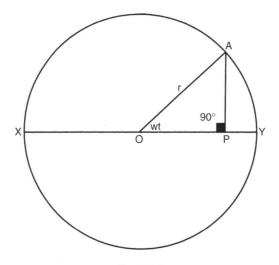

Fig. 40.1

Let the radius vector rotate from OY to OA in 't' seconds, then angle AOY is equal to 'wt'. Let the time taken for the radius vector to rotate through one complete revolution (2π radians) be equal to 'T' seconds, then

$$2\pi = wT$$

or

$$T = 2\pi/w$$

Let

$$OP = x$$

then

$$x = r \cos wt$$

$$\frac{dx}{dt} = -rw \sin wt$$

$$\frac{d^2x}{dt^2} = -rw^2 \cos wt$$

but

$$r \cos wt = x$$

$$\therefore \frac{d^2x}{dt^2} = -w^2x$$

or

$$\frac{d^2x}{dt^2} + w^2x = 0$$

The latter equation is the type of differential equation for simple harmonic motion and since $T = 2\pi/w$ and 'w' is the square root of the coefficient of x in the above equation, then

$$T = \frac{2\pi}{\sqrt{\text{coeff. of } x}}$$

When a ship rolls, the axis about which the oscillation takes place cannot be accurately determined but it would appear to be near to the longitudinal axis through the ship's centre of gravity. Hence the ship rotates or rolls about her 'G'.

The mass moment of inertia (I) of the ship about this axis is given by:

$$I = MK^2$$

where
M = the ship's mass
K = the radius of gyration about this axis

But

$$M = \frac{W}{g}$$

where
W = the ship's weight
g = the acceleration due to gravity

$$\therefore I = \frac{W}{g} K^2$$

When a ship is inclined to a small angle (θ) the righting moment is given by:

$$\text{Righting moment} = W \times GZ$$

where
W = the ship's Weight
GZ = the righting lever

But

$$GZ = GM \sin \theta$$
$$\therefore \text{Righting moment} = W \times GM \times \sin \theta$$

And since θ is a small angle, then

$$\text{Righting moment} = W \times GM \times \theta$$
$$\text{The angular acceleration} = \frac{d^2\theta}{dt^2}$$
$$\therefore I \times \frac{d^2\theta}{dt^2} = -W \times GZ$$

or

$$\frac{W}{g} K^2 \times \frac{d^2\theta}{dt^2} = -W \times GM \times \theta$$

$$\frac{W}{g} K^2 \times \frac{d^2\theta}{dt^2} + W \times GM \times \theta = 0$$

$$\frac{d^2\theta}{dt^2} + \frac{gGM\theta}{K^2} = 0$$

This is the equation for a simple harmonic motion having a period 'T' given by the equation:

$$T = \frac{2\pi}{\sqrt{\text{Coeff. of } \theta}}$$

or

$$T = \frac{2\pi K}{\sqrt{gGM}} = \frac{2\pi}{\sqrt{g}} \times \frac{K}{\sqrt{GM}} = \frac{2K}{\sqrt{GM}} \text{ approx}$$

$$T_R = 2 \times \pi \times (K^2/(g \times GM_T)^{0.5}) \text{ seconds}$$
$$\quad = 2(K^2/GM_T)^{0.5}$$
$$\quad = 2 \times K/(GM_T)^{0.5} \text{ sec}$$

The transverse K value can be approximated with acceptable accuracy to be $K = 0.35 \times B$, where B = breadth mld of the ship.

$$T_R = 2 \times \pi \times (K^2/(g \times GM_T)^{0.5})$$
$$T_R = (2 \times 3.142 \times 0.35 \times B)/3.131 \times (GM_T)^{0.5}$$
$$T_R = (0.7 \times B)/(GM_T)^{0.5} \text{ sec approx}$$

Figure 40.2 shows the resulting rolling periods based on the above formula, with the variables of GM_T up to 5 m and breadth B up to 60 m incorporated.

From the above it can be seen that:

1. The time period of roll is completely independent of the actual amplitude of the roll so long as it is a small angle.
2. The time period of roll varies directly as K, the radius of gyration. Hence if the radius of gyration is increased, then the time period is also increased. K may be increased by moving weights away from the axis of oscillation. Average K value is about $0.35 \times$ Br. Mld.
3. The time period of roll varies inversely as the square root of the initial metacentric height. Therefore ships with a large GM will have a short period and those with a small GM will have a long period.
4. The time period of roll will change when weights are loaded, discharged or shifted within a ship, as this usually affects both the radius of gyration and the initial metacentric height.

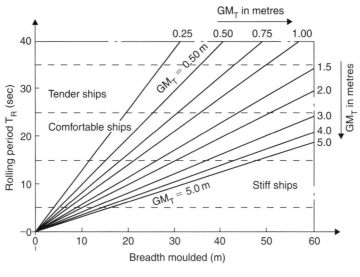

Fig. 40.2

Example 1

Find the still water period of roll for a ship when the radius of gyration is 6 m and the metacentric height is 0.5 m.

$$T = \frac{2\pi K}{\sqrt{gGM}} \qquad = \frac{2K}{\sqrt{GM_T}} \text{ approx}$$

$$T = \frac{2\pi K}{\sqrt{9.81 \times 0.5}} = \frac{2 \times 6}{\sqrt{0.5}} \text{ approx}$$

$$= 16.97\text{s} \qquad \text{Average} = 17\text{s}$$

(99.71 per cent correct giving only 0.29 per cent error!!)

Ans. T = 17.02 s

Note. In the S.I. system of units the value of g to be used in problems is 9.81 m per second per second, unless another specific value is given.

Example 2

A ship of 10 000 tonnes displacement has GM = 0.5 m. The period of roll in still water is 20 seconds. Find the new period of roll if a mass of 50 tonnes is discharged from a position 14 m above the centre of gravity. Assume

$$g = 9.81 \text{ m/sec}^2$$

$$W_2 = W_0 - w = 10\ 000 - 50 = 9950 \text{ tonnes}$$

$$T = \frac{2\pi K}{\sqrt{gGM}} \qquad\qquad GG_1 = \frac{w \times d}{W_2}$$

$$20 = \frac{2\pi K}{\sqrt{9.81 \times 0.5}} \qquad\qquad = \frac{50 \times 14}{9950}$$

$$400 = \frac{4\pi^2 K^2}{9.81 \times 0.5} \qquad\qquad GG_1 = 0.07 \text{ m}$$

$$GM = \underline{0.50 \text{ m}}$$

$$\text{New } GM = 0.57 \text{ m}$$

or

$$K^2 = \frac{400 \times 9.81 \times 0.5}{4 \times \pi^2}$$

$$= 49.69$$

$$\therefore K = 7.05$$

$$I \text{ (originally)} = MK^2$$

$$I_o = 10\,000 \times 49.69$$

$$I_o = 496\,900 \text{ tonnes m}^2$$

$$I \text{ of discharged mass about } G = 50 \times 14^2$$

$$= 9800 \text{ tonnes m}^2$$

$$\text{New } I \text{ of ship about the original C of G} = \text{Original } I - I \text{ of discharged mass}$$

$$= 496\,900 - 9800$$

$$= 487\,100 \text{ tonnes m}^2$$

By the theorem of parallel axes:
  Let

$$I_2 = \frac{\text{New } I \text{ of ship about}}{\text{the new C of G}} = \frac{\text{New } I \text{ of ship about}}{\text{the original C of G}} - W \times CG_1^2$$

$$I_2 = 487\,100 - 9950 \times 0.07^2$$

$$I_2 = 487\,100 - 49$$

$$I_2 = 487\,051 \text{ tonnes m}^2$$

$$I_2 = M_2 K_2^2$$

$$\therefore \text{New } K^2 = \frac{I_2}{M_2}$$

$$\therefore K_2^2 = \frac{487\,051}{9950}$$

Let

$$K_2 = \text{New K} = \sqrt{\frac{487\ 051}{9950}}$$

$$K_2 = 7\ m$$

Let

$$T_2 = \text{New T} = \frac{2\pi K_2}{\sqrt{gGM_2}}$$

$$T_2 = \frac{2\pi 7}{\sqrt{9.81 \times 0.57}}$$

Ans. $\underline{T_2 = 18.6\,s}$

Procedure steps for Example 2
1. Calculate the new displacement in tonnes (W_2).
2. Estimate the original radius of gyration (K).
3. Evaluate the new displacement and new GM (W_2 and GM_2).
4. Calculate the new mass moment of inertia (I_2).
5. Calculate the new radius of gyration (K_2).
6. Finally evaluate the new period of roll (T_2).

For 'stiff ships' the period of roll could be as low as 8 seconds due to a large GM. For 'tender ships' the period of roll will be, say, 25 to 35 seconds, due to a small GM. A good comfortable period of roll for those on board ship will be 15 to 25 seconds, averaging out at 20 seconds (see Figure 40.3).

Pitching motions

Pitching is the movement of the ship's bow, from the lowest position to the highest position and back down to its lowest position. Pitching can be also assumed to be longitudinal rolling motion, where the ship is oscillating around a point at or very near to amidships. This point is known as the 'quiescent point' and is very close to the longitudinal centre of gravity (LCG).

$$T_p = 2 \times \pi \times (K^2/(g \times GM_L)^{0.5})\ \text{seconds}$$
$$= 2(K^2/GM_L)^{0.5}$$
$$= 2 \times K/(GM_L)^{0.5}\ \text{sec}$$

The longitudinal K value can be approximated with acceptable accuracy to be K = 0.25 ×L, where L = LBP of the ship.
  The longitudinal GM value can be approximated to $GM_L = 1.1 \times L$

$$T_p = 2 \times \pi \times (K^2/(g \times GM_L)^{0.5})$$
$$T_p = (2 \times 3.142 \times 0.25 \times L)/3.131 \times (1.1 \times L)^{0.5}$$
$$T_p = \tfrac{1}{2} \times (L)^{0.5}\ \text{sec approx}$$

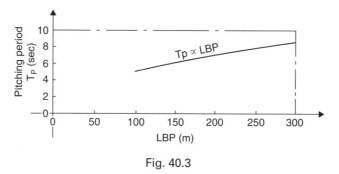

Fig. 40.3

Figure 40.3 shows the resulting pitching periods, based on the above formula, with the variable of LBP up to 300 m incorporated.

Heaving motions

This is the vertical upward or downward movement in the water of the ship's centre of gravity 'G'.

Let the heaving period = T_H seconds.

$$T_H = 2 \times \pi \times (W/(TPC \times 100 \times g))^{0.5} \text{ seconds} \text{ ------------------- (1)}$$

where
W = ship's displacement in tonnes = $L \times B \times d \times C_b \times p$
TPC = tonnes per centimetre immersion = $WPA \times p/100$
WPA = waterplane Area in m^2 = $L \times B \times C_W$
g = gravity = 9.81 m/sec^2
p = density of salt water = 1.025 t/m^3

Substitute all of these values back into Equation (1):

$$T_H = 2 \times 3.142 \times (L \times B \times d \times C_b \times 1.025 \times 100)/(L \times B \times C_W \times 1.025 \times 100 \times 9.81)^{0.5}$$

Hence

$$T_H = 2 \times (d \times C_b/C_W)^{0.5} \text{ sec}$$

Hence the heaving period depends on the draft 'd' and the ratio of the C_b/C_W. This ratio varies with each type of ship, as shown in Table 40.1.

Figure 40.4 shows the resulting heaving periods from Table 40.1, with the type of hull form and the variable of SLWL (up to 22.00 m) incorporated.

Table 40.1   Approximate heaving periods for several merchant ships.

Ship type	Typically fully loaded C_b	Corresponding loaded C_W	Approximate C_b/C_W	Approximate heaving period in seconds
ULCC	0.850	0.900	0.944	$1.94 \times (d)^{0.5}$
VLCC	0.825	0.883	0.934	$1.93 \times (d)^{0.5}$
Bulk carrier	0.775	0.850	0.912	$1.91 \times (d)^{0.5}$
General cargo	0.700	0.800	0.875	$1.87 \times (d)^{0.5}$
Passenger liner	0.600	0.733	0.819	$1.81 \times (d)^{0.5}$
Container ship	0.575	0.717	0.802	$1.79 \times (d)^{0.5}$

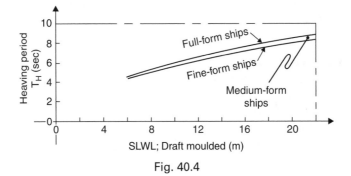

Fig. 40.4

Worked example

For a general cargo ship in a particular loaded condition, LBP = 140 m, B = 19.17 m, C_b = 0.709, C_W = 0.806, draft = 8.22 m and GM_T = 0.45 m. Estimate the rolling period, the pitching period and the heaving period in seconds. From the notes

$$T_R = (0.7 \times B)/(GM_T)^{0.5} \text{ seconds approx}$$

So

$$T_R = (0.7 \times 19.17)/(0.45)^{0.5}$$
$$T_R = 20 \text{ seconds approx}$$

From the notes

$$T_P = \tfrac{1}{2} \times (L)^{0.5} \text{ seconds approx}$$

So

$$T_P = \tfrac{1}{2} \times (L)^{0.5}$$
$$T_P = \tfrac{1}{2} \times (140)^{0.5}$$
$$T_P = 5.92 \text{ sec}$$

From the notes

$$T_H = 2 \times (d \times C_b/C_W)^{0.5} \text{ seconds}$$

$$T_H = 2 \times (8.22 \times 0.709/0.806)^{0.5} \text{ seconds}$$

$$T_H = 5.38 \text{ sec}$$

Note how the rolling period is much greater in value than the pitching period and the heaving period. Figures 40.2 to 40.4 graphically confirm the mathematics of this statement.

In examinations, only use the approximate formulae in the absence of being given more detailed information. Calculations are to be based wherever possible on real stability information.

Exercise 40

1   Find the still water period of roll for a ship when the radius of gyration is 5 m and the initial metacentric height is 0.25 m.

2   A ship of 5000 tonnes displacement has GM = 0.5 m. The still water period of roll is 20 seconds. Find the new period of roll when a mass of 100 tonnes is discharged from a position 14 m above the centre of gravity.

3   A ship of 9900 tonnes displacement has GM = 1 m, and a still water rolling period of 15 seconds. Calculate the new rolling period when a mass of 100 tonnes is loaded at a position 10 m above the ship's centre of gravity.

4   A vessel has the following particulars:
    Displacement is 9000 tonnes, natural rolling period is T_R of 15 seconds, GM is 1.20 m. Determine the new natural rolling period after the following changes in loading have taken place:

    2000 tonnes added at 4.0 m above ship's VCG.
    500 tonnes discharged at 3.0 m below ship's VCG.

    Assume that KM remains at the same value before and after changes of loading have been completed. Discuss if this final condition results in a 'stiff ship' or a 'tender ship'.

5   For a bulk carrier, in a particular loaded condition, the LBP is 217 m, B is 32.26 m, C_b is 0.795, C_W is 0.863, draft is 12.20 m and the GM_T is 0. 885 m. Estimate the rolling period, the pitching period and the heaving period in seconds.

Chapter 41
Synchronous rolling and parametric rolling of ships

Synchronous rolling of ships

Synchronous rolling is caused by the ship's rolling period T_R becoming synchronous or resonant with the wave period. When this occurs, the ship will heel over and, in exceptional circumstances, be rolled further over by the action of the wave.

Consequently, there is a serious danger that the vessel will heel beyond a point or angle of heel from which it cannot return to an upright condition. The ship ends up having negative stability, and will capsize.

Figure 41.1 shows a ship with synchronous rolling problems.

To reduce synchronous rolling:

1. Use water ballast changes to alter the KG of the vessel. This should alter the GM_T and hence the natural rolling period T_R to a non-synchronous value.
2. Change the course heading of the ship so that there will be a change in the approaching wave frequencies. In other words, introduce a yawing effect.

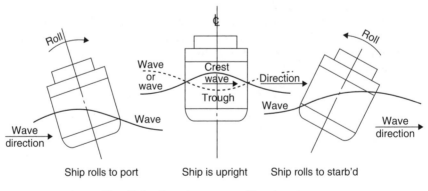

Fig. 41.1   Synchronous rolling in waves.

3. Alter the ship's speed until synchronism or resonance no longer exists with the wave frequency.

Parametric rolling of ships

Parametric rolling is produced by pitching motions on vessels which have very fine bowlines together with very wide and full stern contours. One such ship type is the container ship. Figure 41.2 shows a ship with parametric rolling problems.

The cause depends very much on the parameters of the vessel, hence the name 'parametric rolling'. It is most marked when the pitching period T_P is either equal to, or half that of the vessel's rolling period T_R.

As the stern dips into the waves it produces a rolling action. This remains unchecked as the bow next dips into the waves due to pitching forces. It is worst when $T_P = T_R$ or when $T_P = \frac{1}{2} \times T_R$.

In effect, the rolling characteristics are different at the stern to those at the bow. It causes a twisting or torsioning along the ship leading to extra rolling motions.

If $T_P = T_R$, or $T_P = \frac{1}{2} \times T_R$, then interaction exists and the rolling of the ship is increased. A more dangerous situation develops because of the interplay between the pitching and rolling motions.

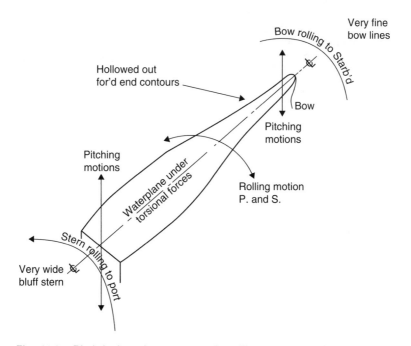

Fig. 41.2   Pitch induced or parametric rolling on a container vessel.

Parametric rolling is worse when a ship is operating at reduced speed in heavy sea conditions. Such condition can cause containers to be lost overboard due to broken deck lashings.

The IMO suggest that parametric rolling is particularly dangerous when the wavelength is 1.0 to 1.5 times the ship's length.

Parametric rolling problems are least on box-shaped vessels or full-form barges where the aft and forward contours are not too dissimilar. Very little transverse and longitudinal interplay occurs.

To reduce parametric rolling:

1. A water ballast could be used to alter the GM_T and hence the natural rolling period T_R, to a non-synchronous value.
2. The ship needs to have an anti-rolling acting stabilising system. Anti-rolling stability tanks that transfer water across the ship or vertically between two tanks are effective for all ship speeds. A quick response time is vital to counteract this type of rolling.
3. Hydraulic fin stabilisers would also help to reduce parametric rolling. They maybe telescopic or hinged into the sides of the vessel at or near to amidships.
4. Alter the ship's forward speed.
5. Alter the ship's course.

Exercise 41

1 With the aid of a sketch, describe *synchronous rolling*. Suggest three methods a master or mate can consider for reducing the effects of *synchronous rolling*.
2 With the aid of sketches, describe *parametric rolling*. Suggest five methods a master or mate can consider for reducing the effects of *parametric rolling*.

Chapter 42
List due to bilging side compartments

When a compartment in a ship is bilged the buoyancy provided by that compartment is lost. This causes the centre of buoyancy of the ship to move directly away from the centre of the lost buoyancy and, unless the centre of gravity of the compartment is on the ship's centreline, a listing moment will be created, b = w.

Let the ship in Figure 42.1 float upright at the waterline, WL. G represents the position of the ship's centre of gravity and B the centre of buoyancy.

Now let a compartment which is divided at the centreline be bilged on the starboard side as shown in the figure. To make good the lost buoyancy

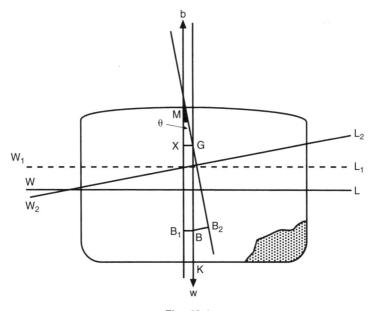

Fig. 42.1

the ship will sink to the waterline W_1L_1. That is, the lost buoyancy is made good by the layer between WL and W_1L_1.

The centre of buoyancy will move from B to B_1, directly away from the centre of gravity of the lost buoyancy, and the distance BB_1 is equal to $\left(\dfrac{w \times d}{W}\right)$, where w represents the lost buoyancy and d represents the distance between the ship's centre of buoyancy and the centre of the lost buoyancy.

The shift in the centre of buoyancy produces a listing moment.

Let θ be the resultant list.

Then

$$\tan\theta = \frac{GX}{XM} = \frac{BB_1}{XM}$$

where XM represents the initial metacentric height for the bilged condition.

Example 1

A box-shaped vessel of length 100 m and breadth 18 m, floats in salt water on an even keel at 7.5 m draft. KG = 4 m. The ship has a continuous centreline bulkhead which is watertight. Find the list if a compartment amidships, which is 15 m long and is empty, is bilged on one side.

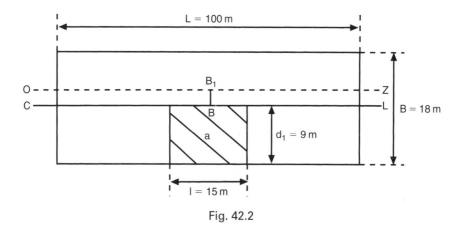

Fig. 42.2

(a) Find the new mean draft

$$\text{Bodily increase in draft} = \frac{\text{Volume of lost buoyancy}}{\text{Area of intact W.P.}}$$

$$= \frac{15 \times 9 \times 7.5}{(100 \times 18) - (15 \times 9)} = 0.61$$

$$\text{New draft} = 7.50 + 0.61$$

$$\therefore \ \text{New draft} = 8.11 \text{ m} = \text{Draft } d_2$$

(b) Find the shift of the centre of buoyancy

$$BB_1 = \frac{a \times B/4}{LB - a}$$

$$= \frac{15 \times 9 \times 18/4}{(100 \times 18) - (15 \times 9)} = \frac{607.5}{1665}$$

$$= 0.37 \text{ m}$$

(c) To find I_{OZ}

$$I_{CL} = \left(\frac{B}{2}\right)^3 \times \frac{L}{3} + \left(\frac{B}{2}\right)^3 \times \frac{(L - l)}{3}$$

$$I_{CL} = \frac{9^3 \times 100}{3} + \frac{9^3 \times 85}{3} = 24\,300 + 20\,655$$

$$= 44\,955 \text{ m}^4$$

$$I_{OZ} = I_{CL} - ABB_1^2$$

$$= 44\,955 - \{(100 \times 18) - (15 \times 9)\} \times 0.365^2$$

$$= 44\,955 - 222$$

$$= 44\,733 \text{ m}^4$$

(d) To find GM

$$BM = \frac{I_{OZ}}{V}$$

$$= \frac{44\,733}{100 \times 18 \times 7.5}$$

$$= 3.31 \text{ m}$$

$$+$$

$$KB = \frac{d_2}{2} \therefore KB = \underline{4.06 \text{ m}}$$

$$KM = 7.37 \text{ m}$$

$$-$$

$$KG = \underline{4.00 \text{ m}} \text{ as before bilging}$$

$$\text{After bilging, GM} = \underline{3.37 \text{ m}}$$

(e) To find the List

$$\tan \text{List} = \frac{BB_1}{GM}$$

$$= \frac{0.37}{3.37} = 0.1098$$

Ans. <u>List = 6° 16′</u>

Example 2

A box-shaped vessel, 50 m long × 10 m wide, floats in salt water on an even keel at a draft of 4 m. A centreline longitudinal watertight bulkhead extends from end to end and for the full depth of the vessel. A compartment amidships on the starboard side is 15 m long and contains cargo with permeability 'μ' of 30 per cent. Calculate the list if this compartment is bilged. KG = 3 m.

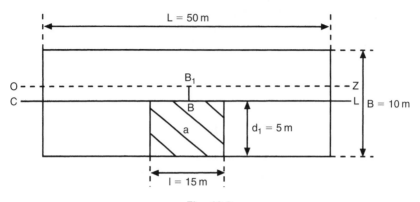

Fig. 42.3

(a) Find the new mean draft

$$\text{Bodily increase in draft} = \frac{\text{Volume of lost buoyancy}}{\text{Area of intact W.P.}}$$

$$= \frac{\dfrac{30}{100} \times 15 \times 5 \times 4}{(50 \times 10) - \left(\dfrac{30}{100} \times 15 \times 5\right)} = \frac{90}{477.5}$$

$$= 0.19 \text{ m}$$

$$\therefore \text{ New draft } = 4.00 + 0.19 = 4.19 \text{ m say draft } d_2$$

(b) Find the shift of the centre of buoyancy

$$BB_1 = \frac{\mu a \times \dfrac{B}{4}}{LB - \mu a}$$

$$= \frac{\dfrac{30}{100} \times 15 \times 5 \times \dfrac{10}{4}}{(50 \times 10) - \left(\dfrac{30}{100} \times 15 \times 5\right)} = \frac{56.25}{477.5}$$

$$= 0.12 \text{ m}$$

(c) To find I_{OZ}

$$I_{CL} = \frac{LB^3}{12} - \frac{\mu l b^3}{3}$$

$$= \left(\frac{50 \times 10^3}{12}\right) - \left(\frac{30}{100} \times \frac{15 \times 5^3}{3}\right)$$

$$= 4166.7 - 187.5$$

$$= 3979 \text{ m}^4$$

$$I_{OZ} = I_{CL} - A \times BB_1^2$$
$$= 3979 - (477.5 \times 0.12^2)$$
$$= 3979 - 7$$
$$= 3972 \text{ m}^4$$

(d) To find GM

$$BM_2 = \frac{I_{OZ}}{V}$$

$$BM_2 = \frac{3972}{50 \times 10 \times 4}$$

$$\therefore BM_2 = 1.99 \text{ m}$$

$$+$$

$$KB_2 = 2.10 \text{ m}$$
$$KM_2 = \overline{4.09 \text{ m}}$$

$$-$$

$$\text{as before bilging, KG} = 3.00 \text{ m}$$
$$GM_2 = 1.09 \text{ m}$$

(e) To find the list

$$\tan \text{List} = \frac{BB_1}{GM}$$
$$= \frac{0.12}{1.09} = 0.1101$$

Ans. List = 6° 17′ to starboard

Note: When μ = 100 per cent then:

$$I_{CL} = \left(\frac{B}{2}\right)^3 \times \frac{L}{3} + \left(\frac{B}{2}\right)^3 \left(\frac{L-1}{3}\right) \text{ m}^4$$

or

$$I_{CL} = \frac{LB^3}{12} - \frac{lb^3}{3} \text{ m}^4$$

Both formulae give the same answer.

Summary
1. Make a sketch from the given information.
2. Calculate the mean bodily increase in draft.
3. Calculate the ship in the centre of buoyancy.
4. Estimate the second moment of area in the bilged condition with the use of the parallel axis theorem.
5. Evaluate the new KB, BM, KM and GM.
6. Finally calculate the requested angle of list.

Exercise 42

1 A box-shaped tanker barge is 100 m long, 15 m wide and floats in salt water on an even keel at 5 m draft. KG = 3 m. The barge is divided longitudinally by a centreline watertight bulkhead. An empty compartment amidships on the starboard side is 10 m long. Find the list if this compartment is bilged.

2 A box-shaped vessel, 80 m long and 10 m wide, is floating on an even keel at 5 m draft. Find the list if a compartment amidships, 15 m long is bilged on one side of the centreline bulkhead. KG = 3 m.

3 A box-shaped vessel, 120 m long and 24 m wide, floats on an even keel in salt water at a draft of 7 m. KG = 7 m. A compartment amidships is 12 m long and is divided at the centreline by a full depth watertight bulkhead. Calculate the list if this compartment is bilged.

4 A box-shaped vessel is 50 m long, 10 m wide and is divided longitudinally at the centreline by a watertight bulkhead. The vessel floats on an even keel in salt water at a draft of 4 m. KG = 3 m. A compartment amidships is 12 m long and contains cargo of permeability 30 per cent. Find the list if this compartment is bilged.

5 A box-shaped vessel 68 m long and 14 m wide has KG = 4.7 m, and floats on an even keel in salt water at a draft of 5 m. A compartment amidships 18 m long is divided longitudinally at the centreline and contains cargo of permeability 30 per cent. Calculate the list if this compartment is bilged.

Chapter 43
Effect of change of density on draft and trim

When a ship passes from water of one density to water of another density the mean draft is changed and if the ship is heavily trimmed, the change in the position of the centre of buoyancy will cause the trim to change.

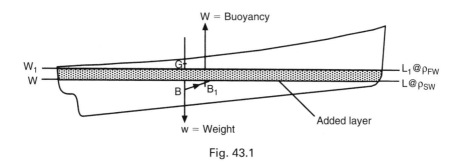

Fig. 43.1

Let the ship in Figure 43.1 float in salt water at the waterline WL. B represents the position of the centre of buoyancy and G the centre of gravity. For equilibrium, B and G must lie in the same vertical line.

If the ship now passes into fresh water, the mean draft will increase. Let W_1L_1 represent the new waterline and b the centre of gravity of the extra volume of the water displaced. The centre of buoyancy of the ship, being the centre of gravity of the displaced water, will move from B to B_1 in a direction directly towards b. The force of buoyancy now acts vertically upwards through B_1 and the ship's weight acts vertically downwards through G, giving a trimming moment equal to the product of the displacement and the longitudinal distance between the centres of gravity and buoyancy. The ship will then change trim to bring the centres of gravity and buoyancy back in to the same vertical line.

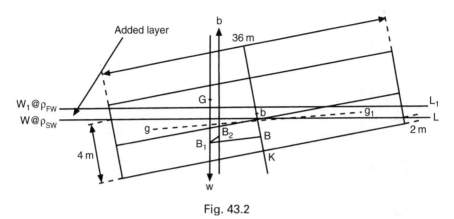

Fig. 43.2

Example

A box-shaped pontoon is 36 m long, 4 m wide and floats in salt water at drafts F 2.00 m, A 4.00 m. Find the new drafts if the pontoon now passes into fresh water. Assume salt water density is $1.025\,t/m^3$. Assume fresh water density $= 1.000\,t/m^3$.

(a) *To find the position of B_1*

$$BB_1 = \frac{v \times gg_1}{V}$$

$$v = \frac{1}{2} \times 1 \times \frac{36}{2} \times 4 \qquad gg_1 = \frac{2}{3} \times 36 \qquad V = 36 \times 4 \times 3$$

$$= 36 \text{ cu. m} \qquad gg_1 = 24 \text{ m} \qquad = 432 \text{ cu.m}$$

$$\therefore BB_1 = \frac{36 \times 24}{432}$$

$$= 2 \text{ m}$$

Because the angle of trim is small, BB_1 is considered to be the horizontal component of the shift of the centre of buoyancy.

Now let the pontoon enter fresh water, i.e. from ρ_{SW} into ρ_{FW}. Pontoon will develop mean bodily sinkage.

(b) *To find the new draft*
In salt water:

$$\text{Mass} = \text{Volume} \times \text{Density}$$
$$= 36 \times 4 \times 3 \times 1.025$$

In fresh water:

$$\text{Mass} = \text{Volume} \times \text{Density}$$
$$\therefore \text{Volume} = \frac{\text{Mass}}{\text{Density}}$$
$$= \frac{36 \times 4 \times 3 \times 1.025}{1.000} \text{ cu. m}$$
$$(\text{Mass in salt water} = \text{Mass in fresh water})$$

Let

$$MBS = \text{Mean bodily sinkage} \quad \rho_{SW} = \text{Higher density}$$
$$\rho_{FW} = \text{Lower density}$$

$$MBS = \frac{W}{TPC_{SW}} \times \frac{(\rho_{SW} - \rho_{FW})}{\rho_{FW}}$$

$$MBS = \frac{L \cdot B \cdot d \cdot \rho_{SW}}{\dfrac{L \times B}{100} \times \rho_{SW}} \left\{ \frac{\rho_{SW} - \rho_{FW}}{\rho_{FW}} \right\}$$

$$\therefore MBS = \frac{d(\rho_{SW} - \rho_{FW})}{\rho_{FW}} \times 100$$

$$MBS = \frac{3 \times 0.025}{1.000} \times 100 = \underline{0.075 \text{ m}}$$

$$\therefore MBS = 0.075 \text{ m}$$

$$\text{Original mean draft} = 3.000 \text{ m}$$
$$\text{New mean draft} = 3.075 \text{ m say draft } d_2$$

(c) Find the change of trim

Let B_1B_2 be the horizontal component of the shift of the centre of buoyancy. Then

$$B_1B_2 = \frac{v \times d}{V} \qquad\qquad W = LBd_{SW} \times \rho_{SW}$$

$$= \frac{10.8 \times 2}{442.8} \qquad\qquad = 36 \times 4 \times 3 \times 1.025$$

$$\therefore B_1B_2 = 0.0487 \text{ m} \qquad \therefore W = 442.8 \text{ tonnes}$$

$$\text{Trimming moment} = W \times B_1B_2$$

$$= 36 \times 4 \times 3 \times \frac{1.025}{1.000} \text{ t} \times 0.0487 \text{ m} = 21.56 \text{ t m}$$

$$BM_{L(2)} = \frac{L^2}{12d_{(2)}}$$

$$= \frac{36^2}{12 \times 3.075}$$

$$= \frac{36}{1.025} \text{ m} = 35.12 \text{ m}$$

$$MCTC \approx \frac{W \times BM_L}{100 \times L}$$

$$= \frac{442.8 \times 35.12}{100 \times 36}$$

$$= 4.32 \text{ tonnes metres}$$

$$\text{Changes of trim} = \frac{\text{Trimming moment}}{MCTC}$$

$$= \frac{21.56}{4.32} = 5 \text{ cm}$$

Change of trim = 5 cm by the stern

= 0.05 m by the stern

Drafts before trimming	A	4.075 m	F	2.075 m
Change due to trim	A	+0.025 m		−0.025 m
New drafts	A	4.100 m	F	2.050 m

In practice the trimming effects are so small that they are often ignored by shipboard personnel. Note in the above example the trim ratio forward and aft was only $2\frac{1}{2}$ cm.

However, for DfT examinations, they must be studied and fully understood.

Exercise 43

1 A box-shaped vessel is 72 m long, 8 m wide and floats in salt water at drafts F 4.00 m. A 8.00 m. Find the new drafts if the vessel now passes into fresh water.

2 A box-shaped vessel is 36 m long, 5 m wide and floats in fresh water at drafts F 2.50 m. A 4.50 m. Find the new drafts if the vessel now passes into salt water.

3 A ship has a displacement of 9100 tonnes, LBP of 120 m, even-keel draft of 7 m in fresh water of density of $1.000 \, t/m^3$.

From her Hydrostatic Curves it was found that:

$MCTC_{SW}$ is 130 t m/cm
TPC_{SW} is 17.3 t
LCB is 2 m forward ⵂ and
LCF is 1.0 aft ⵂ

Calculate the new end drafts when this vessel moves into water having a density of $1.02 \, t/m^3$ without any change in the ship's displacement of 9100 tonnes.

Chapter 44
List with zero metacentric height

When a weight is shifted transversely in a ship with zero initial metacentric height, the resulting list can be found using the 'Wall sided' formula.

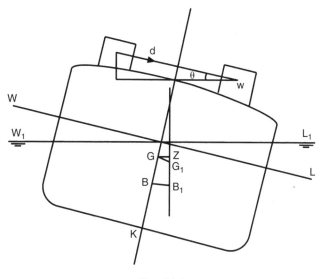

Fig. 44.1

The ship shown in Figure 44.1 has zero initial metacentric height. When a weight of mass 'w' is shifted transversely through a distance 'd', the ship's centre of gravity shifts from G to G_1 where the direction GG_1 is parallel to the shift of the centre of gravity of the weight shifted. The ship will then incline to bring the centres of gravity and buoyancy in the same vertical line.

The horizontal component of the shift of the ship's centre of gravity is equal to GZ and the horizontal component of the shift of the centre of gravity of the weight is equal to d cos θ.

$$\therefore w \times d \times \cos\theta = W \times GZ$$

The length GZ, although not a righting lever in this case can be found using the 'wall-sided' formula, i.e.

$$GZ = (GM + \tfrac{1}{2} \times BM \times \tan^2 \theta)\sin \theta$$

$$\therefore \ w \times d \times \cos \theta = W \times \sin \theta(GM + \tfrac{1}{2} \times BM \times \tan^2 \theta)$$

If

$$GM = 0$$

Then

$$w \times d \times \cos \theta = W \times \sin \theta \times \tfrac{1}{2} BM \times \tan^2 \theta$$

$$\frac{w \times d}{W} = \frac{\sin \theta}{\cos \theta} \times \tfrac{1}{2} \times BM \times \tan^2 \theta$$

$$\frac{2 \times w \times d}{BM \times W} = \tan^3 \theta$$

or

$$\tan \theta = \sqrt[3]{\frac{2 \times w \times d}{BM \times W}}$$

Example
A ship of 12 250 tonnes displacement has KM = 8 m, KB = 3.8 m, KG = 8 m and is floating upright. Find the list if a weight of 2 tonnes, already on board, is shifted transversely through a horizontal distance of 12 m, assuming that the ship is wall-sided.

$$
\begin{aligned}
KM &= 8.0 \text{ m} \\
KB &= -3.8 \text{ m} \\
BM &= \underline{4.2 \text{ m}}
\end{aligned}
$$

The ship has zero GM

$$\therefore \ \tan \text{ list} = \sqrt[3]{\frac{2 \times w \times d}{W \times BM}}$$

$$\tan \text{ list} = \sqrt[3]{\frac{2 \times 2 \times 12}{12\,250 \times 4.2}} = 0.0977$$

Ans. List = 5° 34'

Exercise 44

1 Find the list when a mass of 10 tonnes is shifted transversely through a horizontal distance of 14 m in a ship of 8000 tonnes displacement which has zero initial metacentric height. BM = 2 m.

2 A ship of 8000 tonnes displacement has zero initial metacentric height. BM = 4 m. Find the list if a weight of 20 tonnes is shifted transversely across the deck through a horizontal distance of 10 m.

3 A ship of 10 000 tonnes displacement is floating in water of density 1.025 kg/cu. m and has a zero initial metacentric height. Calculate the list when a mass of 15 tonnes is moved transversely across the deck through a horizontal distance of 18 m. The second moment of the waterplane area about the centreline is $10^5 \, \text{m}^4$.

4 A ship of 12250 tonnes displacement is floating upright. KB = 3.8 m, KM = 8 m and KG = 8 m. Assuming that the ship is wall-sided, find the list if a mass of 2 tonnes, already on board, is shifted transversely through a horizontal distance of 12 m.

Chapter 45
The deadweight scale

The deadweight scale provides a method for estimating the additional draft or for determining the extra load that could be taken onboard when a vessel is being loaded in water of density less than that of salt water. For example, the vessel may be loading in a port where the water density is that of fresh water at 1.000 t/cu.m.

This deadweight scale (see Figure 45.1) displays columns of scale readings for:

Freeboard (f).
Dwt in salt water and in fresh water.
Draft of ship (mean).
Displacement in tonnes in salt water and in fresh water.
Tonnes per cm (TPC) in salt water and in fresh water.
Moment to Change Trim 1 cm (MCTC).

On every dwt scale the following constants must exist:

$$\text{Any freeboard (f)} + \text{Any draft (d)} = \text{Depth of ship (D)}$$

hence $f + d = C1$.

$$\text{Any displacement (W)} - \text{Any Dwt} = \text{Lightweight (Lwt)}$$

hence $W - Dwt = C2$.

The main use of the Dwt scale is to observe Dwt against draft. Weight in tonnes remains the same but the volume of displacement will change with a change in density of the water in which the ship floats. The salt water and fresh water scales relate to these changes.

On many ships this Dwt scale has been replaced by the data being presented in tabular form. The officer onboard only needs to interpolate to obtain the information that is required. Also the Dwt scale can be part of a computer package supplied to the ship. In this case the officer only needs to key in the variables and the printout supplies the required data.

The following worked example shows the use of the Dwt scale.

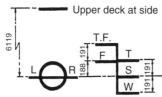

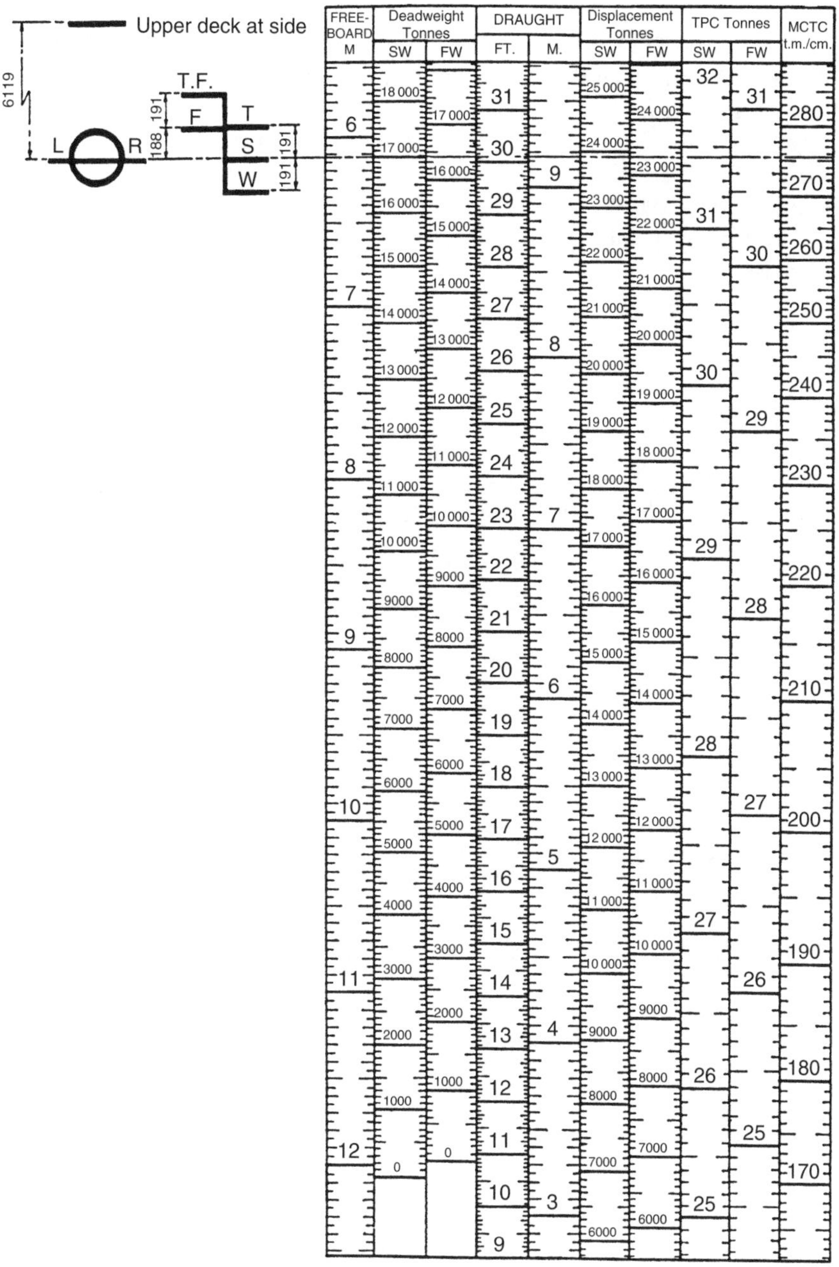

FREE-BOARD	Deadweight Tonnes		DRAUGHT		Displacement Tonnes		TPC Tonnes		MCTC
M	SW	FW	FT.	M.	SW	FW	SW	FW	t.m./cm.

Fig. 45.1 Deadweight scale.

Question:

Determine the TPC at the fully loaded draft from the Dwt scale shown in Figure 45.1 and show the final displacement in tonnes remains similar for fresh and salt water.

From Figure 45.1 TPC is 31.44 and the permitted fresh water sinkage as shown on the freeboard marks is 19 cm with displacement in salt water being almost 23 900 t.

Consequently, the approximate load displacement in fresh water is given by:

$$\text{FW sinkage} = W/\text{TPC} \times 40 \text{ cm}$$

So

$$W = \text{TPC} \times \text{FW sinkage} \times 40 = 31.44 \times 19 \times 40 = 23\,894 \text{ tonnes}$$

Hence this vessel has loaded up an extra 19 cm of draft in fresh water whilst keeping her displacement at 23 894 t (equivalent to salt water draft of 9.17 m).

Chapter 46
The Trim and Stability book

When a new ship is nearing completion, a Trim and Stability book is produced by the shipbuilder and presented to the shipowner. Shipboard officers will use this for the day to day operation of the vessel. In the Trim and Stability book is the following technical data:

1. General particulars of the ship and General Arrangement Plan.
2. Inclining experiment report and its results.
3. Capacity, VCG, LCG particulars for all holds, compartments, tanks etc.
4. Cross curves of stability. These may be GZ curves or KN curves.
5. Deadweight scale data. May be in diagram form or in tabular form.
6. Hydrostatic curves. May be in graphical form or in tabular form.
7. Example conditions of loading such as:

   Lightweight (empty vessel) condition.
   Full-loaded departure and arrival conditions.
   Heavy-ballast departure and arrival conditions.
   Medium-ballast departure and arrival conditions.
   Light-ballast departure and arrival conditions.

   For the arrival conditions, a ship should arrive at the end of the voyage (with cargo and/or passengers as per loaded departure conditions) with at least "10% stores and fuel remaining".

   A mass of 75 kg should be assured for each passenger; but may be reduced to not less than 60 kg where this can be justified.

   On each condition of loading there is a profile and plan view (at upper deck level usually). A colour scheme is adopted for each item of deadweight. Examples could be red for cargo, blue for fresh water, green for water ballast, brown for oil. Hatched lines for this Dwt distribution signify wing tanks P and S.

   For each loaded condition, in the interests of safety, it is necessary to show:

   Deadweight.
   End draughts, thereby signifying a satisfactory and safe trim situation.

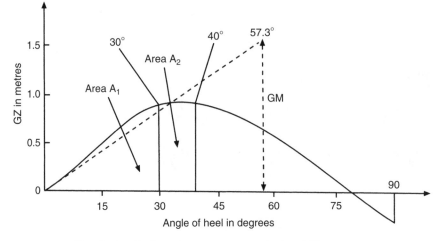

Fig. 46.1   Enclosed areas on a statical stability curve.

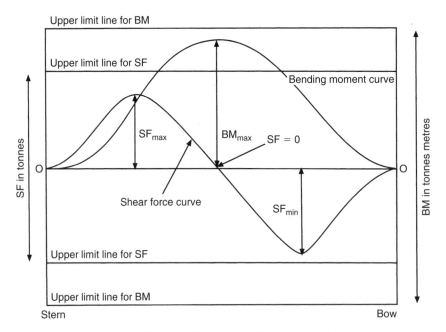

Fig. 46.2   SF and BM curves with upper limit lines.

KG with no free surface effects (FSE), and (KG) with FSE taken into account.

Final transverse metacentric height (GM). This informs the officer if the ship is in stable, unstable or neutral equilibrium. It can also indicate if the ship's stability is approaching a dangerous state.

Total free surface effects of all slack tanks in this condition of loading.

A statical stability curve relevant to the actual loaded condition with the important characteristics clearly indicated. For each S/S curve it is important to observe the following:

Maximum GZ and the angle of heel at which it occurs.

Range of stability.

Area enclosed from zero degrees to thirty degrees (A1) and the area enclosed from thirty degrees to forty degrees (A2) as shown in Figure 46.1.

Shear force and bending moment curves, with upper limit lines clearly superimposed as shown in Figure 46.2.

8. Metric equivalents. For example, TPI″ to TPC, MCTI″ to MCTC, or tons to tonnes.

All of this information is sent to a DfT surveyor for appraisal. On many ships today this data is part of a computer package. The deadweight items are keyed in by the officer prior to any weights actually being moved. The computer screen will then indicate if the prescribed condition of loading is indeed safe from the point of view of stability and strength.

Summary

The results and graphs can also be obtained using computer packages supplied to the ships. Each time a ship is loaded or discharged, the computer will give a printout of all the stability information required by the master or mate.

The computer can be programmed to illustrate the information shown in Figures 46.1 and 46.2. Furthermore, if the limiting lines are exceeded for SF and/or BM, a flashing light or audio alarm will signal to warn the officer. Corrective loading arrangements can then be undertaken.

Chapter 47
Simplified stability information

DEPARTMENT FOR TRANSPORT   MERCHANT SHIPPING NOTICE NO. 1122

SIMPLIFIED STABILITY INFORMATION
Notice to shipowners, Masters and Shipbuilders

1. It has become evident that the masters' task of ensuring that his ship complies with the minimum statutory standards of stability is in many instances not being adequately carried out. A feature of this is that undue traditional reliance is being placed on the value of GM alone, while other important criteria which govern the righting lever GZ curve are not being assessed as they should be. For this reason the Department, appreciating that the process of deriving and evaluating GZ curves is often difficult and time-consuming, strongly recommends that in future simplified stability information be incorporated into ships' stability booklets. In this way masters can more readily assure themselves that safe standards of stability are met.

2. Following the loss of the *Lairdsfield*, referred to in Notice M.627, the Court of Inquiry recommended that simplified stability information be provided. This simplified presentation of stability information has been adopted in a large number of small ships and is considered suitable for wider application in order to overcome the difficulties referred to in Paragraph 1.

3. Simplified stability information eliminates the need to use cross curves of stability and develop righting lever GZ curves for varying loading conditions by enabling a ship's stability to be quickly assessed, to show whether or not all statutory criteria are complied with, by means of a single diagram or table. Considerable experience has now been gained and three methods of presentation are in common use. These are:
   (a) The Maximum Deadweight Moment Diagram or Table.
   (b) The Minimum Permissible GM Diagram or Table.
   (c) The Maximum Permissible KG Diagram or Table.

In all three methods the limiting values are related to salt water displacement or draft. Free surface allowances for slack tanks are, however, applied slightly differently.

4. Consultation with the industry has revealed a general preference for the Maximum Permissible KG approach, and graphical presentation also appears to be preferred rather than a tabular format. The Department's view is that any of the methods may be adopted subject to:
   (a) clear guidance notes for their use being provided and
   (b) submission for approval being made in association with all other basic data and sample loading conditions.

   In company fleets it is, however, recommended that a single method be utilised throughout.

5. It is further recommended that the use of a *Simplified Stability Diagram* as an adjunct to the *Deadweight Scale* be adopted to provide a direct means of comparing stability relative to other loading characteristics. Standard work forms for calculating loading conditions should also be provided.

6. It is essential for masters to be aware that the standards of stability obtainable in a vessel are wholly dependent on exposed openings such as hatches, doorways, air pipes and ventilators being securely closed weathertight; or in the case of automatic closing appliances such as air-pipe ball valves that these are properly maintained in order to function as designed.

7. Shipowners bear the responsibility to ensure that adequate, accurate and uptodate stability information for the master's use is provided. It follows that it should be in a form which should enable it to be readily used in the trade in which the vessel is engaged.

Maximum Permissible Deadweight Moment Diagram

This is one form of simplified stability data diagram in which a curve of Maximum Permissible Deadweight Moments is plotted against displacement in tonnes on the vertical axis and Deadweight Moment in tonnes metres on the horizontal axis, the Deadweight Moment being the moment of the Deadweight about the keel.

The total Deadweight Moment at any displacement must not, under any circumstances, exceed the Maximum Permissible Deadweight Moment at that displacement.

Figure 47.1 illustrates this type of diagram. The ship's displacement in tonnes is plotted on the vertical axis from 1500 to 4000 tonnes while the Deadweight Moments in tonnes metres are plotted on the horizontal axis. From this diagram it can be seen that, for example, the Maximum Deadweight Moment for this ship at a displacement of 3000 tonnes is 10 260 tonnes metres (Point 1). If the light displacement for this ship is 1000 tonnes then the Deadweight at this displacement is 2000 tonnes. The maximum kg

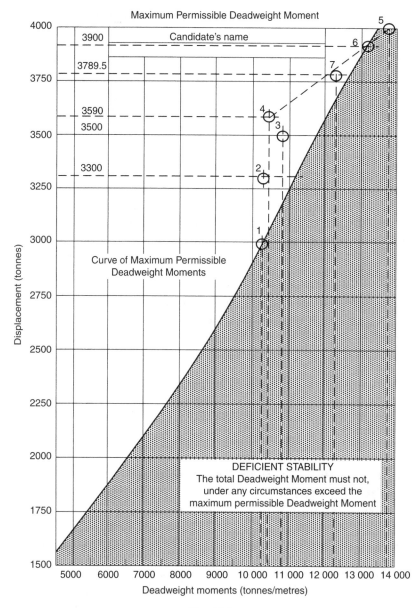

Fig. 47.1

for the Deadweight tonnage is given by:

$$\text{Maximum kg} = \frac{\text{Deadweight Moment}}{\text{Deadweight}}$$

$$= \frac{10\,260}{2000}$$

$$= 5.13\,\text{m}$$

Example 1

Using the Simplified Stability Data shown in Figure 47.1, estimate the amount of cargo (Kg 3 m) which can be loaded so that after completion of loading the ship does not have deficient stability. Prior to loading the cargo the following weights were already on board:

250 t	fuel oil	Kg 0.5 m	Free surface moment 1400 t m
50 t	fresh water	Kg 5.0 m	Free surface moment 500 t m
2000 t	cargo	Kg 4.0 m	

The light displacement is 1000 t, and the loaded Summer displacement is 3500 t.

Item	Weight	Kg	Deadweight Moment
Light disp.	1000 t	–	–
Fuel oil	250 t	0.5 m	125 t m
Free surface	–	–	1400 t m
Fresh water	50 t	5.0 m	250 t m
Fresh surface	–	–	500 t m
Cargo	2000 t	4.0 m	8000 t m
Present cond.	3300 t		10 275 t m – Point 2 (satisfactory)
Maximum balance	200 t	3.0 m	600 t m
Summer displ.	3500		10 875 t m – Point 3 (satisfactory)

Since 10 875 tonnes metres is less than the Maximum Permissible Deadweight Moment at a displacement of 3500 tonnes, the ship will not have deficient stability and may load 200 tonnes of cargo.

Ans. Load 200 tonnes

Example 2

Using the Maximum Permissible Deadweight Moment (Figure 47.1) and the information given below, find the quantity of timber deck cargo (Kg 8.0 m) which can be loaded, allowing 15 per cent for water absorption during the voyage.

Summer displacement 4000 tonnes. Light displacement 1000 tonnes. Weights already on board:

Fuel oil 200 tonnes	Kg 0.5 m, Free surface moment 1400 t m
Fresh water 40 tonnes	Kg 5.0 m, Free surface moment 600 t m
Cargo 2000 tonnes	Kg 4.0 m
Ballast 350 tonnes	Kg 0.5 m

The following weights will be consumed during the voyage:

Fuel oil 150 tonnes   Kg 0.5 m   Free surface moment will be
reduced by 800 t m

Fresh water 30 tonnes   Kg 5.0 m   Free surface moment will be
by 200 t m

Departure condition

Item	Weight	Kg	Deadweight Moment
Light ship	1000	–	–
Fuel oil	200	0.5	100
Free surface			1400
Fresh water	40	5.0	200
Free surface			600
Cargo	2000	4.0	8000
Ballast	350	0.5	175
Departure disp. (without deck cargo)	3590		10 475 – Point 4 (satisfactory)
Maximum deck cargo	410	8.0	3280
Summer disp.	4000		13 755 – Point 5 (deficient stability)

From Figure 47.1, where the line joining Points 4 and 5 cuts the curve of
Maximum Permissible Deadweight Moments (Point 6), the displacement is
3920 tonnes.

Total departure displacement   3920 tonnes
Departure displacement without deck cargo   3590 tonnes
∴ Max deck cargo to load   330 tonnes

$$\text{Absorption during voyage} = \frac{15}{100} \times 330 = 49.5 \text{ tonnes}$$

Arrival Condition

Item	Weight	Kg	Deadweight Moment
Departure disp. (without deck cargo)	3590		10 475
Fuel oil	−150	0.5	−75
Free surface			−800
Fresh water	−30	5.0	−150
Free surface			−200
Arrival disp. (without deck cargo)	3410		9250
Deck cargo	330	8.0	2640
Absorption	49.5	8.0	396
Total arrival disp.	3789.5		12 286 – Point 7 (satisfactory stability)

Ans. Load 330 tonnes of deck cargo

Exercise 47

1   Using the Maximum Permissible Deadweight Moment (Figure 47.1), find the amount of deck cargo (Kg 8.0 m) which can be loaded allowing 15 per cent for water absorption during the voyage given the following data:

Light displacement 1000 tonnes. Loaded displacement 4000 tonnes. Weights already on board:

Item	Weight	Kg	Free surface moment
Cargo	1800	4.0	–
Fuel oil	350	0.5	1200
Fresh water	50	5.0	600
Ballast	250	0.5	–

During the voyage the following will be consumed (tonnes and kg):

Fuel oil   250   0.5   Reduction in free surface moment 850 t m
Fresh water 40   5.0   Reduction in free surface moment 400 t m

Chapter 48
The stability pro-forma

There is now a trend to supply a blank pro-forma for the master or mate to fill in that covers the loading condition for a ship. As the trim and stability calculations are made this information is transferred or filled in on the pro-forma. The following example shows the steps in this procedure.

Worked example

Use the supplied blank 'Trim and Stability' pro-forma and the supplied 'Hydrostatic Particulars Table' to fill in the pro-forma and determine each of the following:

(a) The effective metacentric height GM_T.
(b) The final drafts, forward and aft.

The first steps are to estimate the total displacement, the VCG and the LCG for this displacement. They came to 14 184 t, 7.07 m and 66.80 m forward of aft perpendicular (foap).

The true mean draft is then required. This is calculated by direct interpolation of draft for displacements between 14 115 t and 14 345 t in the 'Hydrostatic Particulars Table.'

Based on the ship's displacement of 14 184 t, this direct interpolation must be between 6.80 m and 6.90 m, a difference of 0.10 m.

$$\text{True mean draft} = 6.80 + \{(14\ 184 - 14\ 115)/(14\ 354 - 14\ 115)\} \times 0.10$$
$$= 6.80 + \{0.3 \times 0.10\} = 6.83 \text{ metres}$$
$$MCTC_{sw} = 181.4 + \{0.3 \times 1.6\} = 181.88 \text{ say } 181.9 \text{ tm/cm}$$
$$KM_T = 8.36 - \{0.3 \times 0.01\} = 8.36 \text{ m}$$
$$LCB = 70.12 - \{0.3 \times 0.04\} = 70.11 \text{ m foap}$$
$$LCF = 67.57 - \{0.3 \times 0.11\} = 67.54 \text{ m foap}$$

These values can now be inserted into the pro-forma sheet.

(a) *To determine the effective metacentric height GM_T.*

$$KG_{solid} = \text{Sum of the vertical moments/W}$$
$$= 100\ 272/14\ 184$$
$$KG_{solid} = 7.07 \text{ m}$$

TS002-72   SHIP STABILITY

WORKSHEET Q.1

(This Worksheet must be returned with your examination answer book)

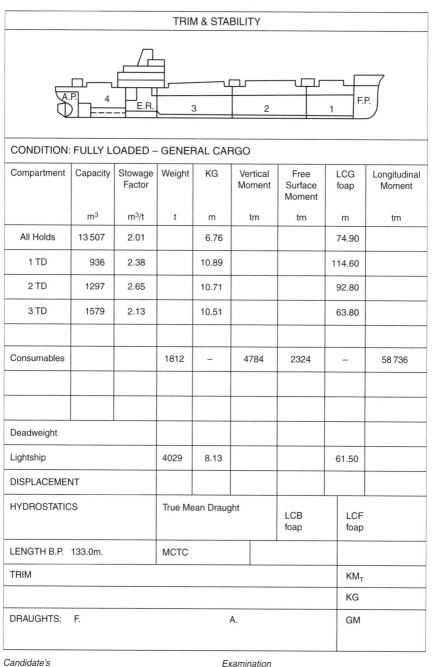

TRIM & STABILITY

CONDITION: FULLY LOADED – GENERAL CARGO

Compartment	Capacity	Stowage Factor	Weight	KG	Vertical Moment	Free Surface Moment	LCG foap	Longitudinal Moment
	m³	m³/t	t	m	tm	tm	m	tm
All Holds	13 507	2.01		6.76			74.90	
1 TD	936	2.38		10.89			114.60	
2 TD	1297	2.65		10.71			92.80	
3 TD	1579	2.13		10.51			63.80	
Consumables			1812	–	4784	2324	–	58 736
Deadweight								
Lightship			4029	8.13			61.50	
DISPLACEMENT								

HYDROSTATICS		True Mean Draught			LCB foap	LCF foap	
LENGTH B.P. 133.0m.		MCTC					
TRIM						KM_T	
						KG	
DRAUGHTS: F.			A.			GM	

Candidate's
Name ....

Examination
Centre ....

TS002-72   SHIP STABILITY

DATASHEET Q.1

(*This Worksheet must be returned with your answer book*)

HYDROSTATIC PARTICULARS 'A'

DRAUGHT	DISPLACEMENT		TPC		MCTC		KM_T	KB	LCB foap	LCF foap
	t		t		tm					
	SW	FW	SW	FW	SW	FW				
m	RD 1.025	RD 1.000	RD 1.025	RD 1.000	RD 1.025	RD 1.000	m	m	m	m
7.00	14576	14220	23.13	22.57	184.6	180.1	8.34	3.64	70.03	67.35
6.90	14345	13996	23.06	22.50	183.0	178.5	8.35	3.58	70.08	67.46
6.80	14115	13771	22.99	22.43	181.4	177.0	8.36	3.53	70.12	67.57
6.70	13886	13548	22.92	22.36	179.9	175.5	8.37	3.48	70.16	67.68
6.60	13657	13324	22.85	22.29	178.3	174.0	8.38	3.43	70.20	67.79
6.50	13429	13102	22.78	22.23	176.8	172.5	8.39	3.38	70.24	67.90
6.40	13201	12879	22.72	22.17	175.3	171.0	8.41	3.33	70.28	68.00
6.30	12975	12658	22.66	22.11	173.9	169.6	8.43	3.28	70.32	68.10
6.20	12748	12437	22.60	22.05	172.5	168.3	8.46	3.22	70.35	68.20
6.10	12523	12217	22.54	21.99	171.1	167.0	8.49	3.17	70.38	68.30
6.00	12297	11997	22.48	21.93	169.8	165.7	8.52	3.11	70.42	68.39
5.90	12073	11778	22.43	21.87	168.5	164.4	8.55	3.06	70.46	68.43
5.80	11848	11559	22.37	21.82	167.3	163.2	8.59	3.01	70.50	68.57
5.70	11625	11342	22.32	21.77	166.1	162.1	8.63	2.95	70.53	68.65
5.60	11402	11124	22.26	21.72	165.0	161.0	8.67	2.90	70.57	68.73
5.50	11180	10908	22.21	21.66	163.9	160.0	8.71	2.85	70.60	68.80
5.40	10958	10691	22.15	21.61	162.9	158.9	8.76	2.80	70.64	68.88
5.30	10737	10476	22.10	21.56	161.8	157.9	8.81	2.74	70.68	68.95
5.20	10516	10260	22.05	21.51	160.8	156.9	8.86	2.69	70.72	69.02
5.10	10296	10045	22.00	21.46	159.8	155.9	8.92	2.63	70.75	69.09
5.00	10076	9830	21.95	21.41	158.8	154.9	8.98	2.58	70.79	69.16
4.90	9857	9616	21.90	21.36	157.9	154.0	9.06	2.53	70.82	69.23
4.80	9638	9403	21.85	21.32	156.9	153.1	9.13	2.48	70.86	69.29
4.70	9420	9190	21.80	21.27	156.0	152.2	9.22	2.43	70.90	69.35
4.60	9202	8978	21.75	21.22	155.1	151.3	9.30	2.38	70.93	69.42
4.50	8985	8766	21.70	21.17	154.2	150.5	9.40	2.32	70.96	69.48
4.40	8768	8554	21.65	21.12	153.3	149.6	9.49	2.27	71.00	69.55
4.30	8552	8344	21.60	21.07	152.4	148.7	9.60	2.22	71.04	69.62
4.20	8336	8133	21.55	21.02	151.5	147.8	9.71	2.17	71.08	69.68
4.10	8121	7923	21.50	20.97	150.6	146.9	9.83	2.12	71.12	69.74
4.00	7906	7713	21.45	20.93	149.7	146.0	9.96	2.07	71.15	69.81
3.90	7692	7505	21.40	20.88	148.7	145.1	10.11	2.01	71.18	69.88
3.80	7478	7296	21.35	20.83	147.8	144.2	10.25	1.96	71.22	69.94
3.70	7265	7088	21.30	20.78	146.8	143.3	10.41	1.91	71.25	70.00
3.60	7052	6880	21.24	20.72	145.9	142.3	10.57	1.86	71.29	70.07
3.50	6840	6673	21.19	20.67	144.9	141.3	10.76	1.81	71.33	70.14
THESE HYDROSTATIC PARTICULARS HAVE BEEN DEVELOPED WITH THE VESSEL FLOATING ON EVEN KEEL										

*Candidate's
Name* ....

*Examination
Centre* ....

TS002-72   SHIP STABILITY

WORKSHEET Q.1

(*This Worksheet must be returned with your examination answer book*)

TRIM & STABILITY								
CONDITION: FULLY LOADED – GENERAL CARGO								
Compartment	Capacity m³	Stowage Factor m³/t	Weight t	KG m	Vertical Moment tm	Free Surface Moment tm	LCG foap m	Longitudinal Moment tm
All Holds	13 507	2.01	**6720**	6.76	**45 427**		74.90	5 03 328
1 TD	936	2.38	**393**	10.89	**4280**		114.60	45 038
2 TD	1297	2.65	**489**	10.71	**5237**		92.80	45 379
3 TD	1579	2.13	**741**	10.51	**7788**		63.80	47 276
Consumables			1812	–	4784	2324	–	58 736
Deadweight			**10 155**					
Lightship			4029	8.13	**32 756**		61.50	2 47 784
DISPLACEMENT			**14 184**	7.07	**100 272**	2324	66.80	9 47 541

HYDROSTATICS	True Mean Draught 6.83 m	LCB foap 70.11 m	LCF foap 67.54 m
LENGTH B.P. 133.0 m	MCTC = 181.9 tm		
TRIM = **2.58 m by the stern**			KM$_T$ = **8.36 m**
			KG$_{FL}$ = **7.23 m**
DRAUGHTS: F. = 5.56 m	A. = 8.14 m		GM$_T$ = **1.13 m** (fluid)

Also see Figure 48.1 and Figure 48.2

*Candidate's
Name* ....

*Examination
Centre* ....

plus

$$\text{FSM/W} = 2324/14\,184 = \underline{0.16 \text{ m}}$$
$$\text{KG}_{\text{fluid}} = \underline{7.23 \text{ m}}$$
$$\text{Effective metacentric height } \text{GM}_T = \text{KM}_T - \text{KG}_{\text{fluid}}$$
$$= 8.36 - 7.23$$
$$\text{Effective metacentric height } \text{GM}_T = 1.13 \text{ m} = \text{GM}_{\text{fluid}}$$

These values can now be inserted into the pro-forma sheet.

(b) *To determine the final end drafts.*

$$\text{Change of trim} = \{W \times (\text{LCB} - \text{LCG})\}/\text{MCTC}_{\text{sw}}$$

Note that

$$\text{LCG} = \text{Sum of the long'l moments/W} = 947\,541/14\,184$$

So

$$\text{LCG} = 66.80 \text{ m foap}$$
$$\text{Change of trim} = \{14\,184\,(70.11 - 66.80)\}/181.9$$
$$= 258 \text{ cm} = 2.58 \text{ m by the stern}$$

(b) Final draft aft = True mean draft + (LCF/LBP) × change of trim
LBP of 133 m for this ship is given on the blank pro-forma.

$$\text{Final draft aft} = 6.83 + (67.54/133) \times 2.58$$
$$\text{Final draft aft} = 8.14 \text{ m}$$

(b)   Final draft ford = True mean draft − (LBP − LCF/LBP) × change of trim
$$= 6.83 - (65.46/133) \times 2.58$$
Final draft forward = 5.56 m

Check:
Aft   draft − forward   draft = 8.14 − 5.56 = 2.58 m   change   of   trim
(as calculated earlier).
These values can be inserted into the pro-forma sheet.
Figure 48.1 shows the positions of all the transverse stability values.
Figure 48.2 shows the ship's final end drafts and final trim by the stern.

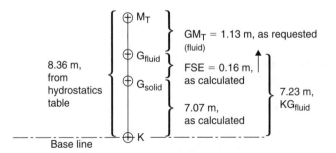

Fig. 48.1   Final transverse stability details.

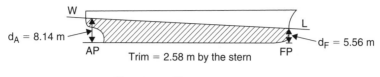

Fig. 48.2   Final trim details.

Exercise 48

1 Use the supplied blank 'Trim and Stability' pro-forma and the supplied 'Hydrostatic Particulars Table' to fill in the pro-forma and determine each of the following:
   (a) Fully loaded deadweight in tonnes.
   (b) Fully-loaded displacement in tonnes.
   (c) Final aft draft in metres.
   (d) Final forward draft in metres.
   (e) Final effective metacentric height GM_T in metres.

TS032-72   SHIP STABILITY

WORKSHEET Q.1

(*This Worksheet must be returned with your answer book*)

TRIM & STABILITY

CONDITION: FULLY LOADED – GENERAL CARGO

Compartment	Capacity	Stowage Factor	Weight	KG	Vertical Moment	Free Surface Moment	LCG foap	Longitudinal Moment
	m³	m³/t	t	m	tm	tm	m	tm
All Holds	14 562	1.86		6.78			73.15	
1 TD	264	2.48		10.71			114.33	
2 TD	1688	2.74		10.60			93.57	
3 TD	1986	2.72		10.51			63.92	
Consumables			1464	–	4112	2560	–	58 675
Deadweight								
Lightship			3831	8.21			61.67	
DISPLACEMENT								

HYDROSTATICS	True Mean Draught		LCB foap	LCF foap
LENGTH B.P. 130.00 m	MCTC			
TRIM				KM$_T$
				KG
DRAUGHTS: F.		A.		GM$_{fluid}$

Candidate's
Name ....

Examination
Centre ....

TS032-72   SHIP STABILITY

DATASHEET Q.1

(This Worksheet must be returned with your answer book)

HYDROSTATIC PARTICULARS

DRAUGHT	DISPLACEMENT		TPC		MCTC		KM$_T$	KB	LCB foap	LCF foap
	t		t		tm					
m	SW RD 1.025	FW RD 1.000	SW RD 1.025	FW RD 1.000	SW RD 1.025	FW RD 1.000	m	m	m	m
7.00	14576	14220	23.13	22.57	184.6	180.1	8.34	3.64	70.03	67.35
6.90	14345	13996	23.06	22.50	183.0	178.5	8.35	3.58	70.08	67.46
6.80	14115	13771	22.99	22.43	181.4	177.0	8.36	3.53	70.12	67.57
6.70	13886	13548	22.92	22.36	179.9	175.5	8.37	3.48	70.16	67.68
6.60	13657	13324	22.85	22.29	178.3	174.0	8.38	3.43	70.20	67.79
6.50	13429	13102	22.78	22.23	176.8	172.5	8.39	3.38	70.24	67.90
6.40	13201	12879	22.72	22.17	175.3	171.0	8.41	3.33	70.28	68.00
6.30	12975	12658	22.66	22.11	173.9	169.6	8.43	3.28	70.32	68.10
6.20	12748	12437	22.60	22.05	172.5	168.3	8.46	3.22	70.35	68.20
6.10	12523	12217	22.54	21.99	171.1	167.0	8.49	3.17	70.38	68.30
6.00	12297	11997	22.48	21.93	169.8	165.7	8.52	3.11	70.42	68.39
5.90	12073	11778	22.43	21.87	168.5	164.4	8.55	3.06	70.46	68.43
5.80	11848	11559	22.37	21.82	167.3	163.2	8.59	3.01	70.50	68.57
5.70	11625	11342	22.32	21.77	166.1	162.1	8.63	2.95	70.53	68.65
5.60	11402	11124	22.26	21.72	165.0	161.0	8.67	2.90	70.57	68.73
5.50	11180	10908	22.21	21.66	163.9	160.0	8.71	2.85	70.60	68.80
5.40	10958	10691	22.15	21.61	162.9	158.9	8.76	2.80	70.64	68.88
5.30	10737	10476	22.10	21.56	161.8	157.9	8.81	2.74	70.68	68.95
5.20	10516	10260	22.05	21.51	160.8	156.9	8.86	2.69	70.72	69.02
5.10	10296	10045	22.00	21.46	159.8	155.9	8.92	2.63	70.75	69.09
5.00	10076	9830	21.95	21.41	158.8	154.9	8.98	2.58	70.79	69.16
4.90	9857	9616	21.90	21.36	157.9	154.0	9.06	2.53	70.82	69.23
4.80	9638	9403	21.85	21.32	156.9	153.1	9.13	2.48	70.86	69.29
4.70	9420	9190	21.80	21.27	156.0	152.2	9.22	2.43	70.90	69.35
4.60	9202	8978	21.75	21.22	155.1	151.3	9.30	2.38	70.93	69.42
4.50	8985	8766	21.70	21.17	154.2	150.5	9.40	2.32	70.96	69.48
4.40	8768	8554	21.65	21.12	153.3	149.6	9.49	2.27	71.00	69.55
4.30	8552	8344	21.60	21.07	152.4	148.7	9.60	2.22	71.04	69.62
4.20	8336	8133	21.55	21.02	151.5	147.8	9.71	2.17	71.08	69.68
4.10	8121	7923	21.50	20.97	150.6	146.9	9.83	2.12	71.12	69.74
4.00	7906	7713	21.45	20.93	149.7	146.0	9.96	2.07	71.15	69.81
3.90	7692	7505	21.40	20.88	148.7	145.1	10.11	2.01	71.18	69.88
3.80	7478	7296	21.35	20.83	147.8	144.2	10.25	1.96	71.22	69.94
3.70	7265	7088	21.30	20.78	146.8	143.3	10.41	1.91	71.25	70.00
3.60	7052	6880	21.24	20.72	145.9	142.3	10.57	1.86	71.29	70.07
3.50	6840	6673	21.19	20.67	144.9	141.3	10.76	1.81	71.33	70.14
THESE HYDROSTATIC PARTICULARS HAVE BEEN DEVELOPED WITH THE VESSEL FLOATING ON EVEN KEEL										

Candidate's Name..

Examination Centre..

TS032-72   **SHIP STABILITY**

WORKSHEET Q.1

(*This Worksheet must be returned with your examination answer book*)

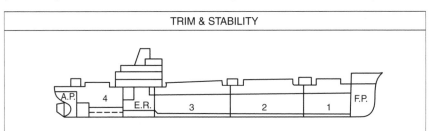

TRIM & STABILITY

CONDITION: FULLY LOADED – GENERAL CARGO

Compartment	Capacity m³	Stowage Factor m³/t	Weight t	KG m	Vertical Moment tm	Free Surface Moment tm	LCG foap m	Longitudinal Moment tm
All Holds	14 562	1.86	**7829**	6.78	**53 081**		73.15	**572 691**
1 TD	264	2.48	**106**	10.71	**1135**		114.33	**12 119**
2 TD	1688	2.74	**616**	10.60	**6530**		93.57	**57 639**
3 TD	1986	2.72	**730**	10.51	**7672**		63.92	**46 662**
Consumables			1464	–	4112	2560	–	58 675
Deadweight			**10 745**					
Lightship			3831	8.21	**31 453**		61.67	**236 258**
DISPLACEMENT			**14 576**	7.13	**103 983**	**2560**	**67.51**	**984 044**

HYDROSTATICS	True Mean Draught 7.00 m	LCB foap 70.03	LCF foap 67.35 m

| LENGTH B.P. 130.00 m | MCTC = **184.6** | | |

$$\text{TRIM} = \frac{14\,576 \times (70.03 - 67.51)}{184.6} = 199 \text{ cm by the STERN} \qquad KM_T = 8.34\,m$$

$$KG_{fluid} = 7.13 + \frac{2560}{14\,576} = 7.13 + 0.18 \qquad KG = 7.31\,m$$

DRAUGHTS: F. = **6.04 m**	A. = **8.03 m**	GM$_{fluid}$ = **1.03 m**

Candidate's
Name ....

Examination
Centre ....

Nomenclature of ship terms

aft terminal	the aftmost point on a ship's floodable length.
air draft	the vertical distance from the waterline to the highest point on the ship.
amidships	a position midway between the aft perpendicular (AP) and the forward perpendicular (FP).
angle of list	an angle of heel where G is transversely offset from the ship's centreline and transverse GM is positive.
angle of loll	an angle of heel where G is on the ship's centreline and transverse GM is zero.
angle of repose	an angle created by a shift of grain.
apparent trim	the difference between the drafts observed at the for'd and aft draft marks.
appendage	a small attachment to a main area or body.
assigned freeboard	the ship's freeboard, after corrections have been made to the DfT Tabular Freeboard value.
blockage factor	area of ship's midship section divided by the cross-sectional area of a river or canal.
boot-topping	the vertical distance between the lightdraft and the SLWL.
bow height	a vertical distance measured at the FP, from the waterline to the uppermost deck exposed to the weather.
bulk carriers	workhorse vessels, built to carry such cargoes as ore, coal, grain, sugar, etc. in large quantities.
calibrated tank	a tank giving volumes or weights at vertical increments of say one centimetre in the tank's depth.

cargo-passenger ship	a vessel that carries cargo and up to 12 paying passengers.
C_b	block coefficient: linking the volume of displacement with LBP, breadth mld and draft.
compressive stress	stress that is forcing the fibres in the material to be crushed together.
computer packages	packages for estimating stability, trim, end drafts, shear forces and bending moments for a condition of loading.
confined channel	a river or canal where there is a nearby presence of banks.
continuous girder	a girder that runs continuously from say main transverse bulkhead to main transverse bulkhead.
C_w	waterplane area coefficient: linking the waterplane area with the LBP and the ship's breadth mld.
deadweight	the weight that a ship carries.
deck camber	transverse curvature of a deck, measured from deck height at centreline to deck height at side, at amidships.
deck sheer	longitudinal curvature of a deck, measured vertically from amidships to the deck, at Aft Perp or Forward Perp.
depth moulded	measured from top of keel to underside of uppermost continuous deck, at amidships.
DfT	Department for Transport.
displacement	for all conditions of loading, it is the lightweight plus any deadweight.
domain of ship	mainly the area in which the pressure bulbs exist on a moving vessel.
draft moulded	distance from the waterline to the top of keel, measured at amidships.
dwt	abbreviation for deadweight, the weight that a ship carries.
dynamical stability	the area under a statical stability curve multiplied by the ship's displacement.
even keel	a vessel with no trim: where the aft draft has the same value as the forward draft.
factor of subdivision F_s	the floodable length ordinates times F_s gives the permissible length ordinates.
F_B	breadth of influence in open water conditions.
forward terminal	the foremost most point on a ship's floodable length.

free surface effects	loss of stability caused by liquids in partially filled tanks.
freeboard ratio	freeboard to the margin line @ amidships/ship's draft moulded.
general particulars	LBP, Br. Mld, depth mld, Draft Mld, Lightweight, Deadweight, Displacement, Cb, Service Speed, etc.
grain cargo	this covers wheat, maize, corn, oats, rye, barley, seeds, rice, etc.
grain heeling moments	transverse moments caused by the transverse movement of grain in a compartment.
heaving motion	the vertical movement of a ship's VCG.
hydrometer	an instrument used to measure the density of a liquid.
hydrostatic curves	used for calculating the trim and stability values for various conditions of loading.
hydrostatic table	a tabular statement of values on hydrostatic curves.
icing allowances	must be considered when dealing with loss of stability and change of trim.
IMO	International Maritime Organisation.
inertia coefficient	used for obtaining moments of inertias of waterplanes.
interaction	action and reaction of ships when they get too close to one another or too close to a river bank.
intercostal girder	a girder that is non-continuous.
knot	1852 m per hour.
lightship draft	draft of ship when ship is empty, with a deadweight of zero.
lightweight	empty ship, with boilers topped up to working level.
LNG ships	liquefied natural gas carrier with cargo at -161 degrees C.
LOA	length overall, from the foremost part of the bow to the aftermost part of the stern.
longitudinal	running from bow to stern in a fore and aft direction.
LPG ships	liquefied petroleum gas carrier with cargo at -42 degrees C.
margin line	a line that is 75 mm below the bulkhead deck @ side.
MCA	Marine Coastguard Agency
moment	an area times a lever or a weight times a lever.

Murray's method	a method for estimating longitudinal bending moments.
NA	data 'not available' or 'not applicable'.
NK	data 'not known'.
open water	a stretch of water where there are no adjacent river or canal banks.
P&S	port and starboard.
Panamax vessel	a vessel having a breadth mld of 32.26 m.
parametric rolling	additional rolling of a ship, caused by having a bluff stern in interacting with a sharp streamlined bow form.
passenger liners	vessels travelling between definite ports, with timetabled departure and arrival dates.
passenger-cargo ship	a vessel that carries cargo and more than 12 paying passengers.
permeability	the amount of water that can enter a bilged compartment, usually expressed as a percentage.
pitching motion	vertical see-saw movement of a ship at the bow and at the stern.
Plimsoll disc	centre of this disc is in line with the SLWL and its centre is spot on amidships.
port	left side of a ship when looking forward.
pressure bulbs	bulbs of pressure that build up around a moving vessel and disappear when vessel stops.
retrofit	a structure later added or deducted after the delivery of the ship to the shipowner.
righting lever	a lever that will bring a stable ship back to the upright after being displaced by temporary external forces.
righting moment	a moment that will bring a stable ship back to the upright.
rolling motion	a ship roll from extreme port to extreme starboard and back again to extreme port.
RO-RO ships	roll-roll vessels that carry cars/lorries and passengers.
shallow water	where the depth of water reduces the ship speed and prop revs, increases squat, reduces rolling motions, etc.
sheer ratio	deck sheer forward or aft/ship's draft moulded.
shifting boards	used to prevent a transverse shift of grain.
ship surgery	lengthening, deepening or widening a ship after cutting her transversely, along or longitudinally.

Simpson's Rules	used for calculating areas, moments, volumes and inertias, without needing to use calculus.
SLWL	summer load waterline similar to draft moulded, in a density of water of 1.025 t/cu. m.
sounding	a vertical distance measured from the bottom of a tank to the surface of a liquid in the tank.
sounding pad	a steel pad at the base of a sounding pipe.
sounding pipe	a pipe of about 37.5 mm that can be used for taking soundings with a steel sounding tape.
sounding tube/glass	a vertical tube used for reading calibrated values of a liquid in a tank.
SQA	Scottish Qualifications Authority.
squat	loss of underkeel clearance as a ship moves forward from rest.
St Lawrence Seaway max	where the ship's breadth mld is 23.8 m.
Starboard	right side of a ship when looking forward.
stiff ship	a vessel with a quick rapid rolling period.
stowage factor	volume per unit weight for a grain cargo.
stringer plate	the line of deck plates nearest to the shearstrake or gunwhale plating.
Summer freeboard	the minimum freeboard for a ship floating in water with a density of salt water of 1.025 t/cu. m.
supertanker	similar to a VLCC: having a dwt of 100 000 t to 300 000 t.
synchronous rolling	the roll of the ship being in phase with the action of the waves.
tabular freeboard	minimum Summer freeboard for a DfT standard ship.
tender ship	a vessel with a long slow lazy rolling period.
tensile stress	stress that is forcing the fibres in the material to be pulled apart.
timber loadlines	marked on ships that carry timber on the main deck.
tonne	equivalent to 1000 kg.
transverse	running from port to starboard across the ship.
transverse squat	squat at the bilge plating, caused by ships overtaking or passing in a river.
trim	the difference between the forward draft and the aft draft.
true trim	the difference between the drafts at for'd perpendicular and after perpendicular.

Type 'A' ship	a ship that carries liquid in bulk.
Type 'B' ship	a ship other than Type 'A' ship.
ukc	underkeel clearance, being depth of water minus a ship's draft.
ULCC	ultra large crude carrier, say over 300 000 t dwt.
ullage	vertical distance from surface of a liquid in the tank to top of the sounding pipe or top of ullage plug.
VLCC	very large crude carrier, say 100 000 t to 300 000 t dwt.
wind heel moments	moments caused by the wind on side of the ship causing ship to have angle of heel.
Young's modulus	the value obtained by stress/strain.

Photographs of merchant ships

The 'HAMANASU', a Ro-Ro vessel
LBP = 208.0 m. Dwt = 5824 t. Vehicles: 158 trailers/65 cars.

The 'BRITISH LIBERTY', a chemical carrier.
LBP = 174.0 m. Dwt = 40 580 t. Built to class 1A ICE strengthening requirements.

The 'DISHA', an LNG vessel.
LBP = 266.0 m. LNG = 138 000 m^3. Dwt = 70 150 t. 17 Officers plus 21 crew.

The 'HELLESPONT ALHAMBRA', an ULCC.
LBP = 366.0 m. Dwt = 407 469 t.
Double hull. Fuel consumption = 140.3 tonnes/day.

The 'Queen Mary 2', passenger liner.
LBP = 301.35 m. Gross tonnage = 150 000 gt.
Number of passengers = 2620. Number of crew = 1254.

The 'MANUKAI', Container ship.
LBP = 202.24 m. Dwt = 30 000 t.
Containers:Total TEU capacity = 3000.

Ships of this millennium

Delivery Date	Name of Vessel	Deadweight tonnes	LBP (L) m	Br. Mld (B) m	L/B value	Depth (D) m	Draft (H) m	H/D value	Speed knots
Gas Carriers									
Oct 2002	Norgas Orinda	6000	115.0	19.80	5.81	11.50	6.70	0.58	16.70
July 2003	Shinju Maru No 1	1781	80.3	15.10	5.32	7.00	4.18	0.60	12.70
Aug 2003	Inigo Tapias	68 200	271.0	42.50	6.38	25.40	11.40	0.45	19.50
Jan 2004	Disha	70 150	266.0	43.40	6.13	26.00	11.40	0.44	19.50
Oct 2004	Polar Viking	36 500	195.0	32.20	6.06	20.80	10.80	0.52	16.00
ULCCs									
Mar 2002	Hellespont Alhambra	407 469	366.0	68.00	5.38	34.00	23.00	0.68	16.00
1980 to 2005	Jahre Viking	564 769	440.0	68.80	6.40	29.80	24.61	0.83	13.00
Passenger Liners									
Nov 2003	Costa Fortuna	8200	230.0	35.50	6.48	19.78	8.20	0.41	20.00
Dec 2003	QM2	15 000	301.4	41.00	7.35	13.70	10.00	0.73	30.00
May 2004	Sapphire Princess	10 343	246.0	37.50	6.56	11.40	8.05	0.71	22.10
Dec 2004	Color fantasy	5600	202.7	35.00	5.79	9.50	6.80	0.72	22.30
June 2005*	Pride of America	8260	259.1	32.21	8.04	10.80	8.00	0.74	22.00

VLCCs									
Oct 2001	Harad	284 000	318.0	58.00	5.48	31.25	21.40	0.68	16.40
Jul 2003	Capricorn star	299 000	319.0	60.00	5.32	30.50	21.50	0.70	14.60
Jan 2004	World Lion	276 840	320.0	58.00	5.52	31.20	20.80	0.67	15.30
Oct 2004	Irene SL	292 205	319.0	60.00	5.32	30.40	21.00	0.69	15.70
Bulk Carriers									
Jan 2004	Tai Progress	64 000	217.0	32.26	6.73	19.50	12.20	0.63	14.50
June 2004	Azzura	52 050	181.0	32.20	5.62	17.30	10.70	0.62	14.70
July 2004	Anangel Innovation	155 200	279.0	45.00	6.20	24.30	16.50	0.68	15.10
Nov 2004	Chin Shan	159 887	281.5	45.00	6.26	24.10	16.50	0.68	14.80
Feb 2005*	Big Glory	47 965	185.0	32.26	5.73	17.80	11.10	0.62	14.60
Cargo Ships									
Apr 2002	Arklow Rally	4500	85.0	14.40	5.90	7.35	5.79	0.79	11.50
Aug 2002	St Apostle Andrew	3670	122.8	16.30	7.53	6.10	4.20	0.69	11.00
XXX	SD14	15 025	137.5	20.40	6.74	11.80	8.90	0.75	15.00
XXX	SD20 (3rd generation)	19 684	152.5	22.80	6.69	12.70	9.20	0.72	15.00
Chemical Carriers									
July 2002	Maritea	34 659	169.0	29.78	5.67	16.81	11.00	0.65	15.00
Jan 2003	Alnoman	30 000	168.0	31.00	5.42	17.20	9.00	0.52	15.00
Nov 2004	British Liberty	40 580	174.0	32.20	5.40	18.80	11.00	0.59	15.20
Apr 2005*	Seychelles Pioneer	37 500	175.2	28.00	6.25	16.80	11.70	0.70	15.20
Oct 2005*	Ottomana	25 000	158.7	27.40	5.79	14.60	9.50	0.65	15.50

Delivery Date	Name of Vessel	Deadweight tonnes	LBP (L) m	Br. Mld (B) m	L/B value	Depth (D) m	Draft (H) m	H/D value	Speed knots
Container ships									
May 2002	Safmarine Convene	27 900	200.0	29.80	6.71	16.70	10.10	0.60	22.00
Apr 2003	OOCL Shenzhen	82 200	308.0	42.80	7.20	24.60	13.00	0.53	25.00
Nov 2003	Thomas Mann	27 930	202.1	32.20	6.28	16.50	10.50	0.64	22.70
Aug 2004	Cap Doukato	27 340	195.4	29.80	6.56	16.40	10.10	0.62	22.30
Jan 2005	Geeststrom	9400	130.0	21.80	5.96	9.50	7.32	0.77	18.00
Mar 2005*	Beluga Revolution	10 700	125.6	21.50	5.84	9.30	7.00	0.75	18.00
Roll-on / roll-off vessels									
Nov 2002	Stena Britannica	9400	197.2	29.30	6.73	9.50	6.30	0.66	22.20
May 2003	Lobo Marinho	1250	98.2	20.00	4.91	7.00	4.50	0.64	21.00
Oct 2003	Tor Magnolia	8780	190.3	26.50	7.18	9.40	6.95	0.74	22.80
June 2004	Nuraghes	3900	192.4	26.40	7.29	10.00	6.90	0.69	29.50
June 2004	Hamanasu	5824	208.0	26.00	8.00	10.00	7.20	0.72	30.50
Sep 2005*	Smyrl	2652	123.0	22.70	5.42	8.10	5.60	0.69	21.55

Reference source: Significant ships of 2002, 2003 and 2004. Published annually by RINA, London.
*signifies "planned publication" for significant ships 2005.
XXX as per Standard General Cargo Ships (Fairplay Publications).

There are several types of smaller and faster ships such as the Hovercraft, the Hydrofoil, the Catamaran and the Trimaran. These vessels have been well documented by E.C. Tupper in his book entitled, 'Introduction to Naval Architecture' published by Elsevier Ltd in 2005.

To put a chapter in this book would amount to unnecessary duplication of effort for the same publisher. Furthermore, consideration of these small vessels do not appear in the new 2004 SQA/MCA syllabuses for Masters and Mates. Because of these reasons it is felt by the Author that these smaller ship types lie outside the remit of this book.

Part 2
Linking Ship Stability and Ship Strength

Chapter 49
Bending of beams

Beam theory

The bending of ships can be likened to the bending of beams in many cases. This chapter shows the procedures employed with beam theory.

The problem of calculating the necessary strength of ships is made difficult by the many and varied forces to which the ship structure is subjected during its lifetime. These forces may be divided into two groups, namely statical forces and dynamical forces.

The statical forces are due to:

1. The weight of the structure which varies throughout the length of the ship.
2. Buoyancy forces, which vary over each unit length of the ship and are constantly varying in a seaway.
3. Direct hydrostatic pressure.
4. Concentrated local weights such as machinery, masts, derricks, winches, etc.

The dynamical forces are due to:

1. Pitching, heaving and rolling.
2. Wind and waves.

These forces cause bending in several planes and local strains are set up due to concentrated loads. The effects are aggravated by structural discontinuities.

The purpose of the present chapter is to consider the cause of longitudinal bending and its effect upon structures.

Stresses

A stress is the mutual actual between the parts of a material to preserve their relative positions when external loads are applied to the material. Thus, whenever external loads are applied to a material stresses are created within the material.

Tensile and compressive stresses

When an external load is applied to a material in such a way as to cause an extension of the material it is called a 'tensile' load, whilst an external load tending to cause compression of the material is a 'compressive' load.

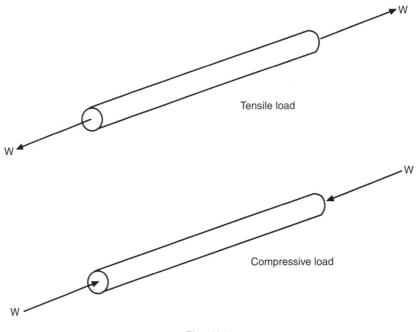

Fig. 49.1

Figure 49.1 shows a piece of solid material of cylindrical section to which an external load W is applied. In the first case the load tends to cause an extension of the material and is therefore a tensile load. The applied load creates stresses within the material and these stresses are called 'tensile' stresses. In the second case the load applied is one of the compression and the consequent stresses within the material are called 'compressive' stresses.

When a tensile or compressive external load is applied to a material, the material will remain in equilibrium only so long as the internal forces can resist the stresses created.

Shearing stresses

A shearing stress is a stress within a material which tends to break or shear the material across.

Figure 49.2(a) and (b) illustrates shearing stresses which act normally to the axis of the material.

In the following text when the direction of a shearing stress is such that the section on the right-hand side of the material tends to move downwards, as shown in Figure 49.2(a), the stress is considered to be positive, and when the direction of a stress is such that the section on the right-hand side tends to move upwards as shown in Figure 49.2(b), the shearing stress is considered to be negative.

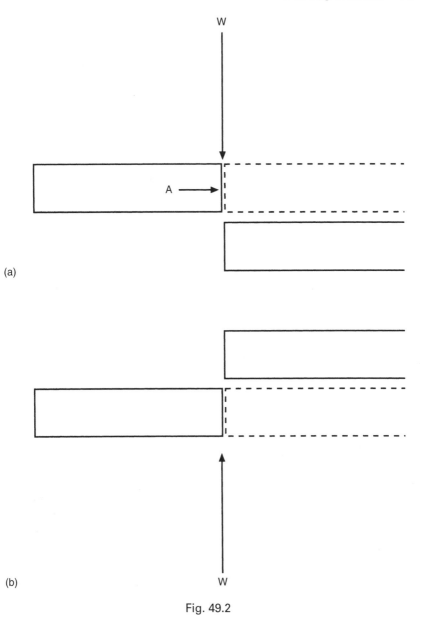

Fig. 49.2

Shearing stresses are resisted by the material but shearing will take place when the shear stress reaches the ultimate shear stress of the material.

Complementary stress

It has already been stated that when a direct load is applied to a material stresses are created within the material and that the material will remain in equilibrium only so long as the internal forces can resist the stresses created.

Let the bar in Figure 49.3(a) be subjected to tensile load W and imagine the bar to be divided into two parts at section AB.

For equilibrium, the force W on the left-hand side of the bar is balanced by an equal force acting to the right at section AB. The balancing force is

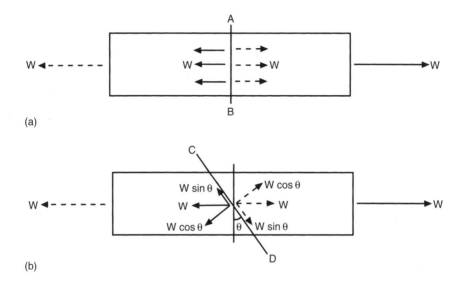

(a)

(b)

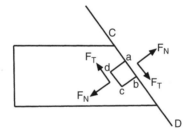

(c)

(d)

Fig. 49.3

supplied by the individual molecular forces which represent the action of the molecules on the right-hand side of the section on those of the left-hand section. Similarly, the force W on the right-hand side of the bar is balanced by the individual molecular forces to the left of the section. Therefore, when an external load W is applied to the bar, at any section normal to the axis of the bar, there are equal and opposite internal forces acting, each of which balances the external force W. The magnitude of the internal forces per unit area of cross-section is called the stress. When the section is well removed from the point of application of the external load then the stress may be considered constant at all parts of the section and may be found by the formula:

$$\text{Stress (f)} = \frac{\text{Load (W)}}{\text{Area (A)}} \quad \therefore f = \frac{W}{A}$$

Let us now consider the stresses created by the external load W in a section which is inclined to the axis of the bar. For example, let section CD in Figure 49.3(b) be inclined at an angle θ to the normal to the axis of the bar and let the section be sufficiently removed from the point of application of the load to give uniform stress across the section.

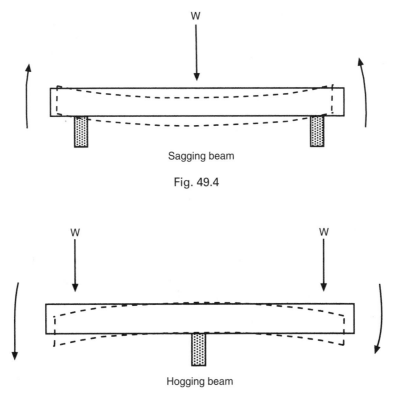

Sagging beam

Fig. 49.4

Hogging beam

Fig. 49.5

The load transmitted by the section CD, for equilibrium, is equal to the external force W. This load can be resolved into two components, one of which is W cos θ and acts normal to the section, and the other is W sin θ and acts tangential to the section. This shows that for direct tensile or compressive loading of the bar stresses other than normal stresses may be created.

Now let us consider the small block of material abcd in the section on the left-hand side of the plane, as shown in Figure 49.3(c). Let the face ab be coincident with the plane CD and let F_N be the internal force normal to this face and F_T the internal force tangential to the face. For the block to be in equilibrium the left-hand side of the section must provide two stresses on the face cd. These are F_N and F_T. Thus the stress F_N normal to the face ab is balanced by the stress F_N normal to the face cd whilst the two tangential stresses (F_T) on these faces tend to produce clockwise rotation in the block. Rotation can only be prevented if an equal and opposite couple is produced by opposing shearing stresses on the faces ad and bc as shown in Figure 49.3(d).

It can therefore be seen that when shear stresses occur at any plane within a material, equal shear stresses are produced on planes at right angles. These equal and opposing shearing stresses are called 'complementary' shearing stresses.

Bending moments in beams

The shear forces and bending moments created within a beam depend on both the way in which the beam is supported and the way in which it is loaded. The bending moment at any section within the beam is the total moment tending to alter the shape of the beam as shown in Figures 49.4 and 49.5, and is equal to the algebraic sum of the moments of all loads acting between the section concerned and either end of the beam.

In the following text, when a bending moment tends to cause sagging or downwards bending of the beam, as shown in Figure 49.4, it is considered to be a negative bending moment and when it tends to cause hogging or convex upwards bending of the beam, as shown in Figure 49.5, it is considered to be positive. Also, when bending moments are plotted on a graph, positive bending moments are measured below the beam and negative bending moments above.

Shear force and bending moment diagrams

The shear forces and bending moments created in a beam which is supported and loaded in a particular way can be illustrated graphically.

Consider first the case of cantilevers which are supported at one end only.

Case I

The beam AB in Figure 49.6 is fixed at one end only and carries a weight 'W' at the other end.

If the weight of the beam is ignored then at any point Y in the beam, which is at distance X from the end B, there is a positive shearing force W

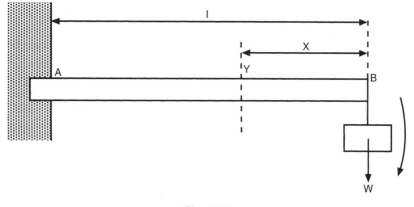

Fig. 49.6

and a positive bending moment W × X. There is thus a positive shearing force W at all sections throughout the length of the beam. This is shown graphically in Figure 49.7 where AB represents the length of the beam (l), and the ordinate AC, which represents the shearing force at A, is equal to the ordinate BD which represents the shearing force at B.

The bending moment at any section of the beam is the algebraic sum of the moments of forces acting on either side of the section. In the present case, the only force to consider is W which acts downwards through the end B. Thus the bending moment at B is zero and from B towards A the bending moment increases, varying directly as the distance from the end B.

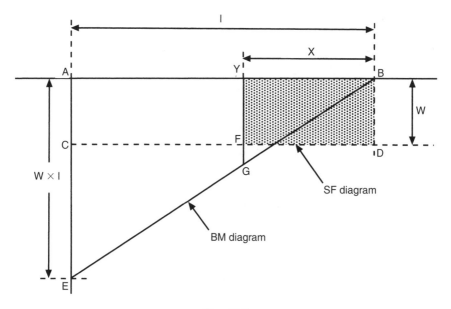

Fig. 49.7

The maximum bending moment, which occurs at A, is equal to W × l. This is shown graphically in Figure 49.7 by the straight line BGE.

The shearing force and bending moment at any point in the length of the beam can be found from the graph by inspection. For example, at Y the shearing force is represented by the ordinate YF and the bending moment by the ordinate YG.

It should be noted that the bending moment at any point in the beam is equal to the area under the shearing force diagram from the end of the beam to that point. For example, in Figure 49.7, the bending moment at Y is equal to W × X and this, in turn, is equal to the area under the shearing force diagram between the ordinates BD and YF.

Case II

Now consider a solid beam of constant cross-section which is supported at one end as shown in Figure 49.8. Let w be the weight per unit length of the beam.

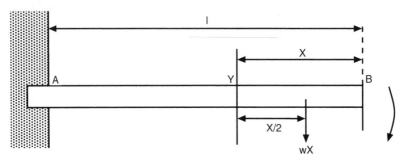

Fig. 49.8

At any section Y in the beam, which is at distance 'X' from B, there is a positive shearing force wX where wX is the weight of the beam up to that section and, since the weight wX may be taken to act half-way along the length X, there is a bending moment wX × X/2 or $\dfrac{wX^2}{2}$.

This is shown graphically in Figure 49.9, where AB represents the length of the beam (l).

The shearing force at B is zero and then increases towards A, varying directly as X, to reach its maximum value at A of wl. This is represented in Figure 49.9 by the straight line BFC.

The bending moment at any point in the beam is equal to $wX^2/2$. It is therefore zero at B and then increases towards A, varying directly as X^2, to reach its maximum value of $wl^2/2$ at A. The curve of bending moments is therefore a parabola and is shown in Figure 49.9 by the curve BGE.

Since the bending moment at any section is equal to the area under the shearing force diagram from the end of the beam to that section, it follows

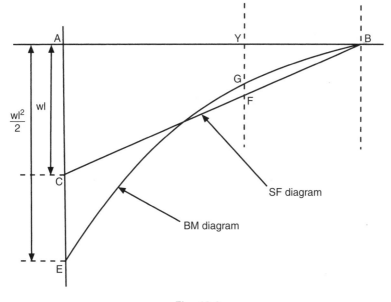

Fig. 49.9

that the bending moment curve may be drawn by first calculating the area under the shearing force diagram from the end of the beam to various points along it and then plotting these values as ordinates of the curve. For example, at section Y in Figure 49.9 the ordinate YF represents the shearing force at this section (wX), and the area under the shearing force diagram between B and the ordinate FY is equal to $\frac{1}{2} \times wX \times X$ or $wX^2/2$. The ordinate YG could now be drawn to scale to represent this value.

Freely supported beams

Case I

Consider now a beam which is simply supported at its ends, and loaded in the middle as shown in Figure 49.10. In this figure AB represents the length of the beam (l) and W represents the load. If the weight of the beam is neglected then the reaction at each support is equal to W/2, denoted by R_A and R_B.

To plot the shearing force diagram first draw two axes of reference as shown in Figure 49.11 with AB representing the length of the beam (l).

Now cover Figure 49.10 with the right hand, fingers pointing to the left, and slowly draw the hand to the right gradually uncovering the figure. At A there is a negative shearing force of W/2 and this is plotted to scale on the graph by the ordinate AC. The shearing force is then constant along the beam to its mid-point O. As the hand is drawn to the right, O is uncovered and a force W downwards appears. This must be considered in addition to the force W/2 upwards at A. The resultant is a shearing force of W/2 downwards

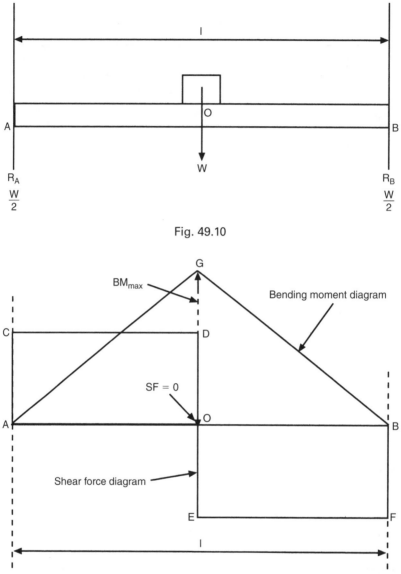

Fig. 49.10

Fig. 49.11

and this force is then constant from O to B as shown in Figure 49.11 by the straight line EF.

The bending moment diagram can now be drawn in the same way or by first calculating the area under the shearing force diagram between the ordinate AC and various points along the beam, and then plotting these values as ordinates on the bending moment diagram. It will be seen that the bending moment is zero at the ends and attains its maximum value at the mid-point in the beam indicated by the ordinate OG. *Note*. BM_{max} occurs when SF = 0.

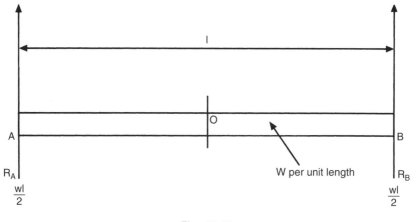

Fig. 49.12

Case II

Now consider a beam of constant cross-sectional area, of length l, and weight W per unit length. Let the beam be simply supported at its ends as shown in Figure 49.12, at reactions R_A and R_B.

The total weight of the beam is wl. The reaction at each end is equal to wl/2, half of the weight of the beam.

The shearing force and bending moment diagrams can now be drawn as in the previous example. In Figure 49.12, let AB represent the length of the beam (l) drawn to scale. At A the shearing force is wl/2 upwards and this is shown on the graph by the ordinate AC. Because the weight of the beam is evenly distributed throughout its length, the shearing force decreases uniformly from A towards B.

At the mid-point (O) of the beam there is a shearing force of wl/2 downwards (half the weight of the beam) and one of wl/2 upwards (the reaction at A) to consider. The resultant shearing force at O is therefore zero. Finally, at B, there is a shearing force of wl downwards (the weight of the beam) and wl/2 upwards (the reaction at A) to consider, giving a resultant shearing force of wl/2 downwards which is represented on the graph by the ordinate BD.

The bending moment diagram can now be drawn in the same way or by first calculating the area under the shearing force diagram between the ordinate AC and other sections along the beam and then plotting these values as ordinates on the bending moment diagram. The bending moment diagram is represented in Figure 49.13 by the curve AEB.

It should be noted that the shearing force at any point Y which is at a distance X from the end A is given by the formula:

$$\text{Shearing force} = \frac{wl}{2} - wX$$
$$= w\left(\frac{l}{2} - X\right)$$

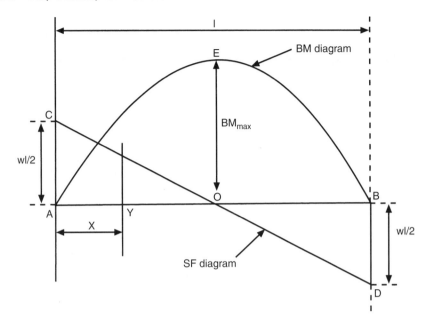

Fig. 49.13

Also, the bending moment at Y is given by the formula:

$$\text{Bending moment} = \frac{wlX}{2} - \frac{wX^2}{2}$$
$$= \frac{wX}{2}(l - X)$$

The maximum bending moment occurs at the mid-point of the beam. Using the above formula:

$$\text{Bending moment} = \frac{wX}{2}(l - X)$$
$$\text{Maximum BM} = \frac{wl}{4}\left(1 - \frac{1}{2}\right)$$
$$\therefore \text{BM}_{max} = \frac{wl^2}{8}$$

Also, the area of the shearing force diagram between the ordinate AC and O is equal to $\frac{1}{2} \times \frac{1}{2} \times \frac{wl}{2}$ or $\frac{wl^2}{8}$.

These principles can now be applied to find the shearing forces and bending moments for any simply supported beam.

Note again how BM_{max} occurs at point 'O', the point at which SF = 0. This must be so for equilibrium or balance of forces to exist.

Example

A uniform beam is 16 m long and has a mass of 10 kg per metre run. The beam is supported on two knife edges, each positioned 3 m from the end of the beam. Sketch the shearing force and bending moment diagrams and state where the bending moment is zero.

$$\text{Mass per metre run} = 10\,\text{kg}$$
$$\text{Total mass of beam} = 160\,\text{kg}$$
$$\text{The reaction at C} = \text{The reaction at B}$$
$$= 80\,\text{kg}$$
$$\text{The shear force at A} = O$$
$$\text{The shear force at L.H. side of B} = +30\,\text{kg}$$
$$\text{The shear force at R.H. side of B} = -50\,\text{kg}$$
$$\text{The shear force at O} = O$$
$$\text{Bending moment at A} = O$$

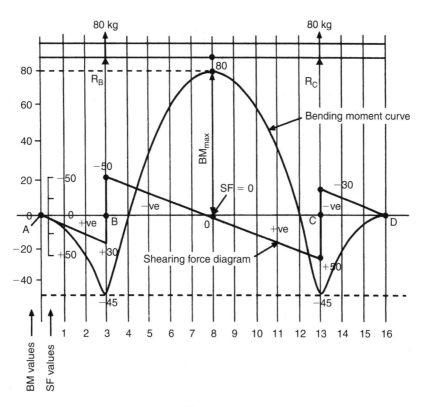

Fig. 49.14

$$\text{Bending moment at 1 m from A} = 1 \times 10 \times \tfrac{1}{2}$$
$$= 5 \text{ kg m (negative)}$$
$$\text{Bending moment at 2 m from A} = 2 \times 10 \times 1$$
$$= 20 \text{ kg m (negative)}$$
$$\text{Bending moment at 3 m from A} = 3 \times 10 \times 1\tfrac{1}{2}$$
$$= 45 \text{ kg m (negative)}$$
$$\text{Bending moment at 4 m from A} = 4 \times 10 \times 2 - 80 \times 1$$
$$= 0$$
$$\text{Bending moment at 5 m from A} = 5 \times 10 \times \tfrac{5}{2} - 80 \times 2$$
$$= 35 \text{ kg m (positive)}$$
$$\text{Bending moment at 6 m from A} = 6 \times 10 \times 3 - 80 \times 3$$
$$= 60 \text{ kg m (positive)}$$
$$\text{Bending moment at 7 m from A} = 7 \times 10 \times \tfrac{7}{2} - 80 \times 4$$
$$= 75 \text{ kg m (positive)}$$
$$\text{Bending moment at 8 m from A} = 8 \times 10 \times 4 - 80 \times 5$$
$$= 80 \text{ kg m (positive)}$$

Ans. Bending moment $= 0$ at 4 m from each end, and at each end of the beam.

The results of the above investigation into the shearing forces and consequent bending moments in simply supported beams will now be applied to find the longitudinal shearing forces and bending moments in floating vessels. Sufficient accuracy of prediction can be obtained.

However, beam theory such as this cannot be used for supertankers and ULCCs. For these very large vessels it is better to use what is known as the finite element theory. This is beyond the remit of this book.

Exercise 49

1   A beam AB of length 10 m is supported at each end and carries a load which increases uniformly from zero at A to 0.75 tonnes per metre run at B. Find the position and magnitude of the maximum bending moment.

2   A beam 15 m long is supported at its ends and carries two point loads. One of 5 tonnes mass is situated 6 m from one end and the other of 7 tonnes mass is 4 m from the other end. If the mass of the beam is neglected, sketch the curves of shearing force and bending moments. Also find (a) the maximum bending moment and where it occurs and (b) the bending moment and shearing force at $\tfrac{1}{3}$ of the length of the beam from each end.

Chapter 50
Bending of ships

Longitudinal stresses in still water

First consider the case of a homogeneous log of rectangular section floating freely at rest in still water as shown in Figure 50.1.

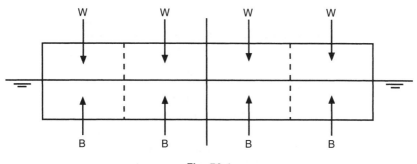

Fig. 50.1

The total weight of the log is balanced by the total force of buoyancy and the weight (W) of any section of the log is balanced by the force of buoyancy (B) provided by that section. There is therefore no bending moment longitudinally which would cause stresses to be set up in the log.

Now consider the case of a ship floating at rest in still water, on an even keel, at the light draft as shown in Figure 50.2.

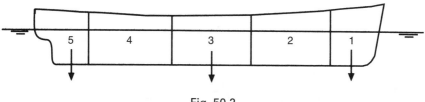

Fig. 50.2

Although the total weight of the ship is balanced by the total force of buoyancy, neither is uniformly distributed throughout the ship's length. Imagine the ship to be cut as shown by a number of transverse sections.

Imagine, too, that each section is watertight and is free to move in a vertical direction until it displaces its own weight of water. The weight of each of the end sections (1 and 5) exceeds the buoyancy which they provide and these sections will therefore sink deeper into the water until equilibrium is reached at which time each will be displacing its own weight of water. If sections 2 and 4 represent the hold sections, these are empty and they therefore provide an excess of buoyancy over weight and will rise to displace their own weight of water. If section 3 represents the engine room then, although a considerable amount of buoyancy is provided by the section, the weight of the engines and other apparatus in the engine room, may exceed the buoyancy and this section will sink deeper into the water. The nett result would be as shown in Figure 50.3 where each of the sections is displacing its own weight of water.

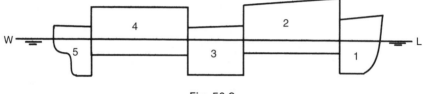

Fig. 50.3

Although the sections in the ship are not free to move in this way, bending moments, and consequently longitudinal stresses, are created by the variation in the longitudinal distribution of weight and buoyancy, and these must be allowed for in the construction of the ship.

Longitudinal stresses in waves

When a ship encounters waves at sea the stresses created differ greatly from those created in still water. The maximum stresses are considered to exist when the wavelength is equal to the ship's length and either a wave crest or trough is situated amidships.

Consider first the effect when the ship is supported by a wave having its crest amidships and its troughs at the bow and the stern, as shown in Figure 50.4.

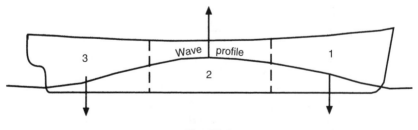

Fig. 50.4

In this case, although once more the total weight of the ship is balanced by the total buoyancy, there is an excess of buoyancy over the weight amidships and an excess of weight over buoyancy at the bow and the stern. This situation creates a tendency for the ends of the ship to move downwards and the section amidships to move upwards as shown in Figure 50.5.

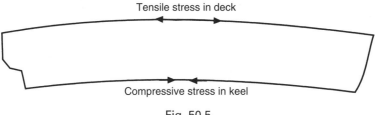

Fig. 50.5

Under these conditions the ship is said to be subjected to a 'hogging' stress.

A similar stress can be produced in a beam by simply supporting it at its mid-point and loading each end as shown in Figure 50.6.

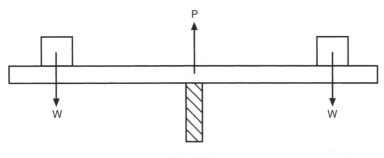

Fig. 50.6

Consider the effect after the wave crest has moved onwards and the ship is now supported by wave crests at the bow and the stern and a trough amidships as shown in Figure 50.7.

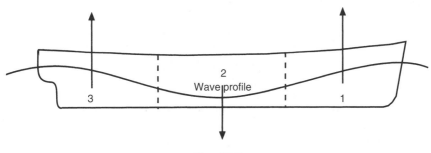

Fig. 50.7

There is now an excess of buoyancy over weight at the ends and an excess of weight over buoyancy amidships. The situation creates a tendency for the bow and the stern to move upwards and the section amidships to move downwards as shown in Figure 50.8.

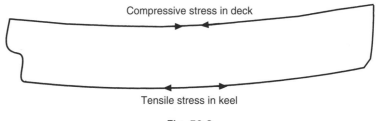

Fig. 50.8

Under these conditions a ship is said to be subjected to a sagging stress. A stress similar to this can be produced in a beam when it is simply supported at its ends and is loaded at the mid-length as shown in Figure 50.9.

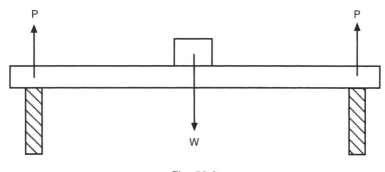

Fig. 50.9

Weight, buoyancy and load diagrams

It has already been shown that the total weight of a ship is balanced by the total buoyancy and that neither the weight nor the buoyancy is evenly distributed throughout the length of the ship.

In still water, the uneven loading which occurs throughout the length of a ship varies considerably with different conditions of loading and leads to longitudinal bending moments which may reach very high values. Care is therefore necessary when loading or ballasting a ship to keep these values within acceptable limits.

In waves, additional bending moments are created, these being brought about by the uneven distribution of buoyancy. The maximum bending moment due to this cause is considered to be created when the ship is moving head-on to waves whose length is the same as that of the ship, and when there is either a wave crest or trough situated amidships.

To calculate the bending moments and consequent shearing stresses created in a ship subjected to longitudinal bending it is first necessary to construct diagrams showing the longitudinal distribution of weight and buoyancy.

The weight diagram

A weight diagram shows the longitudinal distribution of weight. It can be constructed by first drawing a baseline to represent the length of the ship, and then dividing the baseline into a number of sections by equally spaced ordinates as shown in Figure 50.10. The weight of the ship between each pair of ordinates is then calculated and plotted on the diagram. In the case considered it is assumed that the weight is evenly distributed between successive ordinates but is of varying magnitude.

Let

$$CSA = Cross\text{-}sectional\ area$$

Bonjean curves

Bonjean curves are drawn to give the immersed area of transverse sections to any draft and may be used to determine the longitudinal distribution of

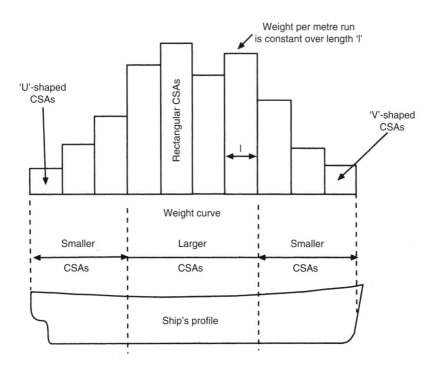

Fig. 50.10   This figure shows the ship divided into 10 elemental strips along its length LOA. In practice the Naval Architect may split the ship into 40 elemental strips in order to obtain greater accuracy of prediction for the weight distribution.

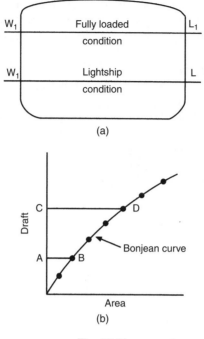

Fig. 50.11

buoyancy. For example, Figure 50.11(a) shows a transverse section of a ship and Figure 50.11(b) shows the Bonjean curve for the same section. The immersed area to the waterline WL is represented on the Bonjean curve by ordinate AB, and the immersed area to waterline W_1L_1 is represented by ordinate CD.

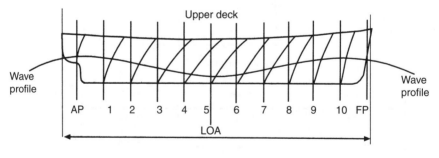

Fig. 50.12

In Figure 50.12 the Bonjean curves are shown for each section throughout the length of the ship. If a wave formation is superimposed on the Bonjean curves and adjusted until the total buoyancy is equal to the total weight of the ship, the immersed transverse area at each section can then be found by inspection and the buoyancy in tonnes per metre run is equal to the immersed area multiplied by 1.025.

Chapter 51
Strength curves for ships

Strength curves consist of five curves that are closely inter-related. The curves are:

1. Weight curve – tonnes/m run or kg/m run.
2. Buoyancy curve – either for hogging or sagging condition – tonnes/m or kg/m run.
3. Load curve – tonnes/m run or kg/m run.
4. Shear force curve – tonnes or kg.
5. Bending moment curve – tonnes m or kg m.

Some firms use units of MN/m run, MN and MN. m.

Buoyancy curves
A buoyancy curve shows the longitudinal distribution of buoyancy and can be constructed for any wave formation using the Bonjean curves in the manner previously described in Chapter 50. In Figure 51.1 the buoyancy

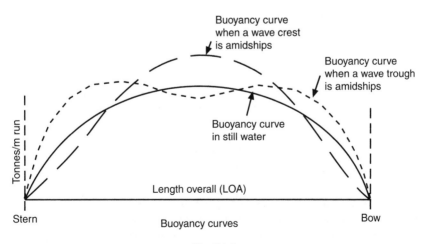

Fig. 51.1

curves for a ship are shown for the still water condition and for the conditions of maximum hogging and sagging. It should be noted that the total area under each curve is the same, i.e. the total buoyancy is the same. Units usually tonnes/m run along the length of the ship.

Load curves

A load curve shows the difference between the weight ordinate and buoyancy ordinate of each section throughout the length of the ship. The curve is drawn as a series of rectangles, the heights of which are obtained by drawing the buoyancy curve (as shown in Figure 51.1) parallel to the weight curve (as shown in Figure 50.10) at the mid-ordinate of a section and measuring the difference between the two curves. Thus the load is considered to be constant over the length of each section. An excess of weight over buoyancy is considered to produce a positive load whilst an excess of buoyancy over weight is considered to produce a negative load. Units are tonnes/m run longitudinally.

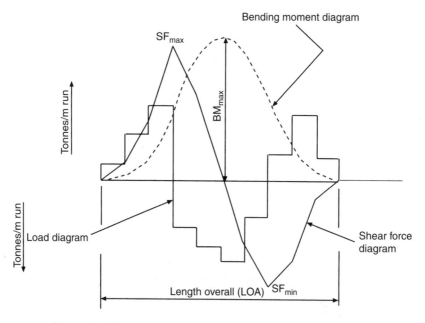

Fig. 51.2   Showing three ship strength curves for a ship in still water conditions.

Shear forces and bending moments of ships

The shear force and bending moment at any section in a ship may be determined from load curve. It has already been shown that the shearing force at any section in a girder is the algebraic sum of the loads acting on either side of the section and that the bending moment acting at any section of the girder is the algebraic sum of the moments acting on either side of the section. It has also been shown that the shearing force at any section is equal to the area under the load curve from one end to the section concerned and that the bending moment at that section is equal to the area under the shearing force curve measured from the same end to that section.

Thus, for the mathematically minded, the shear force curve is the first-order integral curve of the load curve and the bending moment curve is the first-order integral curve of the shearing force curve. Therefore, the bending moment curve is the second-order integral curve of the load curve.

Figure 51.2 shows typical curves of load, shearing force and bending moments for a ship in still water.

After the still water curves have been drawn for a ship, the changes in the distribution of the buoyancy to allow for the conditions of hogging and sagging can be determined and so the resultant shearing force and bending moment curves may be found for the ship in waves.

Example

A box-shaped barge of uniform construction is 32 m long and displaces 352 tonnes when empty, is divided by transverse bulkheads into four equal compartments. Cargo is loaded into each compartment and level stowed as follows:

| No. 1 hold – 192 tonnes | No. 2 hold – 224 tonnes |
| No. 3 hold – 272 tonnes | No. 4 hold – 176 tonnes |

Construct load and shearing force diagrams, before calculating the bending moments at the bulkheads and at the position of maximum value; hence draw the bending moment diagram.

$$\text{Mass of barge per metre run} = \frac{\text{Mass of barge}}{\text{Length of barge}}$$

$$= \frac{352}{32}$$

$$= 11 \text{ tonnes per metre run}$$

$$\text{Mass of barge when empty} = 352 \text{ tonnes}$$

$$\text{Cargo} = 192 + 224 + 272 + 176$$

$$= 864 \text{ tonnes}$$

$$\text{Total mass of barge and cargo} = 352 + 864$$

$$= 1216 \text{ tonnes}$$

$$\text{Buoyancy per metre run} = \frac{\text{Total buoyancy}}{\text{Length of barge}}$$

$$= \frac{1216}{32}$$

$$= 38 \text{ tonnes per metre run}$$

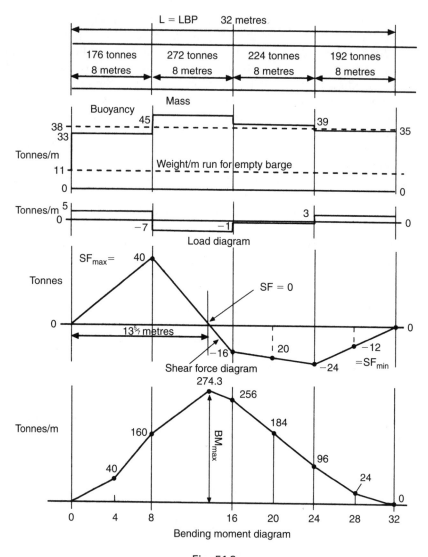

Fig. 51.3

From Figure 51.3:
Bending moments along the barge's length

$$BM_8 = \frac{8 \times 40}{2} = 160 \text{ t m}$$

$$= \underline{160 \text{ tonnes m}}$$

$$BM_0 = 0 \text{ t m}$$

$$BM_4 = \frac{20 \times 4}{2} = 40 \text{ t m}$$

$$BM_8 = \frac{8 \times 40}{2} = 160 \text{ t m}$$

$$BM_{13\frac{5}{7}} = \frac{13\frac{5}{7} \times 40}{2} = 274.3 \text{ t m}$$

$$BM_{16} = \left(\frac{13\frac{5}{7} \times 40}{2}\right) - \left(\frac{2\frac{2}{7} \times 16}{2}\right) = 256 \text{ t m}$$

$$BM_{20} = 256 - \left(\frac{16 + 20}{2}\right) \times 4 = 184 \text{ t m}$$

$$BM_{24} = 184 - \left(\frac{20 + 24}{2}\right) \times 4 = 96 \text{ t m}$$

$$BM_{28} = 96 - \left(\frac{24 + 12}{2}\right) \times 4 = 24 \text{ t m}$$

$$BM_{32} = 24 - \left(\frac{12 \times 4}{2}\right) = 0 \text{ t m}$$

Murray's method

Murray's method is used to find the total longitudinal bending moment amidships on a ship in waves and is based on the division of the Total Bending Moment into two parts:

(a) the Still Water Bending Moment,
(b) the Wave Bending Moment.

The Still Water Bending Moment is the longitudinal bending moment amidships when the ship is floating in still water.

When using Murray's method the Wave Bending Moment amidships is that produced by the waves when the ship is supported on what is called a 'Standard Wave'. A Standard Wave is one whose length is equal to the length of the ship (L), and whose height is equal to $0.607 \sqrt{L}$, where L is measured in metres. See Figure 51.4.

The Wave Bending Moment is then found using the formula:

$$WBM = b \times B \times L^{2.5} \times 10^{-3} \text{ tonnes metres}$$

where B is the beam of the ship in metres and b is a constant based on the ship's block coefficient (C_b) and on whether the ship is hogging or sagging. The value of b can be obtained from the table on the following page.

The Still Water Bending Moment (SWBM)

Let
W_F represent the moment of the weight forward of amidships.
B_F represent the moment of buoyancy forward of amidships.

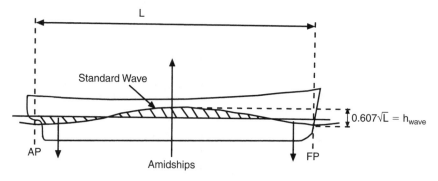

Fig. 51.4

Murray's coefficient 'b' values

	Values of b	
C_b	Hogging	Sagging
0.80	10.555	11.821
0.78	10.238	11.505
0.76	9.943	11.188
0.74	9.647	10.850
0.72	9.329	10.513
0.70	9.014	10.175
0.68	8.716	9.858
0.66	8.402	9.541
0.64	8.106	9.204
0.62	7.790	8.887
0.60	7.494	8.571

Let

W_A represent the moment of the weight aft of amidships.
B_A represent the moment of the buoyancy aft of amidships.
W represent the ship's displacement.

then:

$$\text{Still Water Bending Moment (SWBM)} = W_F - B_F$$
$$= W_A - B_A$$

This equation can be accurately evaluated by resolving in detail the many constituent parts, but Murray's method may be used to give an approximate solution with sufficient accuracy for practical purposes.

The following approximations are then used:

$$\text{Mean Weight Moment } (M_W) = \frac{W_F + W_A}{2}$$

This moment is calculated using the full particulars of the ship in its loaded condition.

$$\text{Mean Buoyancy Moment } (M_B) = \frac{W}{2} \times \text{Mean LCB of fore and aft bodies}$$

An analysis of a large number of ships has shown that the Mean LCB of the fore and aft bodies for a trim not exceeding 0.01 L can be found using the formula:

$$\text{Mean LCB} = L \times C$$

where L is the length of the ship in metres, and the value of C can be found from the following table in terms of the block coefficient (C_b) for the ship at a draft of 0.06 L.

Murray's coefficient 'C' values

Draft	C
0.06 L	$0.179C_b + 0.063$
0.05 L	$0.189C_b + 0.052$
0.04 L	$0.199C_b + 0.041$
0.03 L	$0.209C_b + 0.030$

The Still Water Bending Moment Amidships (SWBM) is then given by the formula:

$$\text{SWBM} = \text{Mean Weight Moment } (M_W)$$
$$- \text{Means Buoyancy Moment } (M_B)$$

or

$$\text{SWBM} = \frac{W_F + W_A}{2} - \frac{W}{2} \times L \times C$$

where the value of C is found from the table above.

If the Mean Weight Moment is greater than the Mean Buoyancy Moment then the ship will be hogged, but if the Mean Buoyancy Moment exceeds the Mean Weight Moment then the ship will sag. So

(i) If $M_W > M_B$ ship hogs $\left.\right\}$ $M_W \backslash M_B$
(ii) If $M_B > M_W$ ship sags

The Wave Bending Moment (WBM)

The actual Wave Bending Moment depends upon the height and the length of the wave and the beam of the ship. If a ship is supported on a Standard Wave, as defined above, then the Wave Bending Moment (WBM) can be calculated using the formula:

$$WBM = b \times B \times L^{2.5} \times 10^{-3} \text{ tonnes metres}$$

where B is the beam of the ship and where the value of b is found from the table previously shown.

Example

The length LBP of a ship is 200 m, the beam is 30 m and the block coefficient is 0.750. The hull weight is 5000 tonnes having LCG 25.5 m from amidships. The mean LCB of the fore and after bodies is 25 m from amidships. Values of the constant b are: hogging 9.795 and sagging 11.02.

Given the following data and using Murray's method, calculate the longitudinal bending moments amidships for the ship on a Standard Wave with: (a) the crest amidships and (b) the trough amidships. Use Figure 51.5 to obtain solution.

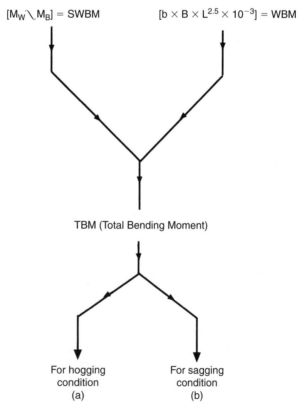

Fig. 51.5   Line diagram for solution using Murray's method.

Data

Item	Weight	LCG from amidships
No. 1 hold	1800 t	55.0 m aft
No. 2 hold	3200 t	25.5 m forward
No. 3 hold	1200 t	5.5 m forward
No. 4 hold	2200 t	24.0 m aft
No. 5 hold	1500 t	50.0 m aft
Machinery	1500 t	7.5 m aft
Fuel oil	400 t	8.0 m aft
Fresh water	150 t	10.0 m forward

Item	Weight	LCG from amidships	Moment
No. 1 hold	1800	55.0 m forward	99 000
No. 2 hold	3200	25.5 m forward	81 600
No. 3 hold	1200	5.5 m forward	6600
No. 4 hold	2200	24.0 m aft	52 800
No. 5 hold	1500	50.0 m aft	75 000
Machinery	1500	7.5 m aft	11 250
Fuel oil	400	8.0 m aft	3200
Fresh water	150	10.0 m forward	1500
Hull	5000	25.5 m	127 500
	16 950		458 450

To find the Still Water Bending Moment (SWBM)

$$\text{Mean Weight Moment } (M_W) = \frac{W_F + W_A}{2}$$

$$= \frac{458\,450}{2}$$

$$M_W = 229\,225 \text{ t m}$$

$$\text{Mean Buoyancy Moment } (M_B) = \frac{W}{2} \times LCB = \frac{16\,950}{2} \times 25$$

$$= 211\,875 \text{ t m}$$

$$\text{Still Water Bending Moment (SWBM)} = M_W - M_B$$

$$= 229\,225 - 211\,875$$

$$\text{SWBM} = 17\,330 \text{ t m (hogging) because } M_W > M_B$$

Wave Bending Moment (WBM)

$$\text{Wave Bending Moment (WBM)} = b \times B \times L^{2.5} \times 10^{-3} \text{ t m}$$
$$\text{WBM hogging} = 9.795 \times 30 \times 200^{2.5} \times 10^{-3} \text{ t m}$$
$$= 166\,228 \text{ t m}$$
$$\text{WBM sagging} = 11.02 \times 30 \times 200^{2.5} \times 10^{-3} \text{ t m}$$
$$= 187\,017 \text{ t m}$$

Total Bending Moment (TBM)

$$\text{TBM hogging} = \text{WBM hogging} + \text{SWBM hogging}$$
$$= 166\,228 + 17\,350$$
$$= \underline{183\,578 \text{ t m}}$$
$$\text{TBM sagging} = \text{WBM sagging} - \text{SWBM hogging}$$
$$= 187\,017 - 17\,350$$
$$= \underline{169\,667 \text{ t m}}$$

Answer (a) with crest amidships, the Total Bending Moment, *TBM is 183 578 tonnes metres.*

Answer (b) with trough amidships, the Total Bending Moment, *TBM is 169 667 tonnes metres.*

The *greatest* danger for a ship to break her back is when the wave crest is at amidships, or when the wave trough is at amidships with the crests at the stem and at the bow.

In the previous example the greatest BM occurs with the crest amidships. Consequently, this ship would fracture across the upper deck if the tensile stress due to hogging condition became too high.

Chapter 52
Bending and shear stresses

The shear forces and bending moments which act upon a ship's structure cause shear and bending stresses to be generated in the structure. We have seen earlier that the shearing forces and bending moments experienced by a ship are similar to those occurring in a simply supported beam. We shall therefore consider the shear and bending stresses created when an ordinary beam of rectangular section is simply supported.

Bending stresses

The beam in Figure 52.1(a) is rectangular in cross-section, is simply supported at each end and has a weight W suspended from its mid-point.

This distribution will tend to cause the beam to bend and sag as shown in Figure 52.1(b).

Consider first the bending stresses created in the beam. Let ab and cd in Figure 52.1(a) be two parallel sections whose planes are perpendicular to the plane AB. Let their distance apart be dx. When the beam bends the

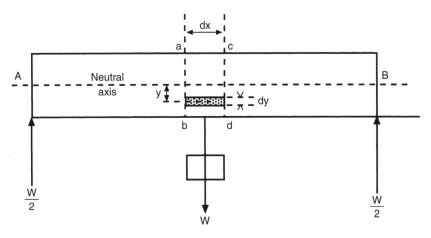

Fig. 52.1(a)

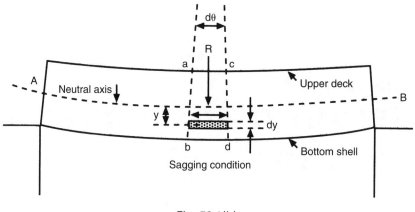

Fig. 52.1(b)

planes of these two sections will remain perpendicular to the plane AB but will now be inclined at an angle dθ to each other. The parts of the beam above the layer AB are in compression and those below the layer AB are in tension. Thus the layer AB is neither in compression or tension and this layer is called the neutral axis.

Let the radius of curvature of the neutral axis layer be R.

Consider a layer of thickness dy which is situated at distance y from the plane of the neutral axis.

$$\text{Original length of layer} = dx$$

After bending, this length $= R\,d\theta$ no stress or strain here.
Length of layer after bending $= (R + y)d\theta$ at a distance y below $A - B$.
But

$$\text{Strain} = \frac{\text{Elongation}}{\text{Original length}}$$

$$= \frac{(R + y)d\theta - R\,d\theta}{R\,d\theta}$$

$$\text{Strain} = y/R$$

This equation indicates that strain varies directly as the distance from the neutral axis. Also, if the modules of elasticity is constant throughout the beam, then:

$$E = \frac{\text{Stress}}{\text{Strain}}$$

or

$$\text{Stress} = E \times \text{Strain}$$

and

$$\text{Stress} = E = \frac{y}{R}$$

So

$$f = E \times \frac{y}{R}$$

This equation indicates that stress is also directly proportional to distance from the neutral axis, stress being zero at the neutral axis and attaining its maximum value at top and bottom of the beam. The fibres at the top of the beam will be at maximum compressive stress whilst those at the bottom will be at maximum tensile stress. Since the beam does not move laterally, the sum of the tensile stresses must equal the sum of the compressive stresses. This is illustrated in Figure 52.2.

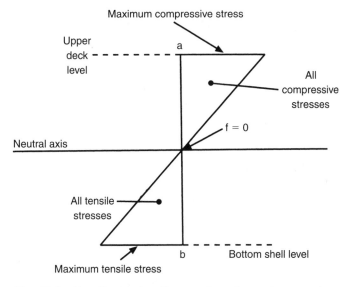

Fig. 52.2   Bending stress diagram for a beam in a sagging condition of loading.

Figure 52.3 shows the cross-section of the beam at ab. NA is a section of the neutral surface at ab, and is called the neutral axis at ab. At NA there is no compressive or tensile stress, so f = 0.

In Figure 52.3, b × dy is an element of area at distance y from the neutral axis. Let b × dy be equal to dA. The force on dA is equal to the product of the stress and the area; i.e.

$$\text{Force on dA} = \frac{Ey}{R} \times dA$$

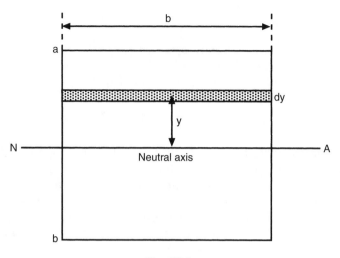

Fig. 52.3

The total or resultant force on the section is equal to the sum of the forces on the section; i.e.

$$\text{Total or resultant force on the section} = \Sigma \frac{Ey}{R} dA$$

But the total or resultant force on the section is zero.

$$\therefore \Sigma \frac{Ey}{R} dA = 0$$

Since E and R are constant, then

$$\Sigma y \, dA = 0$$

But $\Sigma y \, dA$ is the first moment of the area about the neutral axis and if this is to be equal to zero then the neutral axis must pass through the centre of gravity of the section. In the case being considered, the neutral axis is at half-depth of the beam.

The sum of the moments of the internal stresses on the section is equal to M, the external bending moment at the section.

At a distance y from the neutral axis, the moment (m) of the stress acting on dA is given by the formula:

$$m = \frac{Ey}{R} \times y \times dA$$

Also,

$$M = \Sigma \frac{Ey}{R} \times y \times dA$$

$$= \Sigma \frac{E}{R} y^2 \, dA$$

Since E and R are constant, then:

$$M = \frac{E}{R} \Sigma y^2 \, dA$$

But $\Sigma y^2 dA$ is the second moment of the area of the section about the neutral axis. Let $\Sigma y^2 dA$ be equal to I.

$$\therefore M = \frac{E}{R} \times I$$

Now if f is the stress at a distance of y from the neutral axis, then:

$$f = \frac{Ey}{R}$$

or

$$\frac{E}{R} = \frac{f}{y}$$

But

$$\frac{M}{I} = \frac{E}{R}$$

$$\therefore \frac{f}{y} = \frac{M}{I}$$

and

$$f = \frac{M}{I/y}$$

The expression I/y is called the section modulus, designated Z, and attains its minimum value when y has its maximum value. The section modulus is the strength criterion for the beam so far as bending is concerned.

Example 1

A steel beam is 40 cm deep and 5 cm wide. The bending moment at the middle of its length is 15 tonnes metres. Find the maximum stress on the steel.

$$I = \frac{lb^3}{12} \qquad\qquad f = \frac{M}{I} \times y$$

$$= \frac{5 \times 40^3}{12} \qquad\qquad y = \frac{d}{2} = \frac{40}{2} = 20 \text{ cm}$$

$$I = \frac{80\,000}{3} \text{ cm}^4$$

So

$$f = \frac{1500 \times 3 \times 20}{80\,000} \text{ tonnes per sq. cm}$$

$$f = 1.125 \text{ tonnes per sq. cm}$$

Ans. Maximum stress = 1.125 tonnes per sq. cm or 1125 kg per sq. cm.

Example 2

A deck beam is in the form of an H-girder as shown in the accompanying Figure 52.4.

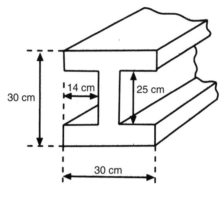

30 cm

14 cm

25 cm

30 cm

Fig. 52.4

If the bending moment at the middle of its length is 15 tonnes metres, find the maximum stress in the steel.

$$I = \frac{BH^3 - 2bh^3}{12}$$

$$f = \frac{M}{I} \times y$$

$$I = \frac{30 \times 30^3 - 2 \times 14 \times 25^3}{12}$$

$$y = \frac{H}{2} = \frac{30}{2} = 15 \text{ cm}$$

$$I = \frac{810\,000 - 437\,500}{12}$$

$$f = \frac{1500 \times 12}{372\,500} \times 15$$

$$I = \frac{372\,500}{12} \text{ cm}^4$$

$$f = 0.725 \text{ tonners per sq. cm}$$

Ans. Maximum stress = 0.725 tonnes per sq. cm or 725 kg per sq. cm.

The above theory can now be used to find the section modulus of the ship. The external bending moment can be calculated, as can the stress at the transverse sections under the conditions of maximum bending moment. The neutral axis passes through the centre of gravity of the section and, because of this, its position can be found. The moments of inertia of all of the continuous longitudinal material about this axis can be found. Then the section modulus is equal to I/y.

Shearing stresses

It has already been shown that a shearing stress across one plane within a material produces another shearing stress of equal intensity on a plane at right angles (see Complementary stresses in Chapter 49).

The mean shearing stress in any section of a structure can be obtained by dividing the shearing force at that section by the cross-sectional area of the section; i.e.

$$\text{Mean shearing stress} = \frac{F}{A}$$

where
F = vertical shearing force and
A = area of cross-section

A more accurate estimation of shear stress distribution can be obtained from a consideration of the bending of beams.

Consider the short length of beam dx in Figure 52.5(a) which lies between the vertical planes ab and cd. Let dy be a layer on this short length of beam which is situated at distance y from the neutral plane.

From the formula for bending moments deduced earlier, the longitudinal stress 'f' on a small area b × dy of section ab can be found from the formula:

$$f = \frac{M_1}{I_1} \times y$$

where
M_1 = the bending moment at this section
I_1 = second moment of the section about the neutral axis
y = the distance of the area from the neutral axis

Let

$$b \times dy = dA$$
$$\text{The force acting on } dA = f \times dA$$
$$= \frac{M_1}{I_1} \times y \times dA$$

Let A_1 be the cross-sectional area at ab, then the total force (P_1) acting on the area A_1 is given by:

$$P_1 = \frac{M_1}{I_1} \times \Sigma y \times dA$$
$$= \frac{M_1}{I_1} \times y \times A_1$$

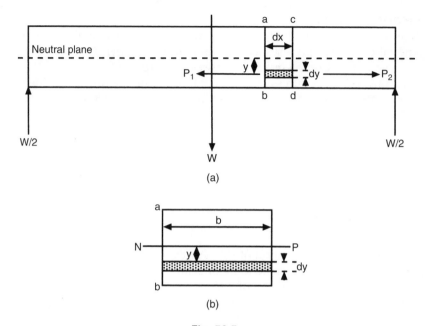

Fig. 52.5

Let P_2 be the force acting on the same area (A_2) at section cd.
Then

$$P_2 = \frac{M_2}{I_2} \times y \times A_2$$

where
M_2 = bending moment at this section

and

I_2 = second moment of the section about the neutral axis

Since dx is small then $A_1 = A_2$ and $I_1 = I_2$. Therefore let A = area and I = second moment of the section about the neutral axis.

$$\text{Shearing force at A} = P_1 - P_2$$
$$= \frac{M_1 - M_2}{I} \times y \times A$$
$$= \frac{M_1 - M_2}{dx} \times \frac{dx \times y \times A}{I}$$

But $\dfrac{M_1 - M_2}{dx}$ is equal to the vertical shearing force at this section of the beam.

$$\therefore P_1 - P_2 = F \times \frac{y \times A \times dx}{I}$$

Now let 'q' be the shearing stress per unit area at A and let 't' be the thickness of the beam, then the shearing force is equal to q × t × dx.

$$\therefore q \times t \times dx = \frac{F \times y \times A \times dx}{I}$$

or

$$q = \frac{F \times A \times y}{I \times t}$$

Example 1

A solid beam of rectangular section has depth 'd' and thickness 't', and at a given section in its length is under a vertical force 'F'. Find where the maximum shearing stress occurs and also determine its magnitude in terms of 'd', 't' and 'F'.

Since the section is rectangular, the I and y about the neutral axis are given by:

$$I = \frac{td^3}{12}$$

and

$$y = \frac{d}{2} \times \frac{1}{2} = \frac{d}{4}$$

and

$$\text{Area } A = \frac{t \times d}{2}$$

In the case being considered, Ay attains its maximum value at the neutral axis and this is therefore where the maximum shearing stress (q) occurs.

$$q = \frac{F \times A \times y}{I \times t}$$

$$\therefore q_{max} = \frac{F \times \dfrac{td}{2} \times \dfrac{d}{4}}{\dfrac{td^3}{12} \times t}$$

Ans. $q_{max} = \dfrac{3F}{2td}$

Consider the worked example for the H-girder shown in Figure 52.6.

When the shear stress distribution for an H-girder is calculated and plotted on a graph it appears similar to that shown in Figure 52.8. It can be seen from this figure that the vertical web of the beam takes the greatest amount of shear.

Example 1.   A worked example showing distribution of shear stress 'q' (with flanges) See Figures 52.6 and 52.7.

Let

$$F = 30 \text{ tonnes}$$

$$I_{NA} = \tfrac{1}{12}(6 \times 12^3 - 5.5 \times 11^3)$$

$$\therefore I_{NA} = 254 \text{ cm}^4$$

At upper edge of top flange and lower edge of lower flange the stress 'q' = zero.

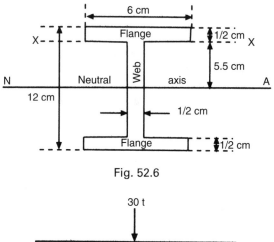

Fig. 52.6

Fig. 52.7

(a)   'q'$_a$ value

$$Ay = m = 6 \times \tfrac{1}{2} \times 5.75 = 17.25 \text{ cm}^3$$

$$b = 6 \text{ cm}$$

$$q = \frac{F \times A \times y}{I_{NA} \times b}$$

$$\therefore \text{'q'}_a = \frac{30 \times 17.25}{254 \times 6} = 0.34 \text{ t/cm}^2$$

(b)   Just below 'x − x'

$$Ay = 17.25 \text{ cm}^2$$

$$b = \tfrac{1}{2}\text{ cm}$$

$$q_b = \frac{30 \times 17.25}{254 \times 0.5} = \underline{4.07 \text{ t/cm}^2}$$

(c)   At 3 cm from neutral axis

$$m = 17.25 + (2.5 \times 0.5) \times 4.25$$

$$= 22.56 \text{ cm}^3 = Ay$$

$$b = \tfrac{1}{2}\text{ cm}$$

$$q_c = \frac{30 \times 22.56}{254 \times 0.5} = \underline{5.33 \text{ t/cm}^2}$$

(d)   At neutral axis

$$m = 17.25 + (5.5 \times \tfrac{1}{2} \times 2.75)$$

$$= 24.81 \text{ cm}^3 = Ay$$

$$b = \tfrac{1}{2}\text{ cm}$$

$$q_d = \frac{30 \times 24.81}{254 \times 0.5} = \underline{5.86 \text{ t/cm}^2}$$

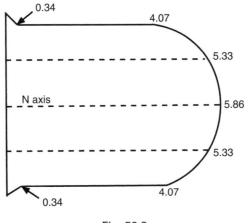

Fig. 52.8

q_{max} occurs at neutral axis.

Load carried by the Web is 28.9 t when force (F) = 30 t. Web gives resistance to shearing stress. Flanges give resistance to bending stress.

Example 2.   **Second worked example showing distribution of shear stress 'q' (with no flanges)** See Figures 52.9 and 52.10.

Let

$$F = 30 \text{ tonnes}$$

$$q = \frac{F \times A \times Y}{I_{NA} \times b}$$

$$I_{NA} = \frac{td^3}{12} = \frac{0.5 \times 12^3}{12} = 72 \text{ cm}^4$$

$$\therefore q = \frac{30 \times Ay}{72 \times 0.5} = \underline{0.833 \times A \times y}$$

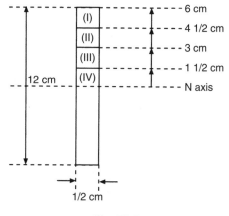

Fig. 52.9

$$q_{(I)} = 0.833 \times 1.5 \times 0.5 \times 5.25 = \underline{3.281 \text{ t/cm}^2}$$
$$q_{(I+II)} = 0.833 \times 3.0 \times 0.5 \times 4.5 = \underline{5.625 \text{ t/cm}^2}$$
$$q_{(I+II+III)} = 0.833 \times 4.5 \times 0.5 \times 3.75 = \underline{7.031 \text{ t/cm}^2}$$
$$q_{(I \text{ to } IV)} = 0.833 \times 6.0 \times 0.5 \times 3 = \underline{7.500 \text{ t/cm}^2}$$

Check:

$$q_{max} = \frac{3 \times F}{2td} = \frac{3 \times 30}{2 \times 0.5 \times 12} = \underline{7.500 \text{ t/cm}^2}$$

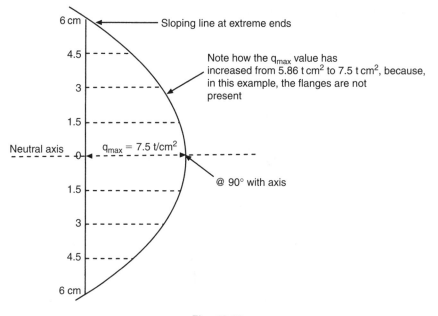

6 cm

4.5

3

1.5

Neutral axis   0

1.5

3

4.5

6 cm

Sloping line at extreme ends

Note how the q_{max} value has increased from 5.86 t cm^2 to 7.5 t cm^2, because, in this example, the flanges are not present

$q_{max} = 7.5$ t/cm^2

@ 90° with axis

Fig. 52.10

Summary sketches for the two worked examples relating to shear stress

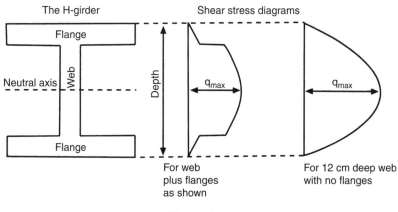

The H-girder

Flange

Web

Neutral axis

Flange

Shear stress diagrams

Depth

q_{max}

q_{max}

For web plus flanges as shown

For 12 cm deep web with no flanges

Fig. 52.11

In a loaded ship with a wave crest or trough amidships, the shearing force attains its maximum value in the vicinity of the neutral axis at about a quarter of the ship's length from each end. The minimum shearing force is exerted at the keel and the top deck.

Bending stresses in the hull girder

Example 1

The effective part of a transverse section of a ship amidships is represented by the steel material shown in Figure 52.12 (also see note at the end of this chapter).

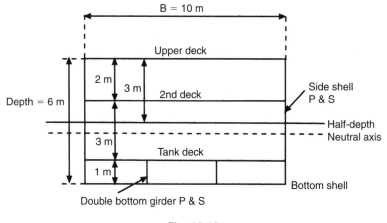

Fig. 52.12

The beam of the ship is 10 m and the depth is 6 m. All plating is 1.5 cm thick. Find the maximum tensile and compressive stresses when the ship is subjected to a sagging moment of 6000 tonnes metres.

Assume initially that neutral axis is at 1/2 depth, i.e. 3 m above base.

Item	Area (sq. m)	Lever	Moment	Lever	I
Upper deck	10 × 0.015 = 0.15	3	0.45	3	1.35
2nd deck	10 × 0.015 = 0.15	1	0.15	1	0.15
Tank top	10 × 0.015 = 0.15	−2	−0.30	−2	0.60
Bottom shell	10 × 0.015 = 0.15	−3	−0.45	−3	1.35
Sideshell P&S	2 × 6 × 0.015 = 0.18	0	0	0	0
Double bottom girders	2 × 1 × 0.015 = 0.03	−2.5	−0.075	−2.5	0.1875
	0.81		−0.225		3.6375

Item	Own inertia $= \frac{1}{12} Ah^2$
Upper deck	$\frac{1}{12} \times 0.15 \times 0.015^2 = 2.8 \times 10^{-6}$
2nd deck	$\frac{1}{12} \times 0.15 \times 0.015^2 = 2.8 \times 10^{-6}$
Tank top	$\frac{1}{12} \times 0.15 \times 0.015^2 = 2.8 \times 10^{-6}$
Bottom shell	$\frac{1}{12} \times 0.15 \times 0.015^2 = 2.8 \times 10^{-6}$

*

Item	Own inertia $= \frac{1}{12} Ah^2$
Sideshell P&S	$\frac{1}{12} \times 0.18 \times 36 = 0.54$
Double bottom girders	$\frac{1}{12} \times 0.03 \times 1 \ = 2500 \times 10^{-6}$
	0.5425

* It can be seen from this table that the second moment of area of *horizontal* members in the structure such as decks, tank tops and outer bottom, about their own neutral axes are small and in practice these values are ignored in the calculation.

$$\text{Depth of neutral axis below half-depth} = \frac{-0.225}{0.81}$$

$$= -0.28 \text{ m}$$

$$\text{Total second moment about half-depth} = 3.6375 + 0.5425$$

$$= 4.18 \text{ m}^4$$

$$\text{Total second moment about neutral axis} = 4.18 - 0.81 \times 0.28^2$$

$$I_{NA} = 4.18 - 0.06$$

$$= 4.12 \text{ m}^4$$

The maximum compressive bending stress is at the upper deck level.

$$\frac{f}{y} = \frac{M}{I}$$

$$\therefore f = \frac{6000}{4.12} \times 3.28$$

$$= 4777 \text{ tonnes per sq. m}$$

The maximum tensile bending stress is at the bottom shell.

$$f = \frac{6000}{4.12} \times 2.72$$

$$= 3961 \text{ tonnes per sq. m}$$

Ans. Maximum tensile bending stress = 3961 tonnes per sq. m.
Maximum compressive bending stress = 4777 tonnes per sq. m.

Summary

The steel material shown in Figure 52.12 is steel that is *continuous* in a longitudinal direction. Intercostal structures are not considered because they contribute very little resistance to longitudinal bending of the ship.

The calculation and table on previous page is known as the 'equivalent girder calculation'. It is the I_{NA} that helps resist hogging and sagging motions of a ship. In doing so, I_{NA} helps reduce the tensile and compressive bending stresses.

Lloyds suggest maximum value for these bending stresses in conjunction with a factor of safety of 4. If mild steel structure, f_{max} is about $110\,MN/m^2$ or about 11 000 tonnes per sq. m for medium-sized ships. If high tensile steel is used, f_{max} is about $150\,MN/m^2$ or about 15 000 tonnes per sq. m for medium-sized ships.

Take medium-sized ships as being 100 m to 150 m LBP. If there is any danger of these bending stresses being too high in value for a ship in service, a change in the loading arrangement could help make this loaded condition safer.

Exercise 52

1 The hull of a box-shaped vessel is 50 m long and has a mass of 600 tonnes. The mass of the hull is uniformly distributed over its length. Machinery of 200 tonnes mass extends uniformly over the quarter length amidships. Two holds extending over the fore and aft quarter length each have 140 tonnes of cargo stowed uniformly over their lengths. Draw the curves of shearing force and bending moments for this condition and state their maximum values.

2 A uniform box-shaped barge, 40 m × 12 m beam, is divided into four cargo compartments each 10 m long. The barge is loaded with 600 tonnes of iron ore, level stowed, as follows:

> No. 1 hold – 135 tonnes   No. 2 hold – 165 tonnes
> No. 3 hold – 165 tonnes   No. 4 hold – 135 tonnes

The loaded barge floats in fresh water on an even keel at a draft of 1.75 m. Construct the curves of shearing force and bending moment for this condition and also find the position and value of the maximum bending moment for the still water condition.

3 A box-shaped vessel, 100 m long, floats on an even keel displacing 2000 tonnes. The mass of the vessel alone is 1000 tonnes evenly distributed and she is loaded at each end for a length of 25 m with 500 tonnes of cargo, also evenly distributed. Sketch the curve of loads, shearing force and bending moments. Also state the maximum shearing force and bending moment and state where these occur.

4 Describe in detail Murray's method for ascertaining the longitudinal bending moments in a ship when she is supported on a standard wave with: (a) the crest amidships and (b) the trough amidships. Include in your answer details of what is meant by a 'standard wave'.

5 A supertanker is 300 m LOA. Second moment of area (I_{NA}) is $752\,m^4$. Neutral axis above the keel is 9.30 m and 9.70 m below the upper deck. Using the information in the table below, proceed to draw the shear force and bending moment curves for this ship. From these two strength curves determine:

(a) Maximum shear force in MN.
(b) Maximum bending moment in MNm.

(c) Position along the ship's length at which this maximum BM occurs.

(d) Bending stress f_{max} in MN/m^2 at the upper deck and at the keel.

Station	Stern	1	2	3	4	5
SF (MN)	0	26.25	73.77	115.14	114.84	21.12
BM (MNm)	0	390	1887	4717	8164	10 199

		6	7	8	9	Bow
SF (MN)	−78.00	−128.74	−97.44	−45.78	0	
BM (MNm)	9342	6238	2842	690	0	

Part 3
Endnotes

Chapter 53
Draft Surveys

When a ship loads up at a port, departs from this port and travels to another port, a Draft Survey is carried out. This is to check that the cargo deadweight or 'constant' is satisfactory for the shipowner at the port of arrival.

It is virtually a check on the amount of cargo that left the first port against that arriving at the second port. This Draft Survey may be carried out by a Master, a Chief Engineer or a Naval Architect.

Prior to starting on a Draft Survey the vessel should be in upright condition and on even keel if possible. If not on even keel then certainly within 1 per cent of her LBP would be advantageous.

When the ship arrives in port ready for this Draft Survey, there are several items of information that have to be known by, say, the Naval Architect. They include:

LBP and C_b relative to the ship's waterline or actual loaded condition.
Lightweight.
Density of the water in which the vessel is floating.
Draft readings port and starboard at the stern, midships and at the bow.
Distance from aft perp to aft draft marks.
Distance from amidships to midship draft marks.
Distance from forward perp to forward draft marks.
Distance of LCF from amidships.
Cargo deadweight or 'constant', for example, say 10 766 t.

Using above data the Naval Architect will modify the actual draft readings to what they would be at AP, amidships and FP. These values are then used to determine the mean draft at the position of the ship's LCF (see chapter on Trim or longitudinal stability).

To take into account any hog or sag a 'mean of means' formula is used. Suppose the drafts at the AP, amidships and FP were 8.994 m, 8.797 m, 8.517 m and LCF was 0.37 m forward of amidships with an LBP of 143.5 m. Then the mean of means draft is:

$$\text{Means of means draft} = (d_{AP} + (6 \times d_{AM}) + d_{FP})/8$$
$$= (8.994 + (6 \times 8.797) + 8.517)/8$$
$$\text{Thus, mean of means draft} = 8.787 \text{ m}$$

When corrected for LCF position, the mean draft is 8.786 m.

The Naval Architect uses this draft of 8.786 m on the ship's hydrostatic curves or tabulated data to obtain a first estimate of the displacement and TPC. These are of course for salt water density of 1.025 t/m³. Assume this displacement was 17 622 t and the TPC was 25.86 t. The Naval Architect can now evaluate the fresh water allowance and proceed to make a correction for density.

$$\text{FWA} = W/(4 \times \text{TPC}) = 17\,662/(4 \times 25.86)$$
$$\text{FWA} = 171\,\text{mm}$$

The density correction is then made (see chapter on Density correction). Assume the water was of 1.019 t/m³. Consequently this correction would be minus 0.041 m.

Thus, the final true draft would be 8.786 − 0.041 = 8.745 m.

Interpolating once more the hydrostatic data the final displacement at the time of the Draft Survey is obtained. Pressume this final displacement is 17 537 tonnes.

From this figure must be deducted the lightweight and all other deadweight items (except the 'constant' or cargo deadweight). Assume this lightweight was 5675 t and the residual dwt of oil, fresh water, stores, crew and effects, water ballast, etc. was 1112 t. This residual dwt would have been estimated after all tanks had been sounded and all compartments been checked for contents. Then the final cargo dwt or 'constant' is:

Cargo deadweight = 17 537 − 5675 − 1112 = 10 750 tonnes

This compares favourably with the cargo dwt given at the port of departure, that is 10 766 tonnes, a difference of 0.15 per cent. In a perfect world the cargo dwt at the arrival port would have been 'constant' at 10 766 t. However, ships are not built to this degree of accuracy and measurement to this standard is very hard to achieve.

Summary

Corrections have been made for:

Draft mark positions	Trim and LCF position
Hog or sag	Density of water

True mean draft is 8.745 m	True displacement is 17 537 t

Cargo deadweight or 'constant' is 10 750 t (−0.15 per cent)

A well-conducted survey is capable of achieving an absolute accuracy of within plus or minus 0.50 per cent of the Cargo dwt. This is as good, if not better, as other systems of direct weighing.

Error can creep in over the years because of the ship's reported Lightweight. This is due to the ship herself gaining about 0.50 per cent Lightweight each year in service. Hence over a period of ten years this ship would have gained about 280 tonnes since her maiden voyage.

Lightweight will also alter slightly depending on the position of the anchors and cables. Obviously, with anchors fully housed the Lightweight will be more. Adjustment may have to be made but it is better if at both ports the anchors are fully housed.

Error can also be made if the draft readings were taken in a tidal current. The speed of the tide would cause the ship to sink slightly in the water (squat effects) and so give draft readings that were too high in value. One reported instance of this occurring resulted in a cargo reduction of over 300 t. Initially this was put down to excessive pilfering until further checks discovered that drafts had been taken in moving water. At departure and arrival ports draft readings must be read when water speed is zero.

One suggestion for improving accuracy of measurement is to have draft marks also at 1/4 L and 3/4 L from aft. In other words at stations 2.5 and 7.5. They would reduce errors where there is an appreciable hog or sag at the time of the Draft Survey.

Chapter 54
Quality control plus the work of ship surveyors

Ship surveyors ensure that stability of motions and stability of ship structures are of the right standards. In the UK, *Lloyds* surveyors look after the *strength* of the ship whilst the *Maritime Department for Transport* surveyors look after the *safety* of ships. Another classification society ensuring the ship is up to acceptable safety standards is the *International Maritime Organization (IMO)*.

With regards to Lloyds surveyors, their responsibilities include the following:

- Inspection of the design drawings with respect to the scantlings. Sometimes plate thickness has to be increased. Sometimes the sectional-modulus of the plating and stiffener has to be modified. When considered satisfactory, the drawing is given the 'Lloyds Approved' stamp.
- Inspection and testing of materials used in the construction of the vessel. Sample tests may be asked for on wood, steel, aluminium or glass reinforced material.
- Examination of shearing forces and bending moments with respect to longitudinal bending of the ship. Racking and buckling stresses will be considered and analysed.
- Examination of jointing within the structure of the vessel. This could mean tests carried out on specimen welds. These may be destructive tests or non-destructive tests that use X-rays and ultrasonics.
- To arrange special surveys to ensure that a vessel has retained its standard of seaworthiness perhaps after maintenance repairs or after a major accident. Special surveys are carried out on the ship every 4 or 5 years and are more severe with each successive special survey. If in doubt, it is a case of take it out and renew.

- Examining to see if corrosion or wear and tear over years of service have decreased scantlings to a state where the structure is no longer of adequate strength. There have been cases where the shell plating thickness (over a period of 10 years in service) has decreased from 18 mm at time of build to only 9 mm!!
- Each shipyard will usually provide an office for use of the Lloyds surveyor, who will ensure that the ship is built as per drawing and that inferior material is not being substituted. The surveyor will also check the standard and quality of workmanship. For obvious reasons, the surveyor often undertakes random testing.
- Final positioning of the load line and freeboard marks together with the official ship number and the official port of registry. This responsibility is also shared with the Maritime Department for Transport surveyor.

With regards to the Maritime Department for Transport surveyors their responsibilities include the following:

- Inspection of drawings. Care is taken with accommodation plans that the floor area for all cabins and communal spaces are above a certain minimum requisite area.
- Care and maintenance of all life-saving appliances. Inspection of inventory of flares, life-jackets, life-rafts, etc.
- Inspection of fire-fighting appliances. Examining arrangements for prevention, detection and extinguishing arrangements for the vessel.
- Consideration of navigation lights. Provision of emergency lighting.
- Consideration of sound-signalling apparatus. The arrangements for radio contact in the event of incidents such as fire or flooding. Emails are now used for quick communication especially when crew or passengers have to leave the ship for emergency medical assistance.
- Inspection of the stability information supplied by the shipbuilder to the ship.
- Inspection of lighting, heating and ventilation in accommodation, navigation spaces, holds and machinery spaces.
- Attendance at the inclining experiment or stability test. This is to verify that the test had been carried out in a correct and fitting manner. The final stability calculations will also be checked out at the local DfT office.
- Written and oral examinations for masters and mates for Certificates of Competency.
- Final positioning of the load line and freeboard marks together with the official ship number and the official port of registry. This responsibility is also shared with the Lloyds surveyor.
- If the ship has undergone major repair, conversion to another ship type, or the ship has had major update/retrofit, then the surveyors check for strength and safety. If deemed of satisfactory quality, then new certificates will be signed, dated and issued to the ship-owner.

The International Maritime Organization (IMO) is a technical organisation, established in 1958. The organization's chief task was to develop a comprehensive body of international conventions, codes and recommendations that could be implemented by all member governments. The effectiveness of IMO measures depends on how widely they are accepted and how they are implemented.

Apart from Safety Of Life At Sea (SOLAS) regulations, the IMO Maritime Safety Committee deal with the following:

- Stability and load lines and fishing vessel safety.
- Standards of training and watchkeeping.
- Carriage of dangerous goods.
- Safety of navigation.
- Radio communications, search and rescue.
- Fire protection.
- Bulk liquids and gases.
- Solid cargoes and containers.
- Ship design and equipment.
- Flag state implementation.

One major piece of work recently produced by IMO is the International Code for the safe carriage of grain in bulk, abbreviated to *International Grain Code* (see Chapter 30). IMO deals with environment protection and pollution prevention. It also has a legal committee, which was formed following the *Torrey Canyon* pollution disaster in 1967.

For further information regarding the work of these three classification societies contact can be made at:

Lloyds Registry of Shipping, London	www.lr.org
Department for Transport/MCA, Southampton	www.mcga.gov.uk
International Maritime Organization, London	www.imo.org

Chapter 55
Extracts from the 1998 Merchant Shipping (Load Line) Regulations Reference Number MSN 1752 (M)

Part VI

INFORMATION FOR THE MASTER

Information as to stability of ships

32. (1) The owner of every ship to which these Regulations apply shall provide, for the guidance of the master, information relating to the stability of the ship in accordance with this regulation. The information shall be in the form of a book which shall be kept on the ship at all times in the custody of the master.

    (2) In the case of a United Kingdom ship this information shall include all matters specified in Schedule 6 in Merchant Shipping Notice MSN 1701(M), and be in the form required by that Schedule. This information shall also be in accordance with the requirements of paragraphs (3), (4) and (5).

    (3) Subject to paragraph (4), this information shall be based on the determination of stability taken from an inclining test carried out in the presence of a surveyor appointed by the Secretary of State or, for ships listed in paragraph (5), by the Assigning Authority. This information shall be amended whenever any alterations are made to the

ship or changes occur to it that will materially affect this information and, if necessary, the ship shall be re-inclined.

(4) The inclining test may be dispensed with if –

   (a) in the case of any ship basic stability data is available from the inclining test of a sister ship and it is known that reliable stability information can be obtained from such data; and

   (b) in the case of:

      (i) a ship specially designed for the carriage of liquids or ore in bulk, or

      (ii) of any class of such ships,

      the information available in respect of similar ships shows that the ship's proportions and arrangements will ensure more than sufficient stability in all probable loading conditions.

(5) Before this information is issued to the master –

   (a) if it relates to a ship that is –

      (i) an oil tanker over 100 metres in length;

      (ii) a bulk carrier, or an ore carrier, over 150 metres in length;

      (iii) a single deck bulk carrier over 100 metres in length but not exceeding 150 metres in length;

      (iv) a single deck dry cargo ship over 100 metres in length;

      (v) a purpose built container ship over 125 metres in length; or

      (vi) a column stabilised mobile offshore drilling unit; or

      (vii) a column stabilised mobile offshore support unit, it shall be approved either by the Secretary of State or the Assigning Authority which assigned freeboards to the ship; and

   (b) if it relates to any other ship, it shall be approved by the Secretary of State.

Information as to loading and ballasting of ships

33. (1) The owner of any ship of more than 150 metres in length specially designed for the carriage of liquids or ore in bulk shall provide, for the guidance of the master, information relating to the loading and ballasting of the ship.

   (2) This information shall indicate the maximum stresses permissible for the ship and specify the manner in which the ship is to be loaded and ballasted to avoid the creation of unacceptable stresses in its structure.

   (3) In the case of a United Kingdom ship the provisions of regulation 32(5) shall have effect in respect of information required under this regulation, and the information so approved shall be included in the book referred in regulation 32(1).

PART I

SHIPS IN GENERAL

Structural strength and stability

2. (1) The construction of the ship shall be such that its general structural strength is sufficient for the freeboards assigned.

(2) The design and construction of the ship shall be such as to ensure that its stability in all probable loading conditions shall be sufficient for the freeboards assigned, and for this purpose due consideration shall be given to the intended service of the ship and to the following criteria.

(a) The area under the curve of righting levers (GZ curve) shall not be less than –

(i) 0.055 metre-radians up to an angle of 30 degrees;

(ii) 0.09 metre-radians up to an angle of 40 degrees or the angle at which the lower edge of any openings in the hull, super-structures or deckhouses which cannot be closed weather-tight, are immersed if that angle is less; and

(iii) 0.03 metre-radians between the angles of heel of 30 degrees and 40 degrees or such lesser angle as is referred to in sub-paragraph (ii) above.

(b) The righting lever (GZ) shall be at least 0.20 metres at an angle of heel equal to or greater than 30 degrees.

(c) The maximum righting lever shall occur at an angle of heel not less than 30 degrees.

(d) The initial transverse metacentric height shall not be less than 0.15 metres. In the case of a ship carrying a timber deck cargo that complies with subparagraph (a) above by taking into account the volume of timber deck cargo, the initial transverse metacentric height shall not be less than 0.05 metres.

(3) To determine whether the ship complies with the requirements of subparagraph (2) the ship shall, unless otherwise permitted, be subject to an inclining test that shall be carried out in the presence of a surveyor appointed by the Secretary of State or, for the ships listed in regulation 32(5), a surveyor appointed by the Assigning Authority.

Regulation 32   SCHEDULE 6.   STABILITY

PART I

INFORMATION AS TO STABILITY

The information relating to the stability of a ship to be provided for the master shall include the particulars specified below.

1. The ship's name, official number, port of registry, gross and register tonnages, principal dimensions, displacement, deadweight and draught to the Summer load line.
2. A profile view and, if necessary, plan views of the ship drawn to scale showing all compartments, tanks, storerooms and crew and passenger accommodation spaces, with their position relative to mid-ship.
3. (1) The capacity and the longitudinal and vertical centre of gravity of every compartment available for the carriage of cargo, fuel, stores, feed-water, domestic or water ballast.
   (2) In the case of a vehicle ferry, the vertical centre of gravity of compartments designated for the carriage of vehicles shall be based on the estimated centres of gravity of the vehicles and not on the volumetric centres of the compartments.
4. (1) The estimated total weight and the longitudinal and vertical centre of gravity of each such total weight of –
      (a) the passengers and their effects; and
      (b) the crew and their effects.
   (2) In estimating such centres of gravity, passengers and crew shall be assumed to be distributed about the ship in the spaces they will normally occupy, including the highest decks to which either or both have access.
5. (1) The estimated weight and the disposition and centre of gravity of the maximum amount of deck cargo which the ship may reasonably be expected to carry on an exposed deck.
   (2) In the case of deck cargo, the arrival condition shall include the weight of water likely to be absorbed by the cargo. (For timber deck cargo the weight of water absorbed shall be taken as 15 per cent of the weight when loaded.)
6. A diagram or scale showing –
   (a) the load line mark and load lines with particulars of the corresponding freeboards; and
   (b) the displacement, tonnes per centimetre immersion, and deadweight corresponding to a range of mean draughts extending between the waterline representing the deepest load line and the waterline of the ship in light condition.
7. (1) A diagram or tabular statement showing the hydrostatic particulars of the ship, including the heights of the transverse metacentre and the values of the moment to change trim one centimetre. These particulars shall be provided for a range of mean draughts extending at least between the waterline representing the deepest load line and the waterline of the ship in light condition.
   (2) Where a tabular statement is used to comply with subparagraph (1), the intervals between such draughts shall be sufficiently close to permit accurate interpolation.
   (3) In the case of ships having raked keels, the same datum for the heights of centres of buoyancy and metacentres shall be used as for the centres of gravity referred to in paragraphs 3, 4 and 5.

8. The effect on stability of free surface in each tank in the ship in which liquids may be carried, including an example to show how the metacentric height is to be corrected.

9. (1) A diagram or table showing cross curves of stability, covering the range of draughts referred to in paragraph 7(1).

   (2) The information shall indicate the height of the assumed axis from which the righting levers are measured and the trim which has been assumed.

   (3) In the case of ships having raked keels and where a datum other than the top of keel has been used, the position of the assumed axis shall be clearly defined.

   (4) Subject to subparagraph (5), only enclosed superstructures and efficient trunks as defined in paragraph 10 of Schedule 4 shall be taken into account in deriving such curves.

   (5) The following structures may be taken into account in deriving such curves if the Secretary of State is satisfied that their location, integrity and means of closure will contribute to the ship's stability –
      (a) superstructures located above the superstructure deck;
      (b) deckhouses on or above the freeboard deck whether wholly or in part only;
      (c) hatchway structures on or above the freeboard deck.

   (6) Subject to the approval of the Secretary of State in the case of a ship carrying timber deck cargo, the volume of the timber deck cargo, or a part thereof, may be taken into account in deriving a supplementary curve of stability appropriate to the ship when carrying such cargo.

   (7) An example shall be included to show how a curve of righting levers (GZ) may be obtained from the cross curves of stability.

   (8) In the case of a vehicle ferry or a similar ship having bow doors, ship-side doors or stern doors where the buoyancy of a superstructure is taken into account in the calculation of stability information, and the cross curves of stability are based upon the assumption that such doors are secured weather-tight, there shall be a specific warning that such doors must be secured weather-tight before the ship proceeds to sea.

10. (1) The diagram and statements referred to in subparagraph (2) shall be provided separately for each of the following conditions of the ship –
      (a) *light condition.* If the ship has permanent ballast, such diagram and statements shall be provided for the ship in light condition both with and without such ballast;
      (b) *ballast condition both on departure and on arrival.* It is to be assumed that on arrival oil fuel, fresh water, consumable stores and the like are reduced to 10 per cent of their capacity;
      (c) *condition on departure and on arrival when loaded to the Summer load line with cargo filling all spaces available for cargo.* Cargo shall be taken to be homogeneous except where this is clearly inappropriate, for example, in cargo spaces that are intended to be used exclusively for the carriage of vehicles or of containers;

(d) *service loaded conditions both on departure and on arrival.*

(2) (a) A profile diagram of the ship drawn to a suitable small scale showing the disposition of all components of the deadweight.

(b) A statement showing the lightweight, the disposition and the total weights of all components of the deadweight, the displacement, the corresponding positions of the centre of gravity, the metacentre and also the metacentric height (GM).

(c) A diagram showing the curve of righting levers (GZ). Where credit is given for the buoyancy of a timber deck cargo the curve of righting levers (GZ) must be drawn both with and without this credit.

(d) A statement showing the elements of stability in the condition compared to the criteria laid down in Schedule 2 paragraph 2(2).

(3) The metacentric height (GM) and the curve of righting levers (GZ) shall be corrected for liquid free surface.

(4) Where there is a significant amount of trim in any of the conditions referred to in subparagraph (1) the metacentric height and the curve of righting levers (GZ) may be required to be determined from the trimmed waterline.

(5) If in the view of the Assigning Authority the stability characteristics in either or both of the conditions referred to in subparagraph (1)(c) are not satisfactory, such conditions shall be marked accordingly and an appropriate warning to the master shall be inserted.

11. A statement of instructions on appropriate procedures to maintain adequate stability in each case where special procedures are applied such as partial or complete filling of spaces designated for cargo, fuel, fresh water or other purposes.

12. The report on the inclining test and of the calculation derived from it to obtain information of the light condition of the ship.

PART II

SHIPS IN RELATION TO WHICH THE SECRETARY OF STATE'S OR THE ASSIGNING AUTHORITY'S APPROVAL OF THE STABILITY INFORMATION IS REQUIRED.

13. The ships referred to in regulation 32(3), (4)(a) and (5)(a) of the Regulations are as follows:

(a) an oil tanker over 100 metres in length;

(b) a bulk carrier, or an ore carrier, over 150 metres in length;

(c) a single deck bulk carrier over 100 metres in length but not exceeding 150 metres in length;

(d) a single deck dry cargo ship over 100 metres in length;

(e) a purpose built container ship over 125 m metres in length.

Penalties

35. (1) to (8). Where Regulations are contravened, the owner and master of the ship shall be guilty of an offence and liable:
  (a) on summary conviction, by a fine to a fine not exceeding the statutory maximum; and
  (b) on conviction on indictment, by a fine or to a fine.

If the appropriate load line on each side of the ship was submerged when it should not have been according to these 1998 load line Regulations then: The owner and master are liable to an additional fine that shall not exceed *one thousand pounds for each complete centimetre* overloaded.

Exercise 55

1 The *Load Line Regulations* 1998 require the master to be provided with stability particulars for various conditions. Detail the information to be provided for a given service condition, describing how this information may be presented.

Chapter 56
Keeping up to date

This book covers many facets of ship stability, ship handling and ship strength. It is very important to keep up to date with developments within these specialist topics.

To help with this, the reader should become a member of the Institute of Nautical Studies, the Royal Institution of Naval Architects or the Honourable Company of Master Mariners, all based in London, UK. These establishments produce monthly journals, and frequently organise conferences and one-day seminars that deal with many topical maritime subjects. Contact can be made at:

Institute of Nautical Studies	www.nautinst.org
Royal Institution of Naval Architects	www.rina.org.uk/tna
Honourable Co. of Master Mariners	www.hcmm.org.uk

Three maritime papers that are of interest and very helpful in the field of Maritime Studies are:

NuMast Telegraph	www.numast.org
Lloyds List	www.lloydslist.com
Journal of Commerce	www.joc.com

Enrolling on short courses is an ideal way to keep up to date with developments within the shipping industry. Signing onto refresher courses is another way of ... a re-treading of the mind.

Ship handling awareness can be greatly enhanced on ship-simulator courses such as the one at Birkenhead (www.lairdside@livjmc.ac.uk). Another very useful contact is The Marine Society & Sea Cadets (bthomas@ms-sc.org).

There are many computer packages used for calculations and graphics, in designing offices and on ships today. Many books have been written giving details of these marine packages.

Knowledge and skills have to be acquired by all involved in the stability and strength of ships. All procedures have to be planned, monitored and adjusted. The handling of ships at sea and in port is very much about teamwork.

All members of that team must share a common understanding of the principles involved. Communication is essential particularly on board ship. To improve techniques, it is essential to discuss evolutions and their consequences. Experiences of past problems and their solutions are always advantageous.

I hope, that you have found this book has been of interest and of benefit to you in your work.

C.B. Barrass

Part 4
Appendices

Appendix I
Summary of stability formulae*

Form coefficients

$$\text{Area of waterplane} = L \times B \times C_w$$
$$\text{Area of amidships} = B \times d \times C_m$$
$$\text{Volume of displacement} = L \times B \times d \times C_b$$
$$C_b = C_m \times C_p$$

Drafts

When displacement is constant (for box shapes):

$$\frac{\text{New draft}}{\text{Old draft}} = \frac{\text{Old density}}{\text{New density}}$$

When draft is constant:

$$\frac{\text{New displacement}}{\text{Old displacement}} = \frac{\text{New density}}{\text{Old density}}$$

$$\text{TPC}_{SW} = \frac{\text{WPA}}{97.56}$$

$$\text{FWA} = \frac{W}{4 \times \text{TPC}_{SW}}$$

$$\text{Change of draft or dock water allowance} = \frac{\text{FWA}\,(1025 - \rho_{DW})}{25}$$

Homogeneous log:

$$\frac{\text{Draft}}{\text{Depth}} = \frac{\text{Relative density of log}}{\text{Relative density of water}}$$

*See Note at end of the Appendix

Variable immersion hydrometer:

$$\text{Density} = \frac{M_y}{M_y - x\left(\dfrac{M_y - M_x}{L}\right)}$$

Trim

$$\text{MCTC} = \frac{W \times GM_L}{100 \times L}$$

$$\text{Change of trim} = \frac{\text{Trimming moment}}{\text{MCTC}} \quad \text{or}$$

$$\frac{W \times (LCB_{foap} - LCG_{foap})}{\text{MCTC}}$$

$$\text{Change of draft aft} = \frac{1}{L} \times \text{Change of trim}$$

$$\text{Change of draft forward} = \text{Change of trim} - \text{Change of draft aft}$$

Effect of trim on tank soundings:

$$\frac{\text{Head when full}}{\text{Length of tank}} = \frac{\text{Trim}}{\text{Length of ship}}$$

True mean draft:

$$\frac{\text{Correction}}{\text{FY}} = \frac{\text{Trim}}{\text{Length}}$$

To keep the draft aft constant:

$$d = \frac{\text{MCTC} \times L}{\text{TPC} \times 1}$$

To find GM_L:

$$\frac{GM_L}{GG_1} = \frac{L}{t}$$

Simpson's rules
1st rule:

$$\text{Area} = h/3 \,(a + 4b + 2c + 4d + e) \quad \text{or} \quad \tfrac{1}{3} \times CI \times \Sigma_1$$

2nd rule:

Area = 3h/8 (a + 3b + 3c + 2d + 3e + 3f + g)   or $\frac{3}{8} \times CI \times \Sigma_2$

3rd rule:

Area = h/12 (5a + 8b − c)   or $\frac{1}{12} \times CI \times \Sigma_3$

KB and BM

Transverse stability

For rectangular waterplanes:

$$I = \frac{LB^3}{12}$$
$$BM = I/V$$

For box shapes:

$$BM = B^2/12d$$
$$KB = d/2$$
$$KM_{min} = B/\sqrt{6}$$

For triangular prisms:

$$BM = B^2/6d$$
$$KB = 2d/3$$

$$\text{Depth of centre of buoyancy below the waterline} = \frac{1}{3}\left(\frac{d}{2} + \frac{V}{A}\right)$$

Longitudinal stability

For rectangular waterplanes:

$$I_L = \frac{BL^3}{12}$$
$$BM_L = \frac{I_L}{V}$$

For box shapes:

$$BM_L = \frac{L^2}{12d}$$

For triangular prisms:

$$BM_L = \frac{L^2}{6d}$$

Transverse statical stability

Moment of statical stability = $W \times GZ$

At small angles of heel:

$$GZ = GM \times \sin\theta$$

By wall-sided formula:

$$GZ = (GM + \tfrac{1}{2} BM \tan^2\theta)\sin\theta$$

By Attwood's formula:

$$GZ = \frac{v \times hh_1}{V} - BG\,\sin\theta$$

Stability curves:

$$\text{New GZ} = \text{Old GZ} \pm GG_1\sin\text{heel or}$$

$$\text{New GZ} = KN - KG\sin\text{heel}$$

$$\text{Dynamical stability} = W \times \text{Area under stability curve}$$

$$= W\left[\frac{v(gh + g_1h_1)}{V} - BG(1 - \cos\theta)\right]$$

$$\lambda_o = \frac{\text{Total VHM}}{SF \times W} \qquad \lambda_{40} = \lambda_o \times 0.8 \qquad \text{Actual HM} = \frac{\text{Total VHM}}{SF}$$

$$\text{Approx angle of heel} = \frac{\text{Actual HM}}{\text{Maximum permissible HM}} \times 12°$$

Approx angle of θ due to grain shift

$$\text{Reduction in GZ} = (GG_H \times \cos\theta) + (GG_V \times \sin\theta)$$

$$
\begin{aligned}
W &= \text{ship displacement in tonnes} \\
GG_H &= \text{horiz movement of 'G'} \\
GG_V &= \text{vert movement of 'G'} \\
\lambda_o &= \text{righting arm @ } \theta = 0° \text{ (if upright ship)} \\
\lambda_{40} &= \text{righting arm @ } \theta = 40° \\
HM &= \text{heeling moment} \\
SF &= \text{stowage factor} \\
VHM &= \text{volumetric heeling moment}
\end{aligned}
$$

List

$$\text{Final KG} = \frac{\text{Final moment}}{\text{Final displacement}}$$

$$GG_1 = \frac{w \times d}{\text{Final W}}$$

$$\tan \text{list} = \frac{GG_1}{GM}$$

Increase in draft due to list:

New draft $= \frac{1}{2} \times b \times \sin\theta + (d)\cos\theta$ ... Rise of floor is zero put 'r' in if rise of floor exists. (measured at full Br. Mld)

Inclining experiment:

$$\frac{GM}{GG_1} = \frac{\text{Length of plumbline}}{\text{Deflection}}$$

Effect of free surface

$$\text{Virtual loss of GM} = \frac{lb^3}{12} \times \frac{\rho}{W} \times \frac{1}{n^2}$$

Drydocking and grounding
Upthrust at stern:

$$P = \frac{MCTC \times t}{l}$$

or

$$P = \text{Old} - \text{New displacement}$$

$$\text{Virtual loss of GM} = \frac{P \times KM}{W}$$

$$\text{or} = \frac{P \times KG}{W - P}$$

Pressure of liquids

$$\text{Pressure (P)} = Dwg$$

$$\text{Thrust} = P \times \text{Area}$$

$$\text{Depth of centre of pressure} = \frac{I_{WL}}{AZ}$$

Bilging and permeability

$$\text{Permeability} = \frac{BS}{SF} \times 100 \text{ per cent}$$

$$\text{Increase in draft} = \frac{\mu v}{A - \mu a}$$

Strength of ships

$$\text{Stress} = \frac{\text{Load}}{\text{Area}}$$

$$\text{Strain} = \frac{\text{Change in length}}{\text{Original length}} = \frac{y}{R}$$

Young's modulus:

$$E = \frac{\text{Stress}}{\text{Strain}}$$

Bending moment:

$$M = \frac{E}{R} \times I$$

$$\text{Section modulus} = \frac{I}{Y}$$

Shearing stress:

$$q = \frac{F \times A \times y}{I \times t}$$

Stress:

$$f = \frac{E}{R} \times y$$

Freeboard marks

$$\text{Distance Summer LL to Winter LL} = \tfrac{1}{48} \times \text{Summer draft}$$

$$\text{Distance Summer LL to Tropical LL} = \tfrac{1}{48} \times \text{Summer draft}$$

Ship squat

$$\text{Blockage factor} = \frac{b \times T}{B \times H}$$

$$\delta_{max} = \frac{C_b \times S^{0.81} \times V_k^{2.08}}{20}$$

$$y_o = H - T$$
$$y_2 = y_o - \delta_{max}$$
$$A_s = b \times T$$
$$A_c = B \times H$$

In open water:

$$\delta_{max} = \frac{C_b \times V_k^2}{100}$$

In confined channel:

$$\delta_{max} = \frac{C_b \times V_k^2}{50}$$
$$\text{Width of influence} = 7.7 + 20 (1 - C_b)^2$$

Miscellaneous
Angle of loll:

$$\tan loll = \sqrt{\frac{2 \times GM}{BM_T}}$$

$$GM = \frac{2 \times \text{initial GM}}{\cos loll}$$

Heel due to turning:

$$\tan heel = \frac{v^2 \times BG}{g \times r \times GM}$$

Rolling period:

$$T = 2\pi \frac{k}{\sqrt{g \times GM}} = \frac{2k}{\sqrt{GM}} \text{ approx}$$

Zero GM:

$$\tan list = \sqrt[3]{\frac{2 \times w \times d}{W \times BM}}$$

Theorem of parallel axes:

$$I_{CG} = I_{OZ} - Ay^2$$

or

$$I_{NA} = I_{xx} - Ay^2$$

Permeability (μ)

$$\text{Permeability} = \frac{BS}{SF} \times 100 \text{ per cent} \qquad \text{Increase in draft} = \frac{\mu\theta}{A - \mu a}$$

$$\text{Permeability } (\mu) = \frac{\text{Volume available for water}}{\text{Volume available for cargo}} \times 100$$

$$\text{Permeability } (\mu) = \frac{SF \text{ of cargo} - \text{solid factor}}{SF \text{ of cargo}} \times 100$$

$$\text{Solid factor} = \frac{1}{RD} \qquad \text{Effective length} = 1 \times \mu$$

$$\text{Sinkage} = \frac{\text{Volume of bilged compartment} \times \mu}{\text{Intact waterplane area}}$$

$$\text{Tan } \theta = \frac{BB_H}{GM_{bilged}}$$

Drafts and trim considerations

$$\text{Correction to observed drafts} = \frac{l_1}{L_1} \times \text{Trim}$$

$$\text{Midships draft corrected for deflection} = \frac{d_{FP} + (6 \times d_m) + d_{AP}}{8}$$

$$\begin{array}{l}\text{Correction of midships draft} \\ \text{to true mean draft when} \\ \text{LCF is not at amidships}\end{array} = \frac{\text{Distance of LCF from midships} \times \text{Trim}}{LBP}$$

Second trim correction for position of LCF, if trimmed hydrostatics are not supplied a (form correction)

$$= \frac{\text{True trim} \times (MCTC_2 - MCTC_1)}{2 \times TPC \times LBP}$$

$$\text{Alternative form correction} = \frac{50 \times (\text{True trim})^2 \times (MCTC_2 - MCTC_1)}{LBP}$$

Note

These formulae and symbols are for guidance only and other formulae which give equally valid results are acceptable.

$$\rho = \frac{\text{Mass}}{\text{Volume}}$$

$$RD = \frac{\rho_{\text{substance}}}{\rho_{FW}}$$

$$\nabla = (L \times B \times d) \times C_b$$

$$\Delta = \nabla \times \rho$$

$$DWT = \Delta - \Delta_{\text{light}}$$

$$A_W = (L \times B) \times C_W$$

$$TPC = \frac{A_W}{100} \times \rho$$

$$\text{Sinkage/rise} = \frac{W}{TPC}$$

$$FWA = \frac{\Delta_{\text{Summer}}}{4 \times TPC_{SW}}$$

$$DWA = \frac{(1025 - \rho_{\text{dock}})}{25} \times FWA$$

$$MSS = \Delta \times GZ$$

$$GZ = GM \times \sin\theta$$

$$GZ = [GM + \tfrac{1}{2}BM \tan^2\theta]\sin\theta$$

$$GZ = KN - (KG \times \sin\theta)$$

$$\text{Dynamic stability} = \text{Area under GZ curve} \times \Delta$$

$$\text{Area under curve (SR1)} = \frac{1}{3} \times h \times (y_1 + 4y_2 + y_3)$$

$$\text{Area under curve (SR2)} = \frac{3}{8} \times h \times (y_1 + 3y_2 + 3y_3 + y_4)$$

$$\lambda_o = \frac{\text{Total VHM}}{SF \times \Delta}$$

$$\lambda_{40} = \lambda_o \times 0.8$$

$$\text{Actual HM} = \frac{\text{Total VHM}}{SF}$$

$$\text{Approx angle of heel} = \frac{\text{Actual HM}}{\text{Max permissible HM}} \times 12°$$

$$\text{Reduction in GZ} = (GG_H \times \cos \theta) + (GG_v \times \sin \theta)$$

$$\text{Rolling period T (sec)} = \frac{2 \times \pi \times K}{\sqrt{g \times GM}} \quad \text{or} \quad 2\pi \sqrt{\frac{K^2}{gGM_T}}$$

$$\text{or} \quad 2\sqrt{\frac{K^2}{GM_T}} \text{ seconds}$$

$$GG_{H/V} = \frac{w \times s}{\Delta}$$

$$FSC = \frac{i}{\Delta} \times \rho_T$$

$$FSC = \frac{l \times b^3}{12 \times \Delta} \times \rho_T$$

$$FSC = \frac{FSM}{\Delta}$$

$$\tan \theta = \frac{GG_H}{GM}$$

$$KG = \frac{\Sigma \text{ Moments}}{\Sigma \text{ Weights}}$$

$$GG_H = \frac{\Sigma \text{ Moments}}{\Sigma \text{ Weights}}$$

$$GM = \frac{w \times s \times \text{length}}{\Delta \times \text{deflection}}$$

$$\tan \text{ angle of loll} = \sqrt{\frac{-2 \times GM}{BM_T}}$$

$$GM \text{ at angle of loll} = \frac{2 \times \text{initial GM}}{\cos \theta}$$

$$\tan \theta = \sqrt[3]{\frac{2 \times w \times s}{\Delta \times BM_T}}$$

Draft when heeled = (Upright draft $\times \cos\theta$) + ($\frac{1}{2} \times$ beam $\times \sin\theta$)

Position of the metacentre $\qquad KM_T = KB + BM_T$

$$BM_T = \frac{I_T}{\nabla}$$

$$BM_T(\text{box}) = \frac{L \times B^3}{12 \times \nabla}$$

Distance Summer LL to Winter LL = $\dfrac{1}{48} \times$ Summer draft

Distance Summer LL to Tropical LL = $\dfrac{1}{48} \times$ Summer draft

$KM_L = KB + BM_L$

$BM_L = \dfrac{I_L}{\nabla}$

$BM_L(\text{box}) = \dfrac{L^3 \times B}{12 \times \nabla}$

$MCTC = \dfrac{\Delta \times GM_L}{100 \times LBP}$

$CoT = \dfrac{\Sigma \text{ Trimming moment}}{MCTC}$

Change of trim aft = Change of trim $\times \dfrac{LCF}{LBP}$

Change of trim fwd = Change of trim $\times \dfrac{LBP - LCF}{LBP}$

Turn mean draft = Draft aft $\pm \left(\text{trim} \times \dfrac{LCF}{LBP} \right)$

$\text{Trim} = \dfrac{\Delta \times (LCG \sim LCB)}{MCTC}$

$P = \dfrac{\text{Trim} \times MCTC}{LCF}$

P = Reduction in TMD $\times$ TPC

Loss of GM $= \dfrac{P \times KM_T}{\Delta} \quad$ or $\quad \dfrac{P \times KG}{\Delta - P}$

$$\tan \theta = \frac{v^2 \times BG}{g \times R \times GM}$$

$$\text{Permeability } (\mu) = \frac{\text{Volume available for water}}{\text{Volume available for cargo}} \times 100$$

$$\text{Solid factor} = \frac{1}{RD}$$

$$\text{Permeability } (\mu) = \frac{\text{SF of cargo} - \text{solid factor}}{\text{SF of cargo}} \times 100$$

$$\text{Effective length} = 1 \times \mu$$

$$\text{Sinkage} = \frac{\text{Volume of bilged compartment} \times \text{permeability } (\mu)}{\text{Intact water plane area}}$$

$$I_{\text{parallel axis}} = I_{\text{centroidal axis}} + As^2$$

$$\text{Tan } \theta = \frac{BB_H}{GM_{\text{bilged}}}$$

$$\text{Correction to observed drafts} = \frac{l_1}{L_1} \times \text{Trim}$$

$$\text{Midships draft corrected for deflection} = \frac{d_{FP} + (6 \times d_M) + d_{AP}}{8}$$

Correction of midships draft to true
mean draft when CF not midships

$$= \frac{\text{Distance of CF from midships} \times \text{trim (true trim at perp's)}}{\text{LBP}}$$

Second trim correction for position of CF if trimmed
hydrostatics are not supplied (form correction)

$$= \frac{\text{True trim} \times (MCTC_2 - MCTC_1)}{2 \times TPC \times LBP}$$

$$\text{Alternative form correction} = \frac{50 \times \text{true trim}^2 \times (MCTC_2 - MCTC_1)}{LBP}$$

Summary

Always write your formula first in letters. If you then make a mathematical error you will at least obtain some marks for a correct formula.

Appendix II
SQA/MCA 2004 syllabuses for masters and mates

(a) Chief mate unlimited stability and structure syllabus

This MCA approved syllabus was prepared by the IAMI Deck Subgroup and subsequently amended, following consultation with all IAMI colleges, in November 2002 through to June 2004.

Notes.

1. The syllabus is based on the HND in Nautical Science. It covers Outcomes 1 and 2 of Unit 24 (Ship Stability), and part of Outcomes 1 and 3 of Unit 25 (Structures and Maintenance).
2. Calculations are to be based wherever possible on the use of real stability information.
3. Longitudinal stability calculations are to be based on taking moments about the after perpendicular and using formula:

$$\text{Trim} = \text{Displacement} \times (\text{LCB} \sim \text{LCG})/\text{MCTC}$$

4. Formula sheets will be provided to candidates for the examination.

1. **Stability information carried on board ship. The inclining experiment.**
   (a) Explains the use of stability information to be carried on board ship.
   (b) Explains the purpose of the inclining experiment.
   (c) Identifies the occasions when the inclining experiment must be undertaken.
   (d) Describes the procedure and precautions to be taken before and during the inclining experiment.
   (e) Calculates the lightship KG and determines the lightship displacement for specified experiment conditions.

(f) Explains why a vessel's lightship displacement and KG will change over a period of time.

2. **Application of 'free surface effect'.**
   (a) Describes free surface effect (FSE) as a virtual loss of GM and relates it to the free surface correction (FSC).
   (b) Calculates FSC given rectangular area tank dimensions and tank liquid density.
   (c) Describes the effect on FSC of longitudinal subdivisions in tanks.
   (d) Calculates FSC given free surface moments (FSM).
   (e) Applies FSC or FSM to all calculations as necessary.

3. **The effect on vessel's centre of gravity of loading, discharging or shifting weights. Final list. Requirements to bring vessel upright.**
   (a) Calculates the final position of vessel's centre of gravity relative to the keel and centreline taking into account loaded, discharged and shifted weights.
   (b) Calculates the resultant list.
   (c) Calculates the minimum GM required prior to loading/discharging/shifting weights to limit the maximum list.

4. **Stability during dry-docking. Using real ship stability information.**
   (a) Explains the virtual loss of metacentric height during dry-docking and the requirements to ensure adequate stability.
   (b) Calculates the virtual loss of metacentric height and hence effective GM during dry-docking.
   (c) Determines the maximum trim at which a vessel can enter dry-dock to maintain a specified GM.
   (d) Calculates the draft at which the vessel takes the blocks fore and aft.
   (e) Describes the practical procedures that can be taken to improve stability prior to dry-docking if it is found to be inadequate.
   (f) Explains why it is beneficial to have a small stern trim when entering dry-dock.

5. **Increase in draft due to list/heel. Angle of heel when turning.**
   (a) Explains increase in draft due to list/heel.
   (b) Calculates increase in draft due to list/heel.
   (c) Explains angle of heel due to turning and the effect on stability.
   (d) Calculates angle of heel due to turning.

6. **The effect of loading/discharging/shifting weights on trim, draft and stability. Using real ship stability information.**
   (a) Defines 'centre of flotation' with respect to waterplane area.
   (b) Defines 'longitudinal centre of flotation' (LCF) with respect to the aft perpendicular and explains change in LCF with change of draft.
   (c) Defines 'true mean draft' (TMD).

(d) Calculates TMD.
(e) Calculates final drafts and effective GM for various conditions of loading.
(f) Calculates where to load/discharge a weight to produce a required trim or draft aft.
(g) Calculates the weight to load/discharge at a given position to produce a required trim or draft aft.
(h) Calculates final drafts when vessel moves from one water density to a different water density.
(i) Calculates the maximum cargo to discharge to pass safely under a bridge.
(j) Calculates the minimum ballast to load to safely pass under a bridge.
(k) Calculates the final drafts in (i) and (j).

7. **Draft Survey.**
   (a) Calculates the correction to the observed forward and after drafts to forward perpendicular and after perpendicular, respectively.
   (b) Calculates the correction to the observed midship draft to amidship.
   (c) Calculates the correction of the amidship draft for hull deflection.
   (d) Calculates the correction of the amidship draft to true mean draft (TMD) when CF is not amidship.
   (e) Calculates the correction for the position of the CF if trimmed hydrostatics are not supplied.

8. **Curves of righting levers (GZ), using real ship stability information. Determine compliance with 'intact stability' requirements for the current load line regulations.**
   (a) Constructs a curve of righting levers (GZ), for a given condition.
   (b) Defines 'righting moment' (moment of statical stability) and 'dynamical stability'.
   (c) Extracts stability information from a curve of righting levers (GZ).
   (d) Calculates appropriate areas under a curve of righting levers (GZ), using Simpson's Rules.
   (e) Assesses whether vessel complies with the 'intact stability' requirements of the current load line regulations.

9. **Simplified stability. Using real ship stability information.**
   (a) Describes the appropriate use of 'simplified stability' information.
   (b) Assesses whether a vessel complies with 'maximum permissible KG' requirements for a given condition.

10. **Angle of loll and effective GM at angle of loll.**
    (a) Describes the stability at an angle of loll and shows the existence of an effective GM.
    (b) Calculates the angle of loll for vessel with a negative initial GM.

(c) Calculates the effective GM at an angle of loll.

(d) Describes the dangers to a vessel with an angle of loll.

(e) Distinguishes between an angle of loll and an angle of list.

(f) Describes the correct procedure for correcting an angle of loll.

11. Factors affecting a curve of righting levers (GZ).

(a) Describes the effects of variations in beam and freeboard on the curve of righting levers (GZ).

(b) Describes the effect of trim on KN values and resultant curve of righting levers (GZ).

(c) Describes the terms 'fixed trim' and 'free trim' with respect to KN values and resultant curve of righting levers (GZ).

(d) Explains the effects of being in a seaway on the curve of righting levers (GZ).

(e) Outlines the conditions for a vessel to be in the stiff or tender condition and describes the effects on the curve of righting levers (GZ).

(f) Describes the use of ballast/bunkers to ensure adequate stability throughout the voyage.

(g) Describes icing allowances.

(h) Describes the changes in stability that may take place on a voyage.

(i) Explains the effects on the curve of righting levers (GZ) of the changes described in (h).

(j) Explains the effects of an angle of list on the curve of righting levers (GZ).

(k) Explains the effects of an angle of loll on the curve of righting levers (GZ).

(l) Explains the effects of a zero initial GM on the curve of righting levers (GZ).

12. The effect on the curve of righting levers (GZ) of shift of cargo and wind heeling moments.

(a) Constructs a curve of righting levers (GZ) taking into account shift of cargo/solid ballast and describe the effects on the vessel's stability.

(b) Explains the precautions to be observed when attempting to correct a large angle of list.

(c) Explains how wind heeling moments are calculated.

(d) Constructs a curve of righting moments taking into account wind heeling moments and describes the effects on the vessel's stability.

(e) Describes the minimum stability requirements taking into account wind heeling moments as specified in current 'Load Line – Instructions for the Guidance of Surveyors'.

(f) Determines that a ship's loaded condition complies with the minimum stability requirements specified in (e).

13. **Use of the current IMO Grain Rules to determine if the vessel complies with the specified stability criteria. Real ship stability information to be used.**
    (a) Calculates the 'grain heeling moments' for a specified loading condition.
    (b) Determines the 'grain heeling moment' calculated in (a) whether the vessel complies with the stability requirements by comparison with the 'maximum permissible heeling moments'.
    (c) Calculates the approximate angle of heel in (b).
    (d) Constructs graphs of a righting arm curve and heeling arm curve.
    (e) Assesses whether a grain laden vessel complies with the 'minimum stability requirements' specified in the 'IMO Grain Rules'.
    (f) Discusses factors to be taken into account to minimum grain heeling moments.

14. **Rolling, pitching, synchronous and parametric rolling.**
    (a) Describes rolling and pitching.
    (b) Defines rolling period.
    (c) Explains factors affecting rolling period.
    (d) Describes synchronous rolling and the associated dangers.
    (e) Describes parametric rolling and the associated dangers.
    (f) Describes actions to be taken by Ship's Officer in the event of synchronous rolling or parametric rolling.

15. **The effect of damage and flooding on stability.**
    (a) Calculates, for box-shaped vessel, the effect on draft, trim, list, freeboard and metacentric height if the following compartments are bilged:
        (i) Symmetrical amidships compartment with permeability.
        (ii) Symmetrical amidships compartment with watertight flat below initial waterline with permeability.
        (iii) Symmetrical amidships compartment with watertight flat above initial waterline with permeability.
        (iv) Extreme end compartment with 100 per cent permeability.
        (v) Extreme end compartment with watertight flat below the initial waterline with 100 per cent permeability.
        (vi) Amidships compartment off the centreline with 100 per cent permeability.
    (b) Describes countermeasures, which may be taken in event of flooding.

16. **Damage stability requirements for passenger vessels and Type A and Type B vessels.**
    (a) Defines for passenger vessels . . . 'bulkhead deck', 'margin line', 'floodable length', 'permissible length', 'factor of subdivision'.
    (b) Describes subdivision load lines for passenger vessels.
    (c) Identifies 'assumed damage' for passenger vessels.

    (d) Identifies 'assumed flooding' for passenger vessels.

    (e) Identifies 'minimum damage stability requirements' for passenger vessels.

    (f) Describes the 'Stockholm Agreement 1996' with respect to stability requirements for passenger vessels.

    (g) Identifies damage stability flooding criteria for Type A, B-60, B-100 vessels.

    (h) Identifies minimum equilibrium stability condition after flooding for vessels specified in (g).

17. Load line terminology and definitions for new builds.

    (a) Defines Type A, B, B-60 and B-100 vessels.

18. Conditions of assignment of load lines.

    (a) Describes 'conditions of assignment' for vessels specified in 17 (a).

    (b) Describes 'tabular freeboard' with respect to vessels specified in 17 (a).

    (c) Explains the corrections to be applied to tabular freeboard to obtain 'statutory assigned freeboard'.

19. Assignment of special load lines, for example 'timber load lines'.

    (a) Describes the special factors affecting the assignment of timber load lines.

    (b) Describes the intact stability requirements for vessels assigned timber load lines.

20. Requirements and codes relating to the stability of specialised vessels.

    (a) Identifies the stability problems associated with Ro-Ro vessels, offshore supply vessels and vessels when towing.

21. The preparations required for surveys.

    (a) Lists surveys required by the load line rules for a vessel to maintain a valid load line certificate.

    (b) Lists the items surveyed at a load line survey and describes the nature of the survey for each item.

(b) OOW unlimited stability and operations syllabus

This MCA approved syllabus was prepared by the IAMI Deck Subgroup and subsequently amended, following consultation with all IAMI colleges, in November 2002 through to June 2004.

Notes.

1. The syllabus is based on the HND in Nautical Science. It covers Outcomes 3 and 4 of Unit 7 (Cargo Work), all Outcomes of Unit 8 (Ship Stability and part of Outcomes 2 of Unit 10 (Maritime Law and Management).

2. Formula sheets will be provided to candidates for the examination.

1. **Hydrostatics.**
   (a) Defines mass, volume, density, relative density, Archimedes' principle FWA, DWA and TPC.
   (b) Determines TPC and displacement at varying drafts using hydrostatic tables.
   (c) Calculates small and large changes in displacement making appropriate use of either TPC or displacement tables.
   (d) Define waterline length, LBP, freeboard, waterplane area, C_W and C_b.
   (e) Calculates the weight to load or discharge to obtain small changes in draft or freeboard.
   (f) Explains the reasons for load lines and load line zones.
   (g) Calculates the weight to load or discharge in relation to load line dimensions, appropriate marks, TPC, FWA and DWA.

2. **Statical stability at small angles.**
   (a) Defines centre of gravity, centre of buoyancy, initial transverse metacentre and initial metacentric height (GM).
   (b) Calculates righting moments given GM and displacement.
   (c) Explains stable, neutral and unstable equilibrium.
   (d) Explains the relationship between equilibrium and the angle of loll.
   (e) Identifies from a given GZ curve, range of stability, initial GM, maximum GZ, angle of vanishing stability, angle of deck edge immersion, angle of loll and angle of list.
   (f) Explains the difference between typical GZ curves for stiff and tender vessels.
   (g) Sketches typical GZ curves for vessels at an angle of loll.

3. **Transverse stability.**
   (a) Calculates the shift of ship's G, vertically and horizontally after the loading/discharging/shifting a weight.
   (b) Calculates the final KG or GM by moments about the keel after loading/discharging/shifting weights including appropriate free surface correction.
   (c) Calculates distance of G horizontally from the centreline after loading/discharging/shifting weights.
   (d) Calculates the effect on stability of loading or discharging a weight using ship's gear.
   (e) Calculates the angle of list resulting from (a) to (d).
   (f) Explains the difference between list and loll, and methods of correction.
   (g) Explains the consequences and dangers of a free surface.
   (h) Explains that the free surface effect can be expresses as virtual rise of G or as a free surface moment.
   (i) Describes the effects on free surface of longitudinal subdivision of a tank.

4. Longitudinal stability.
  (a) Defines LCF, LCG, LCB, AP, trim, trimming moment and MCTC.
  (b) Calculates the effect on drafts of loading, discharging and shifting weights longitudinally by taking moments about the AP.

5. Maintaining a deck watch (alongside or at anchor).

6. Pollution prevention.

7. Legislation.

Note. Parts 5, 6 and 7 are outside the remit of this book on Ship Stability.

Appendix III
Specimen exam questions with marking scheme

Recent Examination questions for Ship Stability @ 032-72 level: (a) STCW 95 Chief Mate/Master Reg.11/2 (unlimited)

Note that all calculations are to be based wherever possible on the use of real ship stability information.　　Marking scheme shown in brackets.

1. (a) Explain why the values of trim and metacentric height in the free-floating condition are important when considering the suitability of a vessel for dry docking. (14)
   (b) Describe how the metacentric height and the trim can be adjusted prior to dry docking so as to improve the vessel's stability at the critical instant. (6)
   (c) Describe *two* methods of determining the upthrust (P) during the critical period. (10)

2. A vessel is floating upright with the following particulars:

   Displacement 12 850 tonnes　　KM 9.30 m.

   A boiler weighing 96 tonnes is to be loaded using the vessel's heavy lift derrick from a position 17.00 m to port of the vessel's centreline.　Kg of derrick head 22.00 m.
   (a) Calculate the maximum allowable KG prior to loading in order to limit the list to a maximum of 7 degrees during the loading operation. (20)
   (b) Calculate the final angle and direction of list if the boiler is to be stowed in a position Kg 2.60 m, 3.90 m to starboard of the vessel's centreline. (15)

3. (a) The *Load Line Regulations* 1998 require the master to be provided with stability particulars for various conditions. Detail the information to be provided for a given service condition, describing how this information may be presented. (20)

(b) State the corrections required when converting *basic* freeboard to *assigned* freeboard, outlining the circumstances when freeboard is increased or decreased in *each* case.   (15)

4. A vessel operating in severe winter conditions may suffer from non-symmetrical ice accretion on decks and superstructure.

   Describe, with the aid of a sketch of the vessel's curve of statical stability, the effects on the overall stability of the vessel.   (35)

5. Describe, with the aid of one or more sketches, the effect on the dynamical stability of a vessel during bad weather, of a transverse and vertical shift of solid bulk cargo originally trimmed level.   (35)

6. With reference to a modern shipboard stability and stress finding instrument:
   (a) state the hydrostatic and stability data already pre-programmed into the instrument;   (8)
   (b) describe the information to be entered into the instrument by the ship's officer;   (12)
   (c) describe the output information.   (15)

7. A vessel with a high deck cargo will experience adverse effects on its stability due to strong beam winds on lateral windage areas.

   Describe, with the aid of a sketch, the minimum stability requirements with respect to wind heeling under current regulations.   (35)

8. (a) A vessel loads a packed timber cargo on deck such that there is an increase in the vessel's KG and an effective increase in freeboard.

   Using a sketch, show the effect of loading this cargo on the vessel's GZ curve.   (14)
   (b) Sketch, showing how the GZ curve for vessel with a zero GM is affected by *each* of the following:
      (i) a rise in the vessel's KG;   (8)
      (ii) a reduction in the vessel's KG.   (8)

9. (a) Discuss how a vessel's still water rolling period may be affected by changes in the distribution of weight aboard the vessel.   (10)
   (b) Explain the term *synchronous rolling*, describing the dangers, if any, associated with it.   (12)
   (c) State the action to be taken by the ship's officer when it becomes apparent that the vessel is experiencing *synchronous rolling*.   (8)

10. A box-shaped vessel floating at an even keel in salt water has the following particulars:

   Length 83.00 m   Breadth 18:00 m   Depth 10.50 m   Draft 6.12 m
   KG 6.86 m

   An empty midship watertight compartment 14.64 m long and extending the full breadth of the vessel is bilged.

Calculate *each* of the following:
(a) the new draft;   (8)
(b) the GM in thee flooded condition;   (12)
(c) the righting moment at an angle of 17 degrees.   (15)

11. Describe the hydrostatic stability and stress data required to be supplied
to ships under the current Load Line Regulations.   (30)

12. (a) Explain why a vessel laden to the same draft on different voyages
may have different rolling periods.   (10)
(b) Sketch on one set of axes, specimen GZ curves showing *stiff* and
tender conditions for a vessel laden to the same draft on different
voyages.   (4)
(c) Describe the different rolling characteristics of a vessel in a *stiff* condition and a vessel in a *tender* condition.   (6)
(d) Sketch the GZ curve for a vessel with a small negative GM.   (4)
(e) Describe the rolling characteristics for a vessel with a small negative
GM.   (6)

13. A box-shaped vessel floating on an even keel in salt water has the following particulars:

Length 146.00 m   Breadth 29.40 m   Draft 6.52 m   KG 9.90 m.

The vessel has a centreline watertight bulkhead with an empty amidships compartment of length 24.00 m on *each* side of the vessel.
Calculate the angle of heel if *one* of these side compartments is
bilges.   (35)

14. A vessel is trimmed 1.50 m by the stern and heeled 8° to starboard and
has the following particulars:

Displacement 15 000 tonnes   MCTC 188   KG 7.25 m   KM 7.52 m

No. 1 D.B. tanks, port and starboard, are empty and have centroids
112.00 m foap and 3.80 m each side of the centreline.
No. 5 D.B. tanks, port and starboard, are full of oil and have centroids
53.00 m foap and 3.60 m each side of the centreline.
Calculate the quantities to transfer into *each* of the No. 1 tanks from
No. 5 tanks to bring this vessel to even keel and upright using equal
quantities from No. 5 port and starboard tanks.   (35)

15. (a) State the formula to determine the virtual loss of GM due to a free
liquid surface within a rectangular tank, explaining *each* of the
terms used.   (8)
(b) Explain the effects on the virtual loss of transverse GM due to the
free surface effects when the slack tank is equally divided in *each* of
the following situations:
(i) by a longitudinal bulkhead;
(ii) by a transverse bulkhead.   (10)

(c) A double bottom tank, initially empty, is to be ballasted full of salt water. Sketch a graph to show the way in which the effective KG of the ship will change from the instant of starting to fill the tank until it is full. (12)

16. (a) Describe, with the aid of sketches, how a vessel overtaking another vessel of a similar size is likely to be affected by interaction in *each* of the following situations:
    (i) when the bows of the two vessels are abreast; (6)
    (ii) when the stern of the overtaking vessel is close to the bow of the vessel being overtaken. (6)
    (b) Describe the measures *each* vessel can take to reduce the danger of collision due to interaction when one vessel is overtaking another in a channel. (18)

17. (a) State the purpose of the inclining experiment. (5)
    (b) Describe the precautions to be taken, by the *ship's officer*, before and during the experiment. (16)
    (c) List the circumstances when the inclining experiment is required to take place on passenger vessels. (9)

18. With reference to the current Passenger Ship Construction Regulations:
    (a) state the assumed hull damage used when calculating the ship's ability to remain afloat in a stable condition after flooding; (10)
    (b) describe the minimum stability requirements in the damaged condition for passenger ships (other than Ro–Ro) built after 1990, including details of the heeling moment assumed. (20)

19. A vessel initially upright, is to carry out an inclining test. Present displacement 5600 tonnes. KM 5.83 m.
    Total weights on board during the experiment:

    Ballast 720 m, Kg 2.32 m. Tank full.
    Bunkers 156 t, Kg 1.94 m. Free surface moment 438 tm.
    Water 87 t, Kg 4.25 m. Slack tank. Free surface moment 600 tm.
    Boiler water 18 t, Kg 4.2 m. Free surface moment 104 tm.
    Deck crane (yet to be fitted) 17 t, Kg 8.74 m.
    Inclining weights 56 t, Kg 8.44 m.

    The plumblines have an effective vertical length of 7.72 m. The inclining weights are shifted transversely 6.0 m on *each* occasion and the mean horizontal deflection of the plumbline is 0.82 m.
    Calculate the vessel's lightship KG. (35)

20. A vessel has the following particulars:

    LBP 177 m   MCTC 258   TPC 24   LCF 82 m foap
    Initial drafts: forward 5.86 m; aft 6.81 m.

The vessel has a fore peak tank, Lcg 162 m foap, which is to be ballasted in order to reduce the draft aft to 6.58 m.

Calculate *each* of the following:

(a) the minimum weight of ballast required;   (20)

(b) the final draft forward.   (15)

21. A vessel has loaded grain, stowage factor 1.65 m³/tonne to a displacement of 13 000 tonnes. In the loaded condition the effective KG is 7.18 m.

All grain spaces are full, except No. 2 Tween Deck, which is partially full.

The tabulated transverse volumetric heeling moments are as follows:

No. 1 Hold	1008 m⁴
No. 2 Hold	1211 m⁴
No. 3 Hold	1298 m⁴
No. 4 Hold	1332 m⁴
No. 1 TD	794 m⁴
No. 2 TD	784 m⁴
No. 3 TD	532 m⁴

The value of the Kg used in the calculation of the vessel's effective KG were as follows:

For lower holds, the centroid of the space.

For Tween Decks, the actual Kg of the cargo.

(a) Using Datasheet Q.21 *Maximum Permissible Grain Heeling Moments Table*, determine the vessel's ability to comply with the statutory Grain Regulations.   (20)

(b) Calculate the vessel's approximate angle of heel in the event of a shift of grain assumed in the Grain Regulations.   (5)

(c) State the stability criteria in the current International Grain Code. (10)

General Ship Knowledge @ 035–22 level: (b) STCW 95 Officer in Charge of Navigational Watch <500 gt Reg.11/3 (Near Coastal)

Note that all calculations are to be based wherever possible on the use of real ship stability information.   Marking scheme shown in brackets.

1. A ship has the following particulars:

Summer freeboard	1.65 m
FWA	75 mm
TPC	6.8

The ship is currently in dock water RD 1.010 with freeboards as follows:

Port	1.85 m.
Starboard	1.79 m.

Calculate the weight of cargo to be loaded in order that the ship will be at her Summer load draft in salt water (RD 1.025).   (20)

2. Describe, with the aid of sketches, how a virtual rise of a vessel's centre of gravity occurs due to slack tanks and the consequent reduction in the righting lever.   (20)

3. Define and explain the purpose for *each* of the following:
   (a) sheer;   (5)
   (b) camber;   (5)
   (c) flare;   (5)
   (d) rise of floor.   (5)

4. Explain, with the aid of a sketch, *each* of the following:
   (a) transverse metacentre;   (5)
   (b) initial metacentric height;   (5)
   (c) righting lever;   (5)
   (d) righting moment.   (5)

5. A ship has a displacement of 1200 tonnes. An 80 tonne generator is in the lower hold in position Kg 2.8 m, 3.2 m to port of the centreline.
   (a) Calculate the vertical and the horizontal components of the shift of the ship's centre of gravity if the generator is restowed on deck in position Kg 8.4 m, 2.2 m to starboard of the centreline.   (20)
   (b) Draw a sketch to illustrate the shift of the ship's centre of gravity calculated in Q5(a).   (10)

6. (a) Define, with the aid of sketches, each of the following ship stability terms:
   (i) metacentre: (3)
   (ii) BM;   (3)
   (iii) metacentric height;   (3)
   (iv) vertical centre of gravity;   (3)
   (v) righting moment.   (3)
   (b) With the aid of sketches, show a vessel in *each* of the following conditions:
   (i) stable equilibrium;   (5)
   (ii) unstable equilibrium.   (5)

7. A ship in salt water (RD 1.025) has a mean draft of 3.10 m, KG 3.70 m. The Summer load draft is 3.80 m.

   120 tonnes of cargo are loaded under deck at Kg 2.30 m.
   50 tonnes of bunkers are loaded at Kg 0.60 m giving a free surface moment of 160 tonne-metres.
   Cargo is then loaded on deck, Kg 2.50 m, to take the ship to her Summer load draft.

Using Datasheet Q.7, calculate *each* of the following:
(a) the weight of deck cargo to load;   (8)
(b) the load KG;   (17)
(c) the GM$_f$ on completion of loading.   (5)

8. A ship is upright. Using the ship's derrick a heavy item of cargo is moved from a stowage position on the centreline of the lower hold to a new stowage position on the starboard side of the weather deck.

Explain, with the aid of sketches, the successive movements of the ship's centre of gravity when the cargo item is:
(a) raised from the lower hold;   (10)
(b) swung to the starboard side;   (10)
(c) landed at the new stowage position.   (10)

9. A ship in salt water (RD 1.025) has a mean draft of 2.70 m, KG 3.90 m. The Summer load draft is 3.60 m.

160 tonnes of cargo are loaded under deck at Kg 2.10 m.
150 tonnes of bunkers are loaded at Kg 0.60 m giving a free surface moment of 190 tonne-metres.
Cargo is then loaded at Kg 4.50 m, to take the ship to her Summer load draft.

Using Datasheet Q5, calculate *each* of the following:
(a) the weight of cargo to take the ship to her Summer draft;   (8)
(b) the load KG;   (17)
(c) the GM on completion of loading.   (5)

10. (a) Discuss, with the aid of sketches, how free surface effects reduce a ship's metacentric height and list the factors involved.   (19)
(b) List *three* methods for reducing free surface effects.   (6)

11. Draw an outline profile and an outline midship section of a ship. On the outlines clearly show the positions of *each* of the following:

LBP, LOA and Amidships.
Draft moulded and Freeboard.
Upper deck sheer forward and aft.
Upper deck camber.
Bulbous bow.
Rise of floor.   (25)

12. A ship has the following particulars:

Summer freeboard   1.70 m
FWA   90 mm
TPC   8.5

She is currently in dock water RD 1.015 with freeboards as follows:

Port   1.80 m
Starboard   1.73 m

Calculate the weight of cargo to be loaded in order that the ship will be at her Summer load draft in salt water (RD 1.025).   (20)

13. A vessel has a draft of 3.75 m in salt water. The lightweight displacement is 658.9 t.
    Using Datasheet Q.13, calculate each of the following:
    (a) fully loaded displacement;   (5)
    (b) fully loaded deadweight;   (5)
    (c) fresh water allowance in mm;   (5)
    (d) tropical fresh water allowance in mm;   (5)
    (e) fresh water draft in metre.   (5)

14. (a) Define, with the aid of sketches, each of the following terms:
        (i) ullage;   (3)
        (ii) sounding;   (3)
        (iii) ullage plug;   (3)
        (iv) sounding pad (striker plate);   (3)
        (v) strum box.   (3)
    (b) Outline the information contained in and use of a ship's tank calibration Tables.   (10)

15. (a) Define TPC and explain its use on ships.   (8)
    (b) Explain clearly why TPC changes with each draft in a conventional merchant ship.   (8)
    (c) A vessel moves from salt water of density $1025 \, \text{kg/m}^3$ to river water at $1012 \, \text{kg/m}^3$.
        If the displacement remains constant at 12 250 t and the TPC in salt water is 27.65, calculate the mean bodily sinkage after the vessel has moved into the river water.   (9)

16. A ship has an initial displacement of 9300 t, KG 8.20 m, and loads the following cargo:

    3200 tonnes at Kg 7.20 m
    1400 tonnes at Kg 8.00 m
    500 tonnes at Kg 10.50 m

    On passage the ship will use:

    275 tonnes of bunkers at Kg 1.60 m
    40 tonnes of FW at Kg 3.00 m

    causing a free surface moment of 300 tonne-metres.
        Determine, using Datasheet Q.15 (maximum KG), whether the ship has adequate stability in *each* of the following:
    (a) on completion of loading;   (15)
    (b) on completion of the voyage.   (15)

17. A ship in salt water (RD 1.025) has a mean draft of 3.10 m, KG 3.70 m. The Summer load draft is 3.80 m.

    120 tonne of cargo are loaded under deck at Kg 2.30 m.

    50 tonne of bunkers are loaded at Kg 0.60 m giving a free surface moment of 160 tonne-metres.

    Cargo is then loaded on deck, Kg 2.50 m, to take the ship to her Summer load draft.

    Using Datasheet Q.16, calculate each of the following:

    (a) the weight of deck cargo to load;   (8)

    (b) the load KG;   (17)

    (c) the GM_f on completion of loading.   (5)

TS035-22   GENERAL SHIP KNOWLEDGE

DATASHEET Q15

(This Datasheet must be returned with your examination answer book)

MAXIMUM KG (METRES) TO COMPLY WITH MINIMUM STABILITY CRITERIA SPECIFIED
IN THE CURRENT LOAD LINE RULES

Displacement t	KG. m
19 500	7.85
19 000	7.93
18 500	8.02
18 000	8.10
17 500	8.17
17 000	8.20
16 500	8.19
16 000	8.18
15 500	8.18
15 000	8.19
14 500	8.20
14 000	8.22
13 500	8.24
13 000	8.28
12 500	8.34
12 000	8.42
11 500	8.50
11 000	8.60
10 500	8.71
10 000	8.85
9 500	9.03
9 000	9.24
8 500	9.48
8 000	9.73
7 500	9.82
7 000	9.62
6 500	9.43
6 000	9.18
5 500	8.84
5 000	8.40

Candidate's
Name ....

Examination
Centre ....

035-22   GENERAL SHIP KNOWLEDGE

DATASHEET Q16

HYDROSTATIC PARTICULARS 'B'

(in salt water R.D. 1.025)

The hydrostatic particulars have been calculated for the vessel on even keel

Draught	Displacement	TPC	MCTC	LCB from amidships	LCF from amidships	KB	KM_T
m	t	t	tm	m	m	m	m
4.00	1888.8	5.20	23.30	1.45A	2.39A	2.09	4.05
3.90	1836.8	5.20	23.20	1.42A	2.43A	2.04	4.05
3.80	1784.8	5.19	23.10	1.39A	2.46A	1.99	4.05
3.70	1733.0	5.18	22.99	1.36A	2.49A	1.93	4.06
3.60	1681.2	5.17	22.89	1.32A	2.52A	1.88	4.06
3.50	1629.5	5.16	22.77	1.29A	2.55A	1.83	4.07
3.40	1577.9	5.16	22.65	1.24A	2.57A	1.78	4.09
3.30	1526.4	5.15	22.52	1.20A	2.59A	1.72	4.11
3.20	1475.1	5.13	22.39	1.15A	2.61A	1.67	4.13
3.10	1423.8	5.12	22.25	1.10A	2.62A	1.62	4.15
3.00	1372.6	5.11	22.11	1.04A	2.63A	1.56	4.19
2.90	1321.5	5.10	21.96	0.98A	2.64A	1.51	4.22
2.80	1270.6	5.09	21.80	0.91A	2.64A	1.46	4.27
2.70	1211.4	5.07	21.64	0.84A	2.64A	1.40	4.32
2.60	1169.1	5.06	21.47	0.76A	2.63A	1.35	4.37
2.50	1118.6	5.05	21.30	0.68A	2.63A	1.30	4.44
2.40	1068.4	5.01	20.87	0.59A	2.51A	1.24	4.51

Candidate's
Name ....

Examination
Centre ....

Appendix IV
100 Revision one-liners

The following are 100 one-line questions acting as an aid to examination preparation. They are similar in effect to using mental arithmetic when preparing for a mathematics exam. Elements of questions may well appear in the written papers or in the oral exams. Good luck.

1. What is another name for the KG?
2. What is a hydrometer used for?
3. If the angle of heel is less than 10 degrees, what is the equation for GZ?
4. What are the formulae for TPC and MCTC for a ship in salt water?
5. Give two formulae for the metacentre, KM.
6. How may free surface effects be reduced on a ship?
7. What is another name for KB?
8. List four requirements before an inclining experiment can take place.
9. With the aid of a sketch, define LOA and LBP.
10. What are cross curves of stability used for?
11. What is the longitudinal centre of a waterplane called?
12. Adding a weight to a ship usually causes two changes. What are these changes?
13. What is Simpson's First Rule for a parabolic shape with seven equally spaced ordinates?
14. What is KB for (a) box-shaped vessel and (b) triangular-shaped vessel?
15. What are hydrostatic curves used for onboard a ship?
16. Using sketches, define the block, the waterplane and midship form coefficients.
17. Sketch a statical stability curve and label six important points on it.
18. What are the minimum values allowed by D.Tp. for GZ and for transverse GM?
19. List three ways in which a ship's end drafts may be changed.
20. GM is 0.45 m. Radius of gyration is 7 m. Estimate the natural rolling period in seconds.
21. What is a deadweight scale used for?
22. What is the formula for bending stress in terms of M, I and y?
23. Sketch a set of hydrostatic curves.

24. List three characteristics of an angle of loll.
25. Define (a) a moment and (b) a moment of inertia.
26. Sketch the first three curves for a set of ship's strength curves.
27. What is the 'theory of parallel axis' formula?
28. What are the effects on a statical stability curve for increased breadth and increased freeboard?
29. Sketch a metacentric diagram for a box-shaped vessel and a triangular-shaped vessel.
30. Block coefficient is 0.715. Midship coefficient is 0.988. Calculate prismatic coefficient.
31. Describe the use of Simpson's Third Rule.
32. What is the wall-sided formula for GZ?
33. Define 'permeability'. Give two examples relating to contents in a hold or tank.
34. Give the equations for BM, box-shaped vessels and triangular-shaped vessels?
35. List three characteristics of an angle of list.
36. Sketch the shear force and bending moment curves. Show their inter-relation.
37. For a curve of seven equally spaced ordinates give Simpson's Second Rule.
38. What is the formula for pressure of water on a lockgate situation?
39. When a weight is lifted from a jetty by a ship's derrick whereabouts does its CG act?
40. Sketch a set of freeboard marks and label dimensions as specified by D.Tp.
41. Sketch a displacement curve.
42. What is Morrish's formula for VCB?
43. For an inclining experiment how is tangent of the angle of list obtained?
44. What do 'a moment of statical stability' and 'dynamical stability' mean?
45. Show the range of stability on an S/S curve having a very small initial negative GM.
46. Breadth is 45 m. Draft is 15 m. What is the increase in draft at a list of 2 degrees?
47. What is the formula for loss of GM due to free surface effects in a slack tank?
48. For what purpose is the inclining experiment made on ships?
49. What is the 'true mean draft' on a ship?
50. When drydocking a ship there is a virtual loss in GM. Give two formulae for this loss.
51. With Simpson's Rules, give formulae for M of I about (a) amidships and (b) centre line.
52. Discuss the components involved for estimating an angle of heel whilst turning a ship.
53. What is a 'stiff ship' and a 'tender ship'. Give typical GM values.

54. With the lost buoyancy method, how does VCG change, after bilging has occurred?
55. Sketch a deadweight moment curve and label the important parts.
56. Sketch a bending stress diagram for a vessel that is in a sagging condition.
57. What are 'Bonjean curves' and for what purpose are they used?
58. Define 'ship squat' and 'blockage factor'.
59. Draw the line diagram for Murray's method for maximum bending moment.
60. What is the formula for shear stress for an H-girder?
61. What happens to cause a vessel to be in unstable equilibrium?
62. What causes hogging in a vessel?
63. Which letters signify the metacentric height?
64. Give typical C_b values for fully loaded VLCC, general cargo ships and passenger liners.
65. What is the air draft and why is it important to know its value?
66. What are Type 'A' ships, Type 'B' ships and 'B-60' ships.
67. What is synchronous rolling?
68. What is DfT tabular freeboard?
69. When dealing with subdivision of passenger vessels, what is the 'margin line'
70. Sketch a set of freeboard marks for a timber ship.
71. Describe how icing can affect the stability of a vessel.
72. List six corrections made to a tabular freeboard to obtain an assigned freeboard.
73. Give the formula for the amidships draft for a vessel with longitudinal deflection.
74. Give the formula for the heaving period T_H in seconds.
75. In grain loading, what is the angle of repose?
76. Sketch a GZ curve together with a wind lever curve (up to 40° angle of heel).
77. Show on the sketch for the previous answer, the angle of heel due to shift of grain.
78. Give a typical range for stowage rates for grain stowed in bulk.
79. When is a RD coefficient used on ships?
80. What is parametric rolling?
81. What is the difference between an ullage and a sounding in a partially filled tank?
82. What are sounding pads and why are they fitted in tanks
83. What is a calibration Book used for?
84. Sketch a gunwhale plate and indicate the placement of the freeboard Dk mark?
85. Seasonal allowances depend on three criteria. What are these three criteria?
86. What is the factor of subdivision Fs?
87. What is the criterion of service numeral Cs?
88. Give a definition of mechanical stability and directional stability.

89. Do transverse bulkheads reduce free surface effects?
90. For a DfT standard ship, what is the depth D in terms of floodable length F_L?
91. For timber ships, LW to LS is LS/n. What is the value of n?
92. The WNA mark is not fitted on ships above a particular length. What is that length?
93. According to DfT, what is standard sheer in teens of LBP?
94. What is a Whessoe gauge used for on ships?
95. List four reasons why a ship's overall G can be raised leading to loss of stability.
96. What exactly defines a bulkhead deck?
97. List four methods for removing ice that has formed on upper structures of ships.
98. Sketch a set of FL and PL curves.
99. Sketch the curves for Type 'A' ships, Type 'B' ships and 'B-60' ships.
100. In the loading of grain, what is the approximate formula for angle of heel relative to 12°, due to shift of grain?

Appendix V
How to pass exams in maritime studies

To pass exams you have to be like a successful football or hockey team. You will need:

Ability, Tenacity, Consistency, Good preparation and Luck!!

The following tips should help you to obtain extra marks that could turn that 36% into a 42%+ pass or an 81% into an Honours 85%+ award. Good luck!!

Before your examination

1. Select 'bankers' for each subject. Certain topics come up very often. Certain topics you will have fully understood. Bank on these appearing on the exam paper.
2. Don't swat 100% of your course notes. Omit about 10% and concentrate on the 90%. In that 10% will be some topics you will never be able to be understand fully.
3. Do all your coursework/homework assignments to the best of your ability. Many a student has passed on a weighted marking scheme, because of a good coursework mark compensating for a bad performance on examination day.
4. In the last days prior to examination, practise final preparation with you revision one-liners.
5. Work through passed exam papers in order to gauge the standard and the time factor to complete the required solution.
6. Write all formulae discussed in each subject on pages at the rear of your notes.
7. In your notes circle each formula in a red outline or use a highlight pen. In this way they will stand out from the rest of your notes. Remember, formulae are like spanners. Some you will use more than others but all can be used to solve a problem. After January 2005, a formulae sheet will be provided in examinations for Masters and Mates.
8. Underline in red important key phrases or words. Examiners will be looking for these in your answers. Oblige them and obtain the marks.

9. Revise each subject in carefully planned sequence so as not to be rusty on a set of notes that you have not read for some time whilst you have been sitting other exams.

10. Be aggressive in your mental approach to do your best. If you have prepared well there will be less nervous approach and like the football team you will gain your goal. Success will come with proper training.

In your examination

1. If the examination is a mixture of descriptive and mathematical questions, select as your first question a descriptive or 'talkie-talkie' question to answer. It will settle you down for when you attempt the later questions to answer.

2. Select mathematical questions to answer if you wish to obtain higher marks for good answers. If they are good answers you are more likely to obtain full marks. For a 'talkie-talkie' answer a lot may be subjective to the feelings and opinion of the examiner. They may not agree with your feelings or opinions.

3. Use big sketches. Small sketches tend to irritate examiners.

4. Use coloured pencils. Drawings look better with a bit of colour.

5. Use a 150 mm rule to make better sketches and a more professional drawing.

6. Have big writing to make it easier to read. Make it neat. Use a pen rather than a biro. Reading a piece of work written in biro is harder to read especially if the quality of the biro is not very good.

7. Use plenty of paragraphs. It makes it easier to read.

8. Write down any data you wish to remember. To write it makes it easier and longer to retain in your memory.

9. Be careful in your answers that you do not suggest things or situations that would endanger the ship or the onboard personnel.

10. Reread your answers near the end of the exam. Omitting the word NOT does make such a difference.

11. Reread your question as you finish each answer. Don't miss for example part (c) of an answer and throw away marks you could have obtained.

12. Treat the exam as an advertisement of your ability rather than an obstacle to be overcome. If you think you will fail, then you probably will fail.

After the examination

1. Switch off. Don't worry. Nothing you can do now will alter your exam mark be it good or bad. Students anyway tend to underestimate their performances in exams.

2. Don't discuss your exam answers with your student colleagues. It can falsely fill you with hope or with despair.

3. Turn your attention to preparing for the next subject in which you are to be examined.

References

No.	Prepared by	Date	Title	Publisher	ISBN
1	Barrass C.B.	2005	*Ship Squat – 33 Years of Research*	Research paper	NA
2	Barrass C.B.	2005	*Ship Squat – A Guide for Masters*	Research paper	NA
3	Barrass C.B.	2003	*Widths and Depths of Influence*	Research paper	NA
4	Barrass C.B.	2004	*Ship Stability for Masters and Mates*	Elsevier Ltd.	0 7506 4101 0
5	Barrass C.B.	2003	*Ship Stability – Notes and Examples*	Elsevier Ltd.	0 7506 4850 3
6	Barrass C.B.	1977	*A Unified Approach to Ship Squat*	Institute of Nautical Studies	NA
7	Barrass C.B.	1978	*Calculating Squat – A Practical Approach*	Safety at Sea Journal	NA
8	Barrass C.B.	2004	*Ship Design and Performance*	Elsevier Ltd.	0 7506 6000 7
9	Cahill R.A.	2002	*Collisions and Their Causes*	Institute of Nautical Studies	1 870077 60 1
10	Cahill R.A.	1985	*Strandings and Their Causes – 1st Edition*	Fairplay Publications	0 9050 4560 2
11	Cahill R.A.	2002	*Strandings and Their Causes – 2nd Edition*	Institute of Nautical Studies	1 870077 61 X
12	Carver A.	2002	*Simple Ship Stability – A Guide for Seafarers*	Fairplay Publications	1 870093 65 8
13	Clark I.C.	2005	*Ship Dynamics for Mariners*	Institute of Nautical Studies	1 870077 68 7
14	Durham C.F.	1982	*Marine Surveys – An Introduction*	Fairplay Publications	0 905045 33 5
15	House D.	2000	*Cargo Work*	Elsevier Ltd.	0 7506 3988 1

No.	Author	Year	Title	Publisher	ISBN
16	IMO	2002	*Load Lines – 2002*	International Maritime Organisation	92-801-5113-4
17	IMO	2004	*Solas Consolidated Edition – 2004*	International Maritime Organisation	92-801-4183-X
18	IMO	2002	*Code of Intact Stability for All Types of Ships*	International Maritime Organisation	92-801-5117-7
19	IMO	1991	*International Grain Code*	International Maritime Organisation	92-801-1275-9
20	Institute of Nautical Studies	2000–2005	*Seaways – Monthly Journals*	Institute of Nautical Studies	NA
21	Merchant Shipping Notice	1998	*M. S. (Load Line) Regulations*	M'Notice MSN 1752 (M)	NA
22	Rawson K.J. & Tupper E.C.	2001	*Basic Ship Theory*	Elsevier Ltd.	0 7506 5397 3
23	RINA	2002–2005	*Significant Ships – Annual Publications*	Royal Institute of Naval Architects	NA
24	RINA	2000–2005	*The Naval Architect – Monthly Journals*	Royal Institute of Naval Architects	NA
25	Stokoe E.A.	1991	*Naval Architecture for Marine Engineers*	Thomas Reed Ltd.	0 94 7637 85 0
26	Tupper E.C.	2004	*Introduction to Naval Architecture*	Elsevier Ltd.	0 7506 6554 8

Answers to exercises

Exercise 1
1. 1800 Nm   **2.** 2 kg m, anti-clockwise   **3.** 0.73 m from the centre towards the 10 kg weight   **4.** 45.83 kg   **5.** 81 kg m

Exercise 2
1. 11.05 m   **2.** 10.41 m   **3.** 4.62 m   **4.** 6.08 m   **5.** 0.25 m

Exercise 3
1. 103.5 t   **2.** 118 t   **3.** 172.8 t   **4.** 133.2 t   **5.** 86.2 t   **6.** 860.3 t

Exercise 4
1. 0.484 m   **2.** 0.256 m   **3.** 1.62 tonnes   **4.** 11.6 tonnes   **5.** 0.04 m, 32 per cent   **6.** 900 kg, S.G. 0.75   **7.** 0.03 m   **8.** 1.02   **9.** 0.75 m, 182.34 tonnes   **10.** (a) 125 kg, (b) 121.4 kg   **11.** 64 per cent   **12.** (a) 607.5 kg, (b) 4.75 cm   **13.** 1.636 m   **14.** (a) 1.2 m, (b) 70 per cent   **15.** 9.4 per cent

Exercise 5
1. 7 612.5 tonnes   **2.** 4352 tonnes   **3.** 1.016 m   **4.** 7.361 m   **5.** 0.4875 m   **6.** 187.5 tonnes   **7.** 13 721.3 tonnes   **8.** 64 06.25 tonnes   **9.** 6.733 m F 6.883 m A   **10.** 228 tonnes   **11.** 285 tonnes   **12.** (a) 8515 tonnes, (b) 11 965 tonnes   **13.** 27 mm   **14.** 83.2 mm

Exercise 6
5. 2604 tonnes metres

Exercise 8
1. (b) 3.78 tonnes 4.42 tonnes, (c) 2.1 m   **2.** (b) 6.46 tonnes 7.8 tonnes, (c) 3.967 m   **3.** 4.53 m   **4.** (b) 920 tonnes, (c) 3.3 m, (d) 6.16 tonnes

5. (a) 2.375 m, (b) 3092 tonnes, (c) 1125 tonnes   **6.** (b) 3230 tonnes,
(c) 1.625 m   **7.** (b) 725 tonnes, (c) 4.48 m, (d) 5 tonnes   **8.** (b) 5150 tonnes,
4.06 m, (c) 5.17 m

Exercise 9

1. (b) 12.3 tonnes   **2.** (b) 8302.5 tonnes   **3.**   12 681.3 tonnes
4.   221 tonnes   **5.**   180 tonnes   **6.**   95   **7.**   53

Exercise 10

1. (a) 508 m^2, (b) 5.2 tonnes, (c) 0.8 m aft of amidships   **2.** (a) 488 m^2,
(b) 5 tonnes, (c) 0.865, (d) 0.86 m aft of amidships   **3.** (a) 122 mm,
(b) 43.4 m from forward   **4.** (a) 30 476.7 tonnes, (b) 371.4 mm, (c) 15.6 m
5.   5062.5 tonnes   **6.** (a) 978.3 m^2, (b) 15.25 cm, (c) 2.03 m aft of amidships
7. (a) 9993$\frac{3}{4}$ tonnes, (b) 97.44 mm, (c) 4.33 m   **8.** (a) 671.83 m^2, (b) 1.57 m
aft of amidships   **9.**   12.125 m^2 **10.**   101 m^2 **11.** (a) 781.67 m^3,
(b) 8.01 tonnes   **12.** (a) 28 93.33 m^3 or 2965.6 tonnes, (b) 3 m

Exercise 11

1.   0.707   **2.**   8a^4/3   **3.**   9:16   **4.**   63 281 cm^4 **5.** (a) 3154 m^4,
(b) 28 283 m^4 **6.** (a) 18 086 m^4, (b) 871 106 m^4 **7.**   BM$_L$ 206.9 m, BM$_T$ 8.45 m
8.   I$_{CL}$ 35 028 m^4, I$_{CF}$ 1 101 540 m^4 **9.**   I$_{CL}$ 20 267 m^4, I$_{CF}$ 795 417 m^4
10.   I$_{CL}$ 13 227 m^4, I$_{CF}$ 396 187 m^4

Exercise 12

1.   5 m   **2.**   1.28 m, 4.56 m   **3.**   1.78 m, 3 m   **4.**   No, unstable when
upright   **5.** (a) 6.2 m, 13.78 m, (b) 4.9 m   **6.** (a) 10.6 m, 5.13 m, (b) 4.9 m at
4.9 m draft   **7.** (a) 6.31 m, 4.11 m, (b) 4.08 m   **8.** (b) GM is + 1.8 m, vessel
is in stable equilibrium, (c) GM is zero, KG = KM, so ship is in neutral
equilibrium

Exercise 13

1.   2.84 m   **2.**   3.03 m   **3.**   3.85 m   **4.**   5.44 m   **5.**   0.063 m
6.   1466.67 tonnes   **7.**   1525 tonnes   **8.**   7031.3 tonnes in L.H. and
2568.7 tonnes in T.D.   **9.**   1.2 m   **10.**   1.3 m   **11.**   55 tonnes
12.   286.3 tonnes   **13.**   1929.67 tonnes

Exercise 14

1.   6° 03 to starboard   **2.**   4.2 m   **3.**   216.5 tonnes to port and 183.5
tonnes to starboard   **4.**   9° 30′   **5.**   5.458 m   **6.**   12° 57′   **7.**   91.9
tonnes   **8.**   282.75 tonnes to port, 217.25 tonnes to starboard
9.   8.52 m to port, GM = 0.864 m   **10.**   14° 04′ to port   **11.**   13° 24′
12.   50 tonnes   **13.**   3.8°

Exercise 15

1. 674.5 tonnes metres   **2.** 7.773 m   **3.** 546.2 m   **4.** 6.027 m,
2000 tonnes metres   **5.** (a) 83.43 tonnes metres, (b) 404.4 tonnes metres
6. (a) 261.6 tonnes metres, (b) 2647 tonnes metres   **7.** 139.5 tonnes
metres, 1366 tonnes metres   **8.** 0.0522 m   **9.** (b) Angle of loll is 14.96°,
KM is 2.67 m, GM is −0.05 m

Exercise 16

1. 218.4 tonnes in No. 1 and 131.6 tonnes in No. 4   **2.** 176.92 tonnes
3. 5.342 mA, 5.152 m F   **4.** 6.726 m A, 6.162 m F   **5.** 668.4 tonnes
from No. 1 and 1331. 6 tonnes from No. 4   **6.** 266.7 tonnes   **7.** 24.4 cm
8. 380 tonnes, 6.785 m F   **9.** 42.9 tonnes in No 1 and 457.1 tonnes in
No. 4, GM is 0.79 m   **10.** 402.1 tonnes from No. 1 and 47.9 tonnes from
No. 4   **11.** 4.340 m A, 3.118 m F   **12.** 5.56 m A, 5.50 m F   **13.** 5.901 m A,
5.679 m F   **14.** 4 metres aft   **15.** 3.78 metres aft   **16.** 4.44 m metres aft
17. 55.556 metres forward   **18.** 276.75 tonnes, 13.6 metres forward
19. 300 tonnes, 6.3 m   **20.** 200 tonnes, 7.6 m   **21.** 405 tonnes in No. 1
and 195 tonnes in No. 4   **22.** 214.3 tonnes   **23.** 215.4 tonnes, 5.96 m F
24. 200 metres   **25.** 240 metres   **26.** 8.23 m A, 7.79 m F, dwt is 9195
tonnes, trim by the stern is 0.44 m

Exercise 17

1. GM is 2 m, range is 0° to 84.5°, max GZ is 2.5 m at 43.5° heel
2. GM is 4.8 m, max moment is 67 860 tonnes metres at 42.25° heel, range
is 0° to 81.5°,   **3.** GM is 3.07 m, max GZ is 2.43 m at 41° heel, range is
0° to 76°, moment at 10° is 16 055 tonnes metres, moment at 40°
is 59 774 tonnes metres   **4.** Moment at 10° is 16 625 tonnes metres, GM is
2 m, max GZ is 2.3 m at 42° heel, range is 0° to 82°   **5.** GM is 3.4 m, range
is 0° to 89.5°, max GZ is 1.93 m at 42° heel   **6.** 6.61 m   **7.** 8.37 m
8. 25.86 m forward   **9.** 39 600 tonnes, 513 tonnes metres, 40.5 tonnes,
9.15 m   **10.** 4.53 m A, 3.15 F   **11.** (a) 0° to 95°, (b) 95°, (c) 3.18 m at 47.5°
12. (a) 0° to 75°, (b) 75°, (c) 2.15 m at 40°   **13.** (a) 1.60 m, (b) 40°, (c) 12°,
(d) 72.5°   **14.** (a) 1.43 m, (b) 39.5°, (c) 44 000 tonnes metres, (d) 70°

Exercise 18

1. 1.19 m   **2.** 1.06 m   **3.** 0.64 m   **4.** 7 m   **5.** 4.62 m

Exercise 19

1. 1.545 m   **2.** 6.15 tonnes/m^2 **3.** 7.175 tonnes/ m^2, 1435 tonnes
4. 6.15 tonnes/m^2, 55.35 tonnes   **5.** 9 tonnes/m^2, 900 tonnes
6. 6.15 tonnes/m^2, 147.6 tonnes   **7.** 1 tonne   **8.** 386.76 tonnes on side
with density 1010 kg/m^3 **9.** 439.4 tonnes on side with density
1016 kg/m^3 **10.** 3075 tonnes   **11.** 128.13 tonnes

Exercise 20

1. Transfer 41.94 tonnes from starboard to port, and 135 tonnes from forward to aft.   **2.** Transfer 125 tonnes from forward to aft, and 61.25 tonnes from port to starboard. Final distribution: No. 1 Port 75 tonnes, No. 1 starboard 200 tonnes, No. 4 Port 63.75 tonnes, No. 4. starboard 61.25 tonnes.   **3.** 13°52′, 3.88 m F, 4.30 m A   **4.** Transfer 133.33 tonnes from each side of No. 5. Put 149.8 tonnes into No 2. port and 116.9 tonnes into No. 2. starboard.

Exercise 21

1. 0.148 m   **2.** 0.431 m   **3.** 1.522 m   **4.** 7° 02′   **5.** Dep. GM is 0.842 m, arr. GM is 0.587 m   **6.** 3° exactly   **7.** 112.4 tonnes **8.** 6.15 m, +0.27 m

Exercise 22

1. (b) 5.12 m   **2.** 0.222 m   **3.** 0.225 m   **4.** 0.558 m   **5.** 0.105 m **6.** 0.109 m   **7.** 0.129 m   **8.** 3.55 m F, 2.01 m A   **9.** 5.267 m A, 7.529 m F   **10.** 2.96 m A, 6.25 m F   **11.** 3.251 m A, 5.598 m F **12.** 4.859 m A, 5.305 mF

Exercise 23

1. 1344 m tonnes   **2.** 2038.4 m tonnes   **3.** 1424 m tonnes **4.** 13.67 m tonnes   **5.** 107.2 m tonnes

Exercise 24

No answers required

Exercise 25

1 and 2. Review chapter notes

Exercise 26

1 and 2. Review chapter notes

Exercise 27

1. Review chapter notes.   **2.** 0.734 m   **3.** Review chapter notes **4.** 53.82

Exercise 28

1 to 4. Review chapter notes   **5.** 556 mm

Exercise 29

1 and 2. Review chapter notes   **3.** 551 mm   **4.** Review chapter notes

Exercise 30

1. 11.5°   **2.** Review chapter notes   **3.** 0.240 m, 0.192 m, 0.216 m
4. Review chapter notes

Exercise 31

1. No, 35° 49′   **2.** 39° 14′   **3.** Probable cause of the list is a negative GM
4. Discharge timber from the high side first

Exercise 32

1. 33 tonnes   **2.** 514 tonnes   **3.** 488 tonnes   **4.** 146.3 tonnes, Sag
0.11 m, 0.843 m

Exercise 33

1. 15.6 cm   **2.** 1962 tonnes, 4.087 m   **3.** 4.576 m   **4.** 5000 tonnes
5. 39.1 cm   **6.** 10.67 m   **7.** 2.92 m   **8.** 4.24 m   **9.** (a) 0.48 m, 1.13°,
(b) 8050 tonnes, 8.51 m

Exercise 34

1. 1.19 m   **2.** 69.12 tonnes, 1.6 m

Exercise 35

1. −0.2 m or −0.25 m   **2.** +0.367 m or +0.385 m   **3.** +0.541 m
or +0.573 m safe to dry-dock vessel   **4.** +0.550 m or +0.564 m safe to
dry-dock vessel   **5.** Maximum trim 0.896 m or 0.938 m by the stern

Exercise 36

1. 658.8 tonnes, 4.74 m   **2.** 17.17 m   **3.** 313.9 tonnes, 2.88 m
4. 309.1 tonnes, 1.74 m   **5.** 5.22 m

Exercise 37

1. (a) 366.5 m, (b) 0.100, (c) 0.65 m at the stern, (d) 0.85 m   **2.** (a) W of I
is 418 m as it is >350 m, (b) Review your notes, (c) 9.05 kt   **3.** 10.56 kt
4. 0.50 m, at the stern

Exercise 38

1. Passenger liner (0.78 m), general cargo ship (0.66 m)   **2.** Review
chapter notes   **3.** Review chapter notes

Exercise 39
1. 4° 34′   2. 6° 04′   3. 5° 44′   4. 6.8° to starboard

Exercise 40
1. 20.05 s   2. 15.6 s   3. 15.87 s   4. T_R is 28.2 s – rather a 'tender ship'.
5. 24 s, 7.37 s, 6.70 s

Exercise 41
1 and 2.   Review chapter notes

Exercise 42
1. 3.6°   2. 11.57°   3. 5.7°   4. 4.9°   5. 8°

Exercise 43
1. 8.20 m A, 4.10 m F   2. 4.39 m A, 2.44 m F   3. 6.91 m A, 6.87 m F

Exercise 44
1. 14° 33′   2. 13° 04′   3. 9° 52′   4. 5° 35′

Exercise 45
No answers required

Exercise 46
No answers required

Exercise 47
1. 420 tonnes

Exercise 48
1. (a) 10 745 t, (b) 14 576 t, (c) 8.03 m, (d) 6.04 m, (e) 1.03 m

Exercise 49
1. Max BM is 2.425 tonnes m at 5.8 m from A   2. (a) Max BM is 29.2
tonnes m between the positions of the two masses,   (b) At 1/3 L from left-
hand side, SF is 4.87 tonnes, BM is 24.34 tonnes m and at 1/3 L from right-
hand side, SF is 0.13 tonnes, BM is 28.65 tonnes m

Exercise 50
No answers required

Exercise 51
No answers required

Exercise 52
1.   Max SF is ±40 tonnes, BM ranges from +145.8 tonnes m to −62.5 tonnes m   **2.**   Max BM is 150 tonnes m, at amidships   **3.**   Max SF is ±250 tonnes at 25 m from each end, max BM is 6250 tonnes m, at amidships   **4.**   Review Chapter notes   **5.**   Max SF is 128.76 MN, max BM is 10 220 MN m at station 5.2, just forward of mid-LOA

Exercise 53
No answers required

Exercise 54
No answers required

Exercise 55
Review chapter notes

Exercise 56
No answers required

Index